Fluvial Hydraulics in the Presence of Vegetation in Channels

Fluvial Hydraulics in the Presence of Vegetation in Channels

Editors

Jueyi Sui
Hossein Afzalimehr

MDPI

Basel • Beijing • Wuhan • Barcelona • Belgrade • Novi Sad • Cluj • Manchester

Editors
Jueyi Sui
School of Engineering
University of Northern British Columbia
Prince George
Canada

Hossein Afzalimehr
Department of Civil Engineering
Iran University of Science and Technology
Tehran
Iran

Editorial Office
MDPI
St. Alban-Anlage 66
4052 Basel, Switzerland

This is a reprint of articles from the Special Issue published online in the open access journal *Water* (ISSN 2073-4441) (available at: www.mdpi.com/journal/water/special_issues/Hydraulics_vegetation_Channels).

For citation purposes, cite each article independently as indicated on the article page online and as indicated below:

Lastname, A.A.; Lastname, B.B. Article Title. *Journal Name* **Year**, *Volume Number*, Page Range.

ISBN 978-3-0365-8745-5 (Hbk)
ISBN 978-3-0365-8744-8 (PDF)
doi.org/10.3390/books978-3-0365-8744-8

Contents

About the Editors

Jueyi Sui

Dr. Jueyi Sui is a Professor at the School of Engineering, University of Northern British Columbia, Prince George, Canada. Jueyi received his Bachelor's and Master's degrees from Hefei University of Technology, China, in 1985 and 1988, respectively. He earned his Dr.-Ing. degree in water resources engineering from Kaiserslautern University in Germany. His research interests focus on fluvial hydraulics in the presence of in-stream infrastructures and vegetation, the interaction between ice accumulation and channel bed deformation, and snow hydrology. He has published over 100 papers in peer-reviewed international journals. Currently, he serves as an associate editor for "International Journal of Sediment Research" and "Journal of Hydrology and Hydromechanics"; and acts as one of members of the editorial board for "Journal of Hydrodynamics" and "Water".

Hossein Afzalimehr

Dr. Hossein Afzalimehr is a Professor at the School of Civil Engineering, Iran University of Science and Technology, Tehran, Iran. He received his Master's and PhD degrees from Laval University, Canada, in 1993 and 1998, respectively. His research interests focus on different aspects of fluvial hydraulics including the presence of vegetation in natural streams, turbulent flow structures in open channels, sediment transport, scouring processes, and channel bed deformation. He has published over 100 papers in peer-reviewed international journals. Currently, he serves as an editorial board member for "International Journal of Sediment Research" and acts as the guest editor for "Water".

Preface

In many semi- and arid regions, water does not flow from the middle of spring to early fall, leading to the development of vegetation patches with irregular distribution on channel beds and banks. These vegetation patches influence the characteristics of flow in rivers and streams. The presence of vegetation patches on channel beds and banks can be either submerged or emergent. The height of an emergent vegetation patch exceeds the water depth, while the height of a submerged vegetation patch is less than the water depth. Vegetation patches on both channel beds and the banks of natural rivers play a significant role in many aspects of the environment, including water quality, sediment transport and bank stability.

The aim of this Special Issue was to bring together research results that improve our understanding of fluvial hydraulics in vegetated channels, including the effects of submerged and emergent vegetation on velocity distribution, validation of the logarithmic law, Reynolds stress, turbulence intensities and bursting events for different aspect ratios (the ratio of width to flow depth) over flat gravel beds and 2D and 3D bedforms. The results showed that the determination of key hydraulic parameters such as the roughness coefficient and drag coefficient is influenced by vegetation arrangements and variation in bedforms.

Jueyi Sui and Hossein Afzalimehr
Editors

water MDPI

Editorial

Fluvial Hydraulics in the Presence of Vegetation in Channels

Hossein Afzalimehr [1] and Jueyi Sui [2,*]

[1] Department of Civil Engineering, Iran University of Science and Technology, Tehran 1311416846, Iran; hafzali@iust.ac.ir
[2] School of Engineering, University of Northern British Columbia, Prince George, BC V2N 4Z9, Canada
* Correspondence: jueyi.sui@unbc.ca

Citation: Afzalimehr, H.; Sui, J. Fluvial Hydraulics in the Presence of Vegetation in Channels. *Water* **2023**, *15*, 2907. https://doi.org/10.3390/w15162907

Received: 3 August 2023
Accepted: 4 August 2023
Published: 11 August 2023

1. Introduction

In many semi- and arid regions, water does not flow from middle spring to early fall, leading to the development of vegetation patches with irregular distribution on channel beds and banks. These vegetation patches influence the characteristics of flow in rivers and streams. The presence of vegetation patches on channel beds and banks can be either submerged or emergent. The height of an emergent vegetation patch exceeds water depth, while the height of a submerged vegetation patch is less than the water depth. Vegetation patches on both channel beds and the banks of natural rivers play a significant role in many aspects of the environment, including water quality, sediment transport and bank stability.

In the presence of a vegetation patch in a channel, the estimation of key hydraulic parameters such velocity, Reynolds stress and turbulence intensities depends on the features of the vegetation patch, such as submergence ratio, vegetation density and configuration. There are numerous key issues for researchers to apply turbulent flow characteristics in bare channels to vegetated channels. In the last three decades, to address issues regarding river restoration and water conservation projects, the effects of vegetation in vegetated rivers have attracted a lot of attention from researchers to conduct studies based on both experimental investigations and numerical simulations. However, our knowledge about the impacts of vegetation patches on the estimation of hydraulic parameters is still limited for engineers in the practice of river restoration projects, especially for cases of vegetated meandering channels and pool–riffle sequences.

The aim of this Special Issue is to bring together research results that improve our understanding of fluvial hydraulics in vegetated channels, including the effects of submerged and emergent vegetation on velocity distribution, validation of the logarithmic law, Reynolds stress, turbulence intensities and bursting events for different aspect ratios (the ratio of width to flow depth) over flat gravel beds and 2D and 3D bedforms. Results showed that the determination of key hydraulic parameters such as the roughness coefficient and drag coefficient is influenced by vegetation arrangements and variation in bedforms.

This Special Issue calls for research work in environmental hydraulics for improving our knowledge of the fluvial hydraulics of vegetated channels. Among the topics of interest in this Special Issue are:

- Turbulent flow characteristics in the presence of submerged and emergent vegetation;
- The water blockage effect of submerged vegetation;
- The influence of vegetation arrangement and density on the flow structure;
- The hydrodynamic effect of emergent vegetation in a U-shape open channel confluence flow with partially rigid emergent vegetation;
- The CDF simulation of flow structures in vegetated channels;
- The influence of bedform and vegetation on flow characteristics;
- The determination of erosion intensity and sedimentation rate from a watershed by means of empirical models.

2. Overview of This Special Issue

Many manuscripts received for this Special Issue show interesting and impressive results. The review process for each manuscript has been completed strictly following the journal requirements. In total, eleven (11) papers have been accepted and published in this Special Issue; they present the last results of fluvial hydraulics in vegetated channels based on laboratory experiments and numerical simulations.

Vegetation in rivers and streams plays an important role in preventing erosion and improving bank stability. Tabesh Mofrad et al. [1] compared the impacts of emergent vegetation on velocity and Reynolds stress distributions to those of submerged vegetation under conditions of the same aspect ratio and flow discharge. It is found that the influence of submerged vegetation on the generation of secondary currents is less than that of emergent vegetation. The application of the logarithmic law is valid for flow up to the depth of $y/h = 0.75$ for emergent vegetation (in which the vegetation cover in banks is partly out of the water) but up to the flow depth of $y/h = 0.25$ for vegetation bank. For submerged vegetation patches, the location of zero-shear stress superposes that of the maximum velocity. However, for emergent vegetation, near the zone of emergent vegetation in channel banks, the maximum velocity is shifted towards the channel bed, and thus the location of the zero-shear stress approaches the bed. The distribution of Reynolds stress shows a convex shape for submerged and emergent vegetation covers in channel beds. However, approaching the bank vegetation (emergent case), the power of secondary currents increases, and thus the maximum Reynolds stress is shifted towards the bed. The results of this study indicate that the estimation of key parameters of fluvial hydraulics such as velocity and Reynolds stress is influenced by the feature of vegetation in a channel including the arrangement patterns and submerged or emergent status.

Vegetation patches are often present in pool–riffle sequences in natural streams. Based on laboratory experiments conducted using two different vegetated pools with various slopes of their entry and exit sections, the impact of the vegetation patch on flow structures in pools has been investigated. The effect of the slopes of both entrance and exit sections of the pools on flow structure has been studied by Tabesh Mofrad et al. [2]. Results show that, for large entrance and exit slopes, the distribution profiles of turbulent kinetic energy (TKE) have no specific shape. However, the maximum TKE value is located near the bed when the slopes of the entrance and exit sections of the pool are small. Both ejections and sweeps govern the turbulence structures and coherent motions at the trailing edge of the vegetation patch. In addition to the location of the pool (entrance, middle pool and exit section), the bedform slopes and features of vegetation patches affect the velocity profile, Reynolds shear stresses, TKE distribution and the contribution of each bursting event on turbulent flow structures. For a pool with a larger entrance and exit slope, the TKE distribution has no specific form. However, for a pool with a small entrance and exit slope, the TKE distribution has a convex shape with the maximum value near the bed. The sweep motion occurs in a narrower zone above the vegetation canopy. The sweep motions of bursting events are the dominant processes directly above the vegetation canopy, while an outward motion with slightly positive values has been observed at the front edge of the vegetation patch. The Kolmogorov $-5/3$ power law rests valid for the 3D vegetation patch. The shedding frequency is affected by the changes in the bedform slopes and the presence of vegetation, resulting in higher values than those reported in the literature.

Based on experiments and numerical simulations, Qiu et al. [3] studied the blockage effect of submerged vegetation. A new numerical simulation model for the flow field in the presence of submerged vegetation in an open channel was proposed. The influence of velocity integration and average treatment on the water-blocking effect generalization model caused by submerged vegetation has been studied. It is found that the relationship between velocity integration and the generalized roughness coefficient is positive, and the relationship between velocity integration and virtual channel elevation is negative. The average treatment and numerical simulation treatment have effects on the generalization of the roughness coefficient and virtual channel elevation value. In addition, the average treatment with or without water-blocking effect generalization makes little difference in the generalization of the roughness coefficient and virtual channel elevation.

The interaction between bedforms and vegetation patches in mountainous streams is a challenging subject for river engineers. Considering the importance of vegetation in stabilizing bedforms, Parvizi et al. [4] studied turbulent flow structures based on a gradual varied shallow flow in a flume with both convective deceleration and acceleration flow sections. The presence of vegetation patches in the pool bed resulted in an increase in the shear velocity, particularly along the entrance section of the pool where the flow decelerates. Results of the quadrant analysis reveal that sweep events play the most important role in different parts of the pool, followed by ejection events. However, near the bed, especially along the pool entrance section, the role of outward and inward events is considerable. The Reynolds shear stress and turbulence intensity distributions over the gravel bed (bare bed) were compared to those in the pool with a vegetation patch. It is found that, in the presence of vegetation in the channel bed of the pool, the velocity profiles show an S-shaped distribution via increasing the aspect ratio. For these velocity profiles, the velocity gradient in the upper part is very small, the velocity gradient is higher in the middle part, and the velocity follows a logarithmic distribution in the region near the bed.

Brahimi and Sui [5] carried out laboratory experiments in a large-scale flume to investigate the effects of submerged vegetation arrangement patterns and density on flow structure. Results of turbulent kinetic energy in the wake zone of the deflected vegetation indicate that the maximum root-mean-square velocity fluctuations of flow occur at the sheath section and the top of the vegetation. In the wake zone behind the vegetation elements, the maximum value of Reynolds shear stress occurred slightly above the interface between deflected vegetation and the non-vegetation layer, showing the Kelvin–Helmholtz instability that is associated with inflectional points of the longitudinal velocity. As the vegetation density increases, the negative and positive values of Reynolds shear stress throughout the flow depth increase. In the presence of non-bending vegetation in a shallow flow bed, a high velocity gradient appears near the channel bed from the depth of $z/H = 0$ to $z/H = 0.1$, reaching a peak velocity at the depth of $z/H = 0.1$, and a decreasing trend of velocity toward the water surface is noticeable. However, for the deeper flow, the peak velocity occurs at a higher location close to the water surface.

By means of the RNG k-ε numerical model coupled with the volume of fluid method, Shi and Jin [6] simulated the hydrodynamics of a U-shape open-channel confluence flow with partially rigid emergent vegetation. Compared to non-vegetated cases, the separation zones in vegetated cases are smaller in both length and width. With a higher vegetation solid volume fraction, the separation zone is divided into two parts, a smaller one right after the confluence point and a larger one on the second half of the curved reach downstream of the confluence. With the increase in the solid volume fraction, the differences in velocities and bed shear stress between the convex and concave banks become larger. With the same solid volume fraction, a larger vegetation density caused more disturbance on the tributary than that resulting from a larger stem diameter.

Fortes et al. [7] simulated the 2019 Typhoon Hagibis at the Nanakita River using a dynamic roughness model. The model estimates the roughness of the river on a pixel level from the relationship between the Manning roughness coefficient and the degree of submergence of vegetation. The dynamic roughness model showed that the water level profile increased by 7.03% on average. It is found that seasonal variations in vegetation area and height have a clear effect on the flow dynamics of the river. It could be concluded that the water level is proportional to the amount of vegetation in the riparian zones. The highest water level was obtained in the summer, when the vegetation volume is at its peak, and the lowest value during the winter, when there is less vegetation. The vegetation area and average height were demonstrated to have a good correlation with the simulated water levels, with average height achieving the strongest relationship. Therefore, considering only the vegetation area to estimate the water level, as does the static roughness model, would be less efficient. The method to obtain the vegetation location and its distributed height, with the consideration of the parameters in the dynamic roughness model proposed in this study, was proven to be applicable for the purpose of river management in the Nanakita River.

Zhang et al. [8] optimized the ratio of the vegetation height-averaged velocity to the water depth-averaged velocity. In the vegetated area, two depth-averaged velocities which included the water depth-averaged velocity and the vegetation height-averaged velocity were defined. The N–S equation with an optimized velocity ratio can predict the lateral distribution of the water depth-averaged streamwise velocity in the open channel partially covered by the submerged vegetation. In both the non-vegetated area and submerged vegetation area, different parameters, including the transverse eddy viscosity coefficient, friction coefficient, porosity and the drag force coefficient, were introduced to determine the analytical solution. Additionally, the methods for calculating the parameters in different zones were discussed. Results showed that a relatively satisfactory prediction of the transverse distribution of water depth-averaged streamwise velocity in the channel flow with submerged vegetation was obtained using the proposed model.

Yu et al. [9] conducted two-dimensional numerical simulations and compared the simulation results to those of laboratory experiments. In their numerical simulations, six different Reynolds-averaged Navier–Stokes (RANS) turbulence models were used. It is found that the shear stress transport *k-w* model achieves the most consistent simulation results with those of experiments for the longitudinal mean flow velocity distribution at the centerline, and the Reynolds stress model provides the poorest results. Results showed that the turbulence intensity determined using the Reynolds stress model, standard k-w model and the Shear-stress transport *k-w* model is stronger than that calculated using the standard *k-ε* model, RNG *k-ε* model and realizable *k-ε* model. For the prediction of turbulent structure around rigid emerged vegetation via 2D numerical simulation, the Shear-stress transport *k-w* model can achieve better results than the standard *k-ε* model, RNG *k-ε* model, realizable *k-ε* model, standard k-w model and Reynolds stress model.

The interaction of the bedform and vegetation cover significantly affects turbulent flow parameters. Based on laboratory experiments with the presence of submerged rigid vegetation elements in the channel bed of a straight pool with different entry and exit slopes, Nosrati et al. [10] studied the turbulent flow field in pools with vegetation cover. It is found that the suitable method for estimating the shear velocity for the flow in the presence of submerged vegetation in the channel bed is the TKE method. The shape and the location of the maximum value of the Reynolds stress distribution depend on the slopes of the entrance and the exit section of the pool. In addition, the layout pattern of vegetation elements also affects the shape and the location of the maximum value of the Reynolds stress distribution. The maximum Reynolds stress and TKE occur at a larger distance above the bed surface, depending on the area density of vegetation. The irregular distributions of Reynolds stress and TKE result from the secondary circulations associated with the irregular distribution pattern of vegetation elements. The momentum between flow, bedform and vegetation elements is mostly transferred via sweep and ejection events. Toward the water surface, the outward event becomes the dominant event. In the presence of an irregular distribution of submerged vegetation elements in the pool bed, ejections become dominant, and then the outward event becomes stronger as the area density of vegetation decreases.

Shahiri Tabarestani et al. [11] carried out field measurements at seven cross-sections along two reaches of the Babolroad River and calculated the bed load transport rate. Suspended sediment discharge was calculated through applying sediment rating curves. The total sediment load was determined based on field measurements along these two reaches of the Babolroad River. On the other hand, they determined the erosion intensity and sedimentation rate from a watershed by means of empirical models, including the modified Pacific Southwest Inter-Agency Committee (MPSIAC), the erosion potential method (EPM) and Fournier. The results using the EPM and MPSIAC models were compared to those of the field measurements and indicated that both models provided good accuracy, with differences of 22.42% and 20.5% from the field results, respectively. Results showed that the Fournier method is not an efficient method since it is unable to consider the erosion potential.

3. Conclusions and Future Works

The application of the results of this Special Issue may help researchers to better understand some important factors in fluvial hydraulics in the presence of vegetation in channel beds, including the resistance to flow and flow velocity for sediment transport. This may help river engineers to apply results of these research works in engineering practice. A variety of experimental and numerical studies in eco-hydraulics must be carried out in order to improve our knowledge about the interaction of flow and vegetation under different conditions. For example, in the presence of vegetation in the channel bed, the effects of a moving bedform on the turbulent flow structures should be investigated through considering the different geometry of a channel under both open channel flow and ice-covered flow conditions. The impact of different types and densities of vegetation on the flow structure in hydraulic models needs to be clarified.

Author Contributions: Methodology H.A. and J.S.; Investigation, H.A. and J.S.; Writing—original draft, H.A. and J.S.; Writing—review & editing, H.A. and J.S. All authors have read and agreed to the published version of the manuscript.

Acknowledgments: We would like to thank all authors who contributed to this Special Issue. We would sincerely appreciate all reviewers for their fruitful and constructive comments, which greatly enrich the quality of the published papers.

Conflicts of Interest: The authors declare no conflict of interest.

References

1. Tabesh Mofrad, M.R.; Afzalimehr, H.; Parvizi, P.; Ahmad, S. Comparison of Velocity and Reynolds Stress Distributions in a Straight Rectangular Channel with Submerged and Emergent Vegetation. *Water* **2023**, *15*, 2435. [CrossRef]
2. Tabesh Mofrad, M.R.; Parvizi, P.; Afzalimehr, H.; Sui, J. Turbulence Kinetic Energy and High-Order Moments of Velocity Fluctuations of Flows in the Presence of Submerged Vegetation in Pools. *Water* **2023**, *15*, 2170. [CrossRef]
3. Qiu, C.; Huang, J.; Liu, S.; Pan, W. Double Parameters Generalization of Water-Blocking Effect of Submerged Vegetation. *Water* **2023**, *15*, 764. [CrossRef]
4. Parvizi, P.; Afzalimehr, H.; Sui, J.; Raeisifar, H.R.; Eftekhari, A.R. Characteristics of Shallow Flows in a Vegetated Pool—An Experimental Study. *Water* **2023**, *15*, 205. [CrossRef]
5. Barahimi, M.; Sui, J. Effects of Submerged Vegetation Arrangement Patterns and Density on Flow Structure. *Water* **2023**, *15*, 176. [CrossRef]
6. Shi, Z.; Jin, S. Numerical Investigation of Hydrodynamics in a U-Shaped Open Channel Confluence Flow with Partially Emergent Rigid Vegetation. *Water* **2022**, *14*, 4027. [CrossRef]
7. Fortes, A.A.; Hashimoto, M.; Udo, K.; Ichikawa, K.; Sato, S. Dynamic Roughness Modeling of Seasonal Vegetation Effect: Case Study of the Nanakita River. *Water* **2022**, *14*, 3649. [CrossRef]
8. Zhang, J.; Mi, Z.; Wang, W.; Li, Z.; Wang, H.; Wang, Q.; Zhang, X.; Du, X. An Analytical Solution to Predict the Distribution of Streamwise Flow Velocity in an Ecological River with Submerged Vegetation. *Water* **2022**, *14*, 3562. [CrossRef]
9. Yu, S.; Dai, H.; Zhai, Y.; Liu, M.; Huai, W. A Comparative Study on 2D CFD Simulation of Flow Structure in an Open Channel with an Emerged Vegetation Patch Based on Different RANS Turbulence Models. *Water* **2022**, *14*, 2873. [CrossRef]
10. Nosrati, K.; Afzalimehr, H.; Sui, J. Interaction of Irregular Distribution of Submerged Rigid Vegetation and Flow within a Straight Pool. *Water* **2022**, *14*, 2036. [CrossRef]
11. Shahiri Tabarestani, E.; Afzalimehr, H.; Sui, J. Assessment of Annual Erosion and Sediment Yield Using Empirical Methods and Validating with Field Measurements—A Case Study. *Water* **2022**, *14*, 1602. [CrossRef]

water

MDPI

Article

Comparison of Velocity and Reynolds Stress Distributions in a Straight Rectangular Channel with Submerged and Emergent Vegetation

Mohammad Reza Tabesh Mofrad [1], Hossein Afzalimehr [1,*], Parsa Parvizi [2] and Sajjad Ahmad [3,*]

[1] Civil Engineering School, Iran University of Science and Technology, Tehran 16846-13114, Iran; mohamadreza_tabesh@yahoo.com

[2] Water Engineering Department, Isfahan University of Technology, Isfahan 84156-83111, Iran; p.parvizi@alumni.iut.ac.ir

[3] Department of Civil and Environmental Engineering and Construction, University of Nevada, Las Vegas, NV 89154, USA

* Correspondence: hafzali@iust.ac.ir (H.A.); sajjad.ahmad@unlv.edu (S.A.)

Abstract: Vegetation in rivers and streams plays an important role in preventing erosion and improving bank stability. Comparison between emergent vegetation (bank vegetation) and submerged vegetation, in terms of velocity and Reynolds stress distributions, for the same aspect ratio and flow discharge, has received limited attention in the literature. This study investigates the velocity and Reynolds stress, as well as the log law for submerged and emergent vegetation in a laboratory flume and compares the results for a different set up with different sediment size and aspect ratio but the same discharge. The results indicate that the influence of submerged vegetation on the secondary currents generation is less than emergent vegetation. In addition, the log law application is valid for both submerged and emergent vegetation cases, however, it is valid up to $y/h = 0.75$ for emergent vegetation (in which the vegetation cover in banks is partly out of the water) but up to $y/h = 0.25$ for vegetation bank. For both submerged and emergent vegetation, Reynolds stress distribution presents a convex form but with a different turning point. Comparison of the results with those in an artificial pool over submerged vegetation and low aspect ratio (<5) keeps almost the same form for velocity and Reynolds stress distributions but decreases the turning point in Reynolds stress distribution. For the submerged vegetation cover, the location of zero shear stress superposes that of maximum velocity, but for the emergent vegetation approaching the bank vegetation and shifting the maximum velocity towards the bed, the location of zero shear stress approaches the bed.

Keywords: bank vegetation; bed vegetation; aspect ratio; Reynolds stress; inflection point

Citation: Mofrad, M.R.T.; Afzalimehr, H.; Parvizi, P.; Ahmad, S. Comparison of Velocity and Reynolds Stress Distributions in a Straight Rectangular Channel with Submerged and Emergent Vegetation. *Water* **2023**, *15*, 2435. https://doi.org/10.3390/w15132435

Academic Editor: Georg Umgiesser

Received: 29 April 2023
Revised: 21 June 2023
Accepted: 27 June 2023
Published: 1 July 2023

1. Introduction

Environmental aquatic currents are rarely free of vegetative effects. Moreover, vegetation over the bed or banks exercises a dominant influence on the fluvial hydraulic studies, including drag coefficient, stable channel design, river plan and bed-forms characteristics [1–7]. Despite considerable investigations on the influence of vegetation in hydraulic projects, there are not many comparative studies for submerged and emergent vegetation in literature. The effect of vegetation can be better considered if one knows that about $1 billion was spent annually during 1990–2003 in USA to manage degraded streams [8,9]. Knowledge of the interaction of flow and vegetation enhances our understanding of sediment transport and resistance to flow because it influences the velocity and turbulent flow characteristics, including bed shear stress estimation. Stream flows over a vegetated bed channel can be observed in river floodplains during floods, or during drought when weeds grow on the bare bed [6,10,11]. Therefore, it is important to understand the impact of vegetation on the velocity and Reynolds stress distributions, which are the key parameters in almost all fluvial hydraulic studies. In fact, floodplain vegetation plays an important role in preventing erosion and improving bank stability. The influence of vegetation depends on many complex interacting factors, including distance from the bank vegetation to the measurement point and spacing of the vegetation, as well as the physical properties including the space of vegetation [12,13]. A survey of the literature suggests that many studies have focused on different aspects of velocity turbulent characteristics of vegetated channels [3]. There are two ways to investigate the effect of vegetation on flow structure: one is an application of rigid and cylindrical roughness and the other is natural vegetation application to simulate at the laboratory scale. Huai et al. investigated a vegetated channel using submerged rigid vegetation, while Stephan and Gutknecht used flexible vegetation in their investigation in a laboratory [14,15].

Kummu performed their laboratory experiments to investigate the relation between vegetation and flow resistance [16]. They found that the vertical velocity profile is logarithmically distributed between 0.1–0.6 water depths over vegetation. This means that some points near the top of the vegetation do not follow the log law. Stephan and Gutknecht presented a modified logarithmic velocity profile for flow over aquatic vegetation. They did not observe a decrease in turbulent intensity with decreasing relative submergence (h/y_p) where h is water depth and y_p is the deflected plant height in vegetated flow [15]. However, Shucksmith et al. carried out their experiments in various submerged and emergent natural vegetation. They found that vegetation causes uniform velocity distribution in vertical direction, and the velocity shear in submerged vegetation is more than an emergent one due to faster flow conditions over submerged vegetation [17]. Kumar and Sharma showed that the vertical velocity reduces in regions of free stream, but at the central axis of channel cross-section and Reynolds shear stress and turbulent intensity are weaker in magnitude at vegetation region. Accordingly, vegetation decreases the flow velocity, Reynolds shear stress and turbulence intensities, showing that vegetation plays an effective role to reduce the resistance in the flow [18]. Much less is understood on the comparison between vegetation in banks and vegetation over the bed for the same aspect ratio (the ratio of the width to flow depth) and flow discharge. Barahimi and Sui [3] found that over submerged vegetation, a high velocity gradient is observed near the bed up to $z/h = 0.1$, and a decreasing trend towards the water surface. In addition, the Reynolds stress distribution is influenced by the aspect ratio (W/h), showing an irregular distribution in the vertical direction [3].

Afzalimehr and Dey observed that the maximum velocity is located at $y/h = 0.2$, near the vegetated bank and gravel-bed stream at $y/h = 0.56$ in center of flume under uniform flow and aspect ratio of ($W/h = 3$) [19]. The position of maximum velocity in the presence of vegetation cover on the walls was located deeper from the water surface, and the dip distance (the location of the maximum flow velocity from the water surface) was observed to be up to 35%. Further, they found that the Reynolds stress distribution was nonlinear, even if the flow depth was constant along the flume with the maximum value near the bed [19]. Setayesh and Afzalimehr showed that Reynolds stress distribution displays a concave form for emergent vegetation and the best method of shear velocity estimation of application of Reynolds stress [20]. For constant flow depth, and equal water surface and bed slopes for vegetation banks over gravel bed, Afzalimehr et al. found that for the aspect ratio of $W/h = 4$, the maximum velocity occurs under water at a water depth of $y/h = 0.46$, and its difference with flow velocity near water surface is 50% [21]. Furthermore, in natural channels and $W/h = 7$, the maximum velocity in the central axis is located at the water surface, and for the near bank it is located on average at a water depth of $y/h = 0.50$. Thus, the vegetation cover on channel/flume banks may considerably affect the location of maximum velocity even for $W/h > 5$, due to development of secondary currents. In addition, Afzalimehr et al. found that the maximum Reynolds stress was near the vegetated banks and close to the channel bed at $y/h = 0.1$ [21].

The influence of aspect ratio was investigated in two parts: less than 5 and greater than 5, showing $W/h = 5$ as critical value, where the flow reveals different patterns. When $W/h < 5$, the flow is 3D and corner flows are developed because the banks produce anisotropy of turbulence and the velocity dip (the occurrence of the maximum velocity under the water surface). However, when $W/h > 5$, the bank effect on the velocity decreases and the maximum velocity occurs at the water surface. Some researchers are pioneers to consider a relation between the aspect ratio and generation of secondary currents [22,23]. They found that nonlinear distribution of shear stress and occurrence of the maximum velocity under the water surface is caused by secondary currents. Accordingly, any change in aspect ratio when it is less than 5 indicates similar patterns. Also, similar patterns are observed when the aspect ratio is larger than 5, indicating that it makes no difference to the velocity and Reynolds stress distributions if W/h is 2 or 4 [24].

In addition, the vegetation covering the bed has small influence on the generation of secondary currents. Huai et al. conducted their experiments over stiff and artificial vegetation cover under uniform flow and found that small aspect ratio did not affect the generation of secondary currents [14]. Carrollo et al. observed an inflection point in velocity profile, which is superposed on the maximum turbulence intensity [25]. Jarvela used wheat stems with a density of 12,000 per square meter and found the occurrence of maximum Reynolds stress (RS) slightly above the top of flexible vegetation [12]. Other investigators have noted that the RS distribution over stiff and flexible vegetation covers is nonlinear in a way that, at the top of vegetation, RS is maximum and then decreases towards the water surface [14,26,27]. Yang et al. found that the Reynolds shear stress in the upper layer of a uniform narrow channel is negative and the zero shear stress would therefore be observed below the free surface [28].

Understanding the difference between submerged and emergent vegetation effects on the velocity and Reynolds stress distributions can improve the estimation of roughness and drag coefficients. The objective of this study is to investigate this comparison for the velocity and the Reynolds stress under similar aspect ratio and discharge, as well as to investigate the limit of the log law application for submerged and bank vegetation (emergent case) in a laboratory flume. To better present the results of this study, the authors compare their results with those of Parvizi et al. [2] in an artificial pool with submerged vegetated cover in laboratory flume.

2. Materials and Methods

Most experimental studies of natural channels and rivers are conducted in straight flumes in a laboratory, so that the flow patterns are studied under controlled conditions (e.g., Nepf [29] and Wang et al. [30]). Moreover, in rivers, a straight reach is considered to study the flow structures ignoring small variations of width (e.g., Julien) [29–31].

In this study, the experiments were conducted in a 20 m long, 0.6 m wide and 0.6 m deep glass-walled rectangular flume, at hydraulic laboratory in University of Science and Technology, Tehran, Iran. A movable weir located at the downstream end of the flume was used to obtain quasi-uniform regime conditions along the vegetated zone. The quasi-uniform flow is considered over rough bed such as gravel-bed streams or submerged vegetation where the depth and velocity are constant alone the stream. The slope of the flume was set to 0.005. A limnimeter was used to record the depth of water along the flume and to determine the water surface profile. Experiments were conducted in four runs. Run 1 and run 2 were carried out over submerged vegetation (wheat stems); run 3 and run 4 were investigated in flow over gravel bed and vegetation banks. The bank and bed vegetation have different characters. The bank vegetation is relatively more moisture-demanding. On the other hand, on a typical bed in rivers, light-demanding and shade-tolerant species (i.e., grasses and vegetation cover such as "Arundo donax") live together.

For runs 1 and 2, sprouts of wheat were planted from 11.5 m to 14.5 m from the entrance of the flume, and the walls were glassy (Figure 1a). The planted wheat over the bed was long enough to have high flexibility and wavy motion. The rest of the flume bed was covered with gravel, which had a median grain size of d_{50} = 21 mm with a standard deviation of less than 1.3. Gravel size can only change the results when relative submergence (h/d_{50}) is less than 3. In these conditions, the log law and the boundary layer theory applications are invalid. This statement was confirmed by other studies (e.g., Afzalimehr, Simons, Graf and Altinakar, and Song and Chiew [23,32–34]. These studies show that for all changes in gravel size for $(h/d_{50}) > 3$, the Reynolds stress and turbulence intensity distributions remain stable, showing a convex distribution decelerating flow and concave distribution for accelerating flow in open channels.

(a)

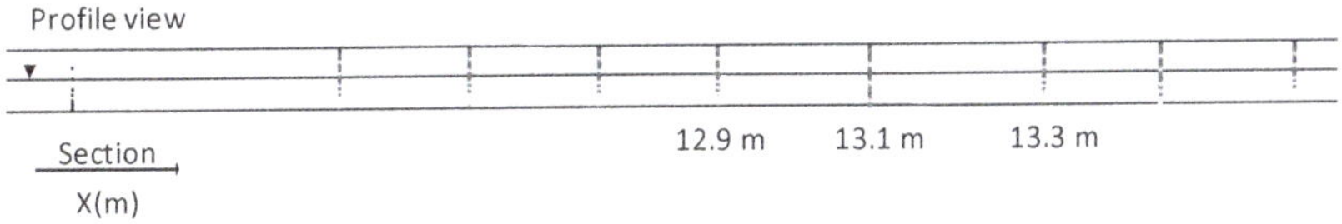

(b)

Figure 1. (**a**). Experimental set up showing: (1 and 2) Wheat sprout canopy; (3 and 4) Wheat stems on bank. (**b**). Measured velocity points of experimental set up, at the distance of 12.9, 13.1, 13.3 m from the entrance of the channel.

The height of submerged wheat was not changed during the tests, and density of wheat stems was kept at 45,604 per square meter in both runs. For runs 3 and 4, "Arundo donax", the selected plant, which was collected from surroundings of a gravel-bed river in Iran, was fixed to the banks of the flume and the bed was covered by gravel (Figure 1). Table 1 shows the hydraulic characteristics of laboratory experiments set up and a summary of experimental data for this study. Flow discharges were selected between 0.037–0.055 m^3/s and water depth was fixed between 11 and 15 cm.

Table 1. Hydraulic conditions of laboratory experiments.

Exp.	Q (m^3/s)	h (cm)	W (cm)	U_{avg} (m/s)	W/h	Covered Area
Run 1	0.037	11	60	0.56	5.45	Bed
Run 2	0.055	15	60	0.61	4	Bed
Run 3	0.037	11	60	0.56	5.45	Bank
Run 4	0.055	15	60	0.61	4	Bank

Notes: Q = flow discharge, h = the flow depth, W = the flume width, U_{avg} = the flow velocity average for each run, covered area by vegetation in each run.

The aspect ratio $W/h \leq 5$ has been considered as a criterion for narrow open channels in which the velocity dip (the maximum velocity occurs under water surface) occurs and the banks produce anisotropy of turbulence. The dip phenomenon has been confirmed by using advanced tools such as the High-power laser Doppler anemometer (LDA) system by Nezu, Nakagawa, and the Acoustic Doppler Velocity Profiler ADVP by Graf and Altimakar for $W/h \leq 5$ or slightly larger. They used $W/h \leq 5$ as a range where the flows are considered 3D [22,23]. Acoustic Doppler Velocimeter (ADV) is a Nortek Vectrino that measures the instantaneous three-dimensional velocity components. The change of flow discharge has no effect on the shape of distribution but leads to higher values of velocity (Afzalimehr and Rennie [35,36]).

Moreover, water depth influences the velocity and Reynolds stress distributions, when the relative submerge changes from $(h/d_{50}) > 3$ to $(h/d_{50}) < 3$, where h is the flow depth and d_{50} the median gravel size (Afzalimehr) [32]. Since the experiments in this study were conducted for $(h/d_{50}) > 5$, the change of flow depth has no effect on the results. In addition, for runs 1 and 2, y_p is the plant height.

A down looking Acoustic Doppler Velocimeter (ADV), 10MHz Nortek Vectrino, was used to measure the instantaneous three-dimensional velocity components. The ADV used in this study has the precision of ± 0.1 mm s^{-1} and a sampling volume with a height of 5.5 mm. After measurements, the data were filtered with a WinADV program. WinADV helps to select suitable series data for velocity to draw reliable results. Accordingly, the coefficiency of determination (R^2) for most data collection was over 95 percent, and signal to noise ratio (SNR) was over 15. Data with SNR lower than or equal to 5 and an average correlation of less than 70% were removed from data set. Collecting data with sampling frequency f = 200 Hz and sampling time 120 sec leads to more than 24,000 observations at each point of velocity and Reynolds stress profile. This shows a great repeatability of data at each point. Accordingly, even if 10% of data were not suitable at each point, the analysis was done with more than 21,000 data points, resulting in high certainty in results.

To investigate the occurrence of fully developed flow along the vegetation patch, the measurements of velocity profiles were made in three cross sections: 12.9, 13.1 and 13.30 m (Figure 2), from the beginning of the channel at three different distances of 5, 12 and 30 cm from the bank (D). About 19 to 23 point velocity values were measured in each velocity profile, and approximately 7 points were measured in the region near the bed (y/h < 0.2); where y is distance of each measuring point from the top of vegetation cover. However, for runs, 1 and 2, ADV was not able to measure velocity inside the vegetation cover and the top of gravel for runs 3 and 4. In each case, two runs with selected discharges and flow depths were carried out.

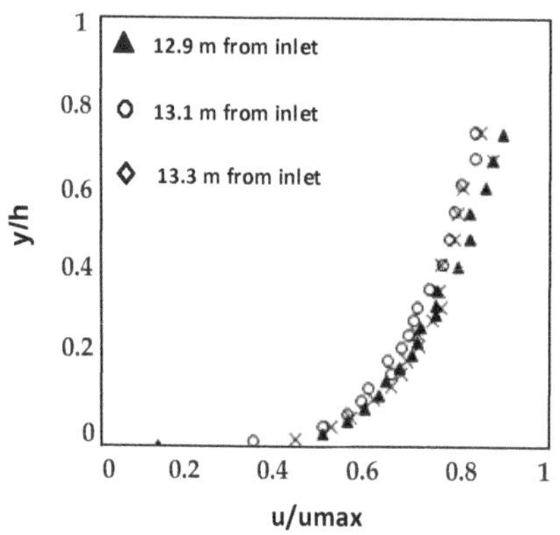

Figure 2. Fully developed flow presentation at the distance of 12.9 m from the flume entrance.

The vertical dotted line in Figure 1b, before 12.9 m, shows the sections where velocity profiles were measured to check the fully developed flow condition. This condition reached after 12.9 m from the flume entrance. Moreover, the vertical dotted lines shown in Figure 1b, after 13.3 m were considered to be sure that the velocity distributions remain similar toward the end of the flume.

The beginning of the channel was at three different distances of 5, 12 and 30 cm from the bank (D) (Figure 1b). About 19 to 23 point velocity values were measured in each velocity profile, and approximately 7 points were measured in the region near the bed (y/h < 0.2); where y is distance of each measuring point from the top of vegetation cover.

However, for runs 1 and 2, ADV was not able to measure velocity inside the vegetation cover and the top of gravel for runs 3 and 4. In each case, two runs with selected discharges and flow depths were carried out. In addition, our results were compared with those of Parvizi et al. [2] who studied the same flow discharge of 0.037 m^3/s over an artificially vegetated and non-vegetated pool, with 5° slope in the entrance and exit section of the pool. Their experiments were conducted in a laboratory flume with W = 0.4 m, h = 0.20 m, an aspect ratio of 2, and a d_{50} of 10.4 mm. The vegetation cover over the examined pool was almost the same as the present grass canopy. The comparison of results of Parvizi et al. [2] with the present study is presented in the discussion section.

The velocity distribution by the log law is presented for flow over vegetation cover as:

$$\frac{u}{u_*} = \frac{1}{k} \ln\left(\frac{y + k_s}{k_s} \right) + C \quad (1)$$

where u is the mean point velocity, u_* is the shear velocity, k is the von Karman constant (equal to 0.4), k_s is the roughness scale, and C is a constant. This study applies the following relation developed by Stephan and Gutknecht [15].

$$\frac{u}{u_*} = \frac{1}{k} \ln\left(\frac{y - y_p}{y_p} \right) + C \quad (2)$$

where y is a distance from the bed, y_p is the thickness of vegetation cover, which was used on the bed in the laboratory. Since ADV cannot collect data inside the dense vegetation, all measurements of velocity are conducted from the top of vegetation. The constant of 8.5 is the universal value for the flow over the rough boundaries. The following relation for the case of bank vegetation and gravel bed:

$$\frac{u}{u_*} = \frac{1}{k} \ln\left(\frac{y + 0.1\,d_{50}}{d_{50}} \right) + C \quad (3)$$

where y is the distance from the top of gravel. Due to a micro variation in gravel bed surface, a reference level is defined as $0.1d_{50}$ under tops of gravel particles where y = 0. Therefore, addition of $0.1d_{50}$ in numerator of (3) causes a better fitness of the log law to the velocity points in the logarithmic region. C is a constant that depends on the flow conditions and roughness size. The value of C has been reported 8.5 by Graf and Altinakar [23], however, there is no universal value for C.

The value of $0.1d_{50}$ for this study is different from the one used in other studies. For example, Graf and Altinakar use $0.2d_{50}$ in order to show a better presentation of data fitness by the log law. The reason for not using any modification factor in Equation (2) such as $0.1y_p$ is the insignificant change in the presentation of the log- law fitness by this modifying factor.

3. Results

3.1. Comparison of Velocity Distributions for Submerged and Emergent Vegetation Covers

Although the gravel bed from the entrance to x = 11.5 m prompts the fully developed flow in the flume (Figure 2), the variation of bed roughness from gravel to vegetation cover (wheat stems) causes the development of a new boundary layer. Figure 2 reveals that after x = 12.9 m from entrance of the flume in this experimental set up, the new boundary layer reaches a fully developed condition where flow velocity profiles are similar at different distances from the leading end of the flume (where water enters to the flume). The horizontal axis shows the mean point velocity (u) normalized by the maximum velocity (u_{max}) in each profile, and y/h shows the flow depth at each point from the bed (y) normalized by total flow depth (h). Therefore, all results for this experimental set up are presented after x = 12.9 m from the channel entrance. Some deviations from unique curves are observed in Figure 2, which may be ascribed to some uncertainty associated with the definition of reference level where u = 0. Since measurements inside the vegetation cover have considerable limitations with ADV, the flow velocity was measured from the top of submerged vegetation (Figure 1a). Figure 2 shows that for the range of $0 < y/h \leq 0.2$, the velocity gradient is different from $y/h > 0.2$ (y is the distance of each point velocity from the top of gravel or vegetation). This is due to a strong shear layer effect on the velocity distribution near the top of vegetation (run 2).

Figure 3 confirms the validity of the log law from the top of vegetation up to $y/h = 0.23$. The mentioned distance of $y/h = 0.23$ shows the deviation of the log law from the measured data in $y/h > 0.23$. This means the data above the bed up to $y/h = 0.23$ follows the log law distribution and from $y/h > 0.23$ toward the water surface this law is invalid. The reason for deviation of data from this law is inapplicability of mixing length theory to quasi-uniform flow, and resistance to flow due to vegetation and gravel. In fact, deviation of measured velocity points is due to the effect of pressure gradient changes, the boundary layer thickness, and the maximum velocity In Figure 3a,b the change of gradient is clearly observed, however, in Figure 3c, the deviation point is not clearly observed. The general rule for choice of deviation point of the inner and outer regions of the boundary layer is to use the coefficient of determination (R^2) between u/u_{max} and $Ln(y/d_{50})$ or $Ln(y/y_p)$. In this study, the value of $R^2 \geq 0.98$ in the logarithmic zone was considered as the deviation point where the data are deviated from the log law zone. The value of R^2 decreases after the selected deviation point towards the water surface.

The maximum velocity occurs under the water surface for 3D flow in which bank effect causes a strong lateral velocity component (v). This component is directed near the water surface from the bank to the flume center, and a down flow ($w < 0$) occurs from the water surface. This difference of velocity components v and w generates the secondary currents. The larger is this difference, the stronger will be deviation of the log law in velocity data far from the bed (Figure 3a). The reason for this deviation is invalidity of two assumptions in the log law application: (1) constant shear stress throughout the fluid, and (2) mixing length approximation $l = ky$, in which k is von Karman constant and y is the distance from the bed. In addition, the choice of $0.1d_{50}$ for the gravel bed has some uncertainties. In other words, if one uses a plot with constant particle diameter, there is less uncertainty in application of the log law compared to gravel bed streams.

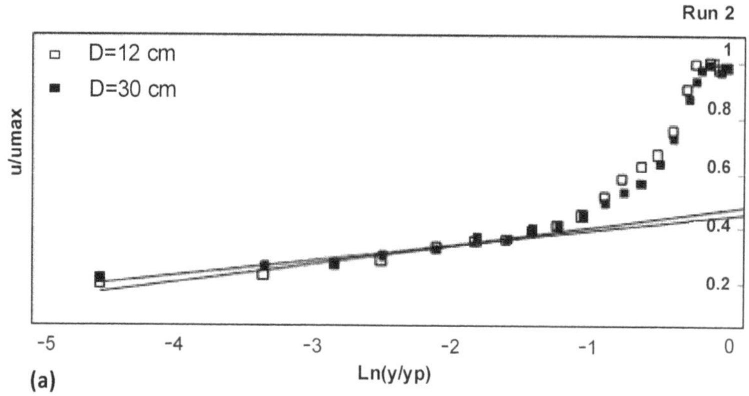

Figure 3. *Cont.*

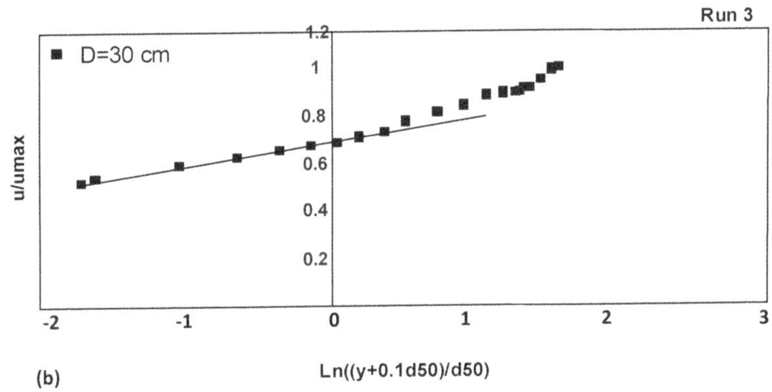

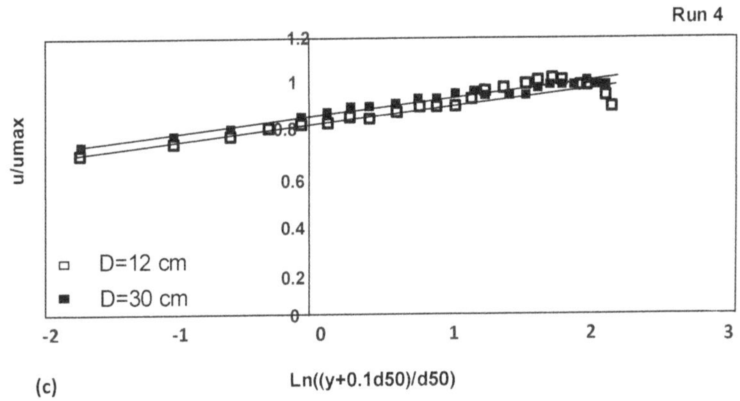

Figure 3. (**a**) The log-law validation for different conditions for submerged vegetation (run 2). (**b**,**c**) The log-law validation for different conditions: (**b**) for the gravel bed without vegetation in bed and banks in the central axis (run 3); (**c**) for the emergent vegetation in banks and gravel bed (run 4).

There are two differences between the case of vegetation on banks and gravel over the bed with the case of vegetation over the bed and bare banks (Figure 4): (1) the former case has lower velocity gradient and the velocity values over vegetation are less than those over gravel with bank vegetation (Figure 4). Since the flow discharge is almost the same for both cases, the reason for decreasing velocity values may be explained by passing some part of flow through the vegetation cover, which causes the diminution of flow velocity. (2) The bank vegetation accelerates the secondary currents, and for a small aspect ratio ($W/h < 5$) the maximum velocity occurs far from the water surface. The dip phenomenon (the location of the maximum velocity from the water surface) is strengthened when approaching the bank vegetation and the difference between the last measuring point velocity with the maximum velocity reaches 50%. It is noted that such a velocity distribution is not observed for the flow in submerged vegetation (Figure 4a), where the velocity profiles show no significant decreasing trend with approaching to the water surface.

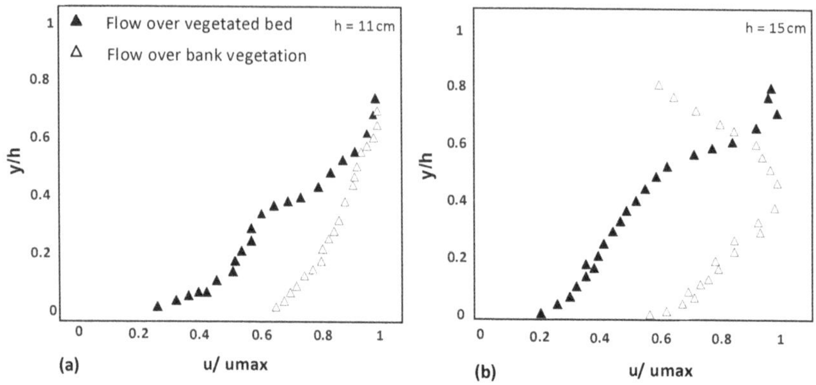

Figure 4. Velocity distributions: (**a**) The flow depth of 11 cm; (**b**) The flow depth of 15 cm.

Considering the equations 2 and 3 for the case W/h = 5.4 (runs 2 and 3), the log-law validity zone is up to y/h = 0.2 for all the velocity profiles at different distances from the vegetation bank. This relative depth (y/h = 0.2) is considered as a classic thickness for validity of log law over gravel and sand bed. However, for run 4, where W/h = 4, the log law can extend up to 0.75 y/h, and for cases near the bank it reaches y/h = 0.25 (Figure 3c).

Figure 4 shows that when aspect ratio is less than 5 the velocity distribution displays different patterns depending on secondary currents power. In Figure 4a,b for submerged vegetation, an irregular velocity profile (flow over vegetated bed) is observed, but for emergent vegetation (bank vegetation) a convex form is observed in which the turning point depends on aspect ratio and pressure gradient. The irregular velocity profile in which a significant change in velocity is observed towards the water surface is due to turbulent structures developed by flow wake around the submerged vegetation and the form-induced stress by vegetation.

3.2. *Comparison of Reynolds Shear Stress for Submerged and Emergent Vegetation*

Figure 5a shows a convex distribution of RS for flow over gravel with vegetation banks (emergent vegetation). The convex form is due to the interaction of small aspect ratio and decelerating flow near the rough bed. Near the rough bed for decelerating flow in which the velocity decreases and pressure increases, a positive pressure gradient develops in the flow direction. Based on Navier Stocks equation, $\partial p/\partial x = \partial \tau/\partial y$ when pressure gradient increases, the vertical Reynolds stress increases as well. Therefore, an increasing trend is observed near the bed for RS distribution. However, RS should reach zero at the water surface, this leads to convex distribution for RS. This convex form occurs for any roughness, vegetation cover and gravel under decelerating flow, however, in Figure 5a the convex form may be observed near the water surface where ADV needs an additional probe to reveal it.

The change in the location of zero RS is due to variations in longitudinal pressure gradient in different distances from the vegetated bank. Figure 5a clearly shows the Reynolds stress distribution changes at different distances from the bank; this is due to a change in longitudinal pressure gradient. Reynolds equation $\partial \tau/\partial y = \partial p/\partial x$ states [36] that a vertical change in Reynolds stress distribution is due to a longitudinal change in pressure gradient. This situation occurs for emergent vegetation due to the considerable effect of bank roughness on the Reynolds stress distribution.

Figure 5b shows the flow over a submerged vegetation. It shows that the RS distribution is nonlinear, but the maximum RS occurs above vegetation at y/h = 0.5. It is noted that above the bed vegetation, there is a strong horizontal shear layer, which diffuses vertically inside of the vegetation cover.

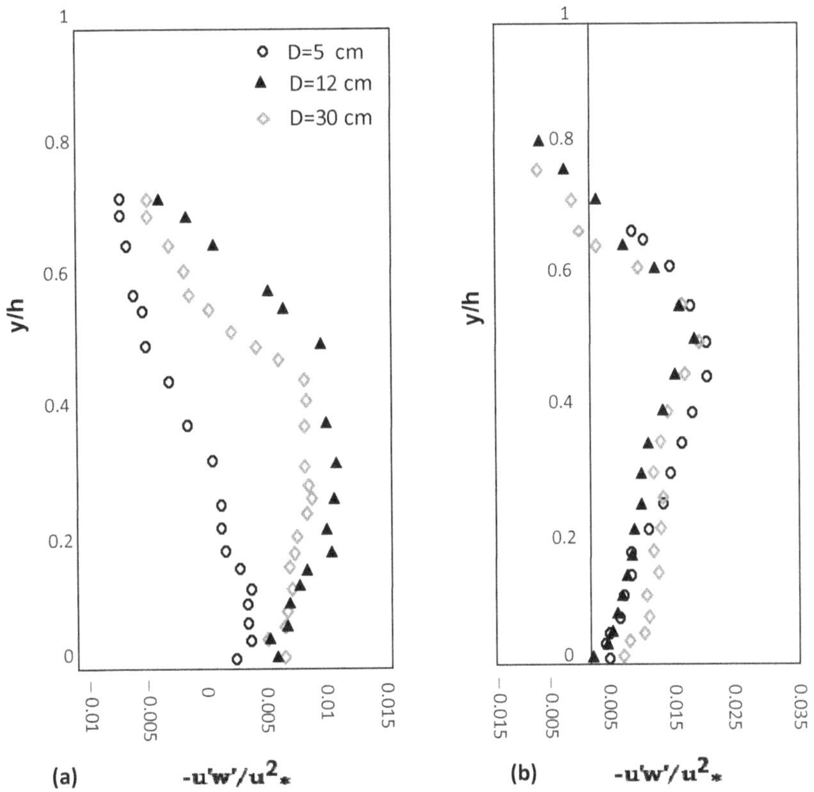

Figure 5. Reynolds stress distributions: (**a**) for emergent vegetation with gravel bed; (**b**) over submerged vegetation. (y) is normalized by the total flow depth (h).

The vegetation over the bed generates high drag force due to additional turbulence, which shifts the maximum turbulent stress to a level above the top of the vegetation cover. In this research, due to high density of vegetation (45,604 wheat stems per square meter), the vertical momentum between flow and vegetation cover decreases and consequently the maximum of RS shifts to the upper flow depth.

The negative values for $y/h > 0.7$ indicate the effect of secondary flows and the occurrence of the maximum flow velocity under the water surface. The vegetation cover on the bed generates a quasi-uniform flow in which there is a non-zero pressure gradient. This non-uniform pressure gradient should reach equilibrium by the vertical shear stress. Since the vegetation bed decelerates the flow, an unfavorable pressure gradient is developed, with a positive sign demanding a positive pressure gradient. Therefore, Reynolds stress increase up to $y/h = 0.5$ and then decreases towards the water surface [37]. As Figure 5b shows, the location of Reynolds stress distribution across the flume does not influence the convex distribution, and for all distances from the bank the maximum RS occurs at $y/h = 0.5$ for submerged vegetation and bare banks.

Comparison of submerged and emergent vegetation covers reveals that for both cases, the Reynolds stress distribution is nonlinear, although the flow depth is constant. This shows that for both submerged and emergent vegetation covers, the change of aspect ratio (W/h) influences the location of the maximum Reynolds stress (Figure 5). In Figure 5a, at a distance of 5 cm from the bank, RS distribution is non-linear, showing the maximum RS at $y/h = 0.15$. There is a difference between RS at 12 cm from the bank in Figure 5a,b. This can be clearly observed by comparing the slope of RS for 12 cm from the bank in y/h < 0.5 for both figures.

In Figure 5a, with a constant aspect ratio, by approaching the bank vegetation the maximum Reynolds stress location is close to the gravel bed. Moving away from the bank vegetation towards the central axis of the flume, the location of maximum Reynolds stress for both submerged and emergent vegetation becomes almost the same.

In addition, although Figure 5b is presented for W/h = 4, it is observed that the change in the location of Maximum Reynolds stress is less affected by aspect ratio due to different roughness of bank for submerged (glass) vs. emergent (vegetation cover). It is observed that the role of secondary currents for the case where the bank is covered with vegetation is stronger than that of the bank with glass. Comparison of Figures 5 and 6 reveals a convex form for RS distribution. This comparison shows that for the submerged vegetation and different distances from the bank, the location of zero or maximum RS is almost the same. However, for emergent vegetation, the distance from the bank plays a significant role in the RS distribution in a manner that at the distance of 5 cm from the vegetation bank its value decreases considerably, reaching near the gravel bed.

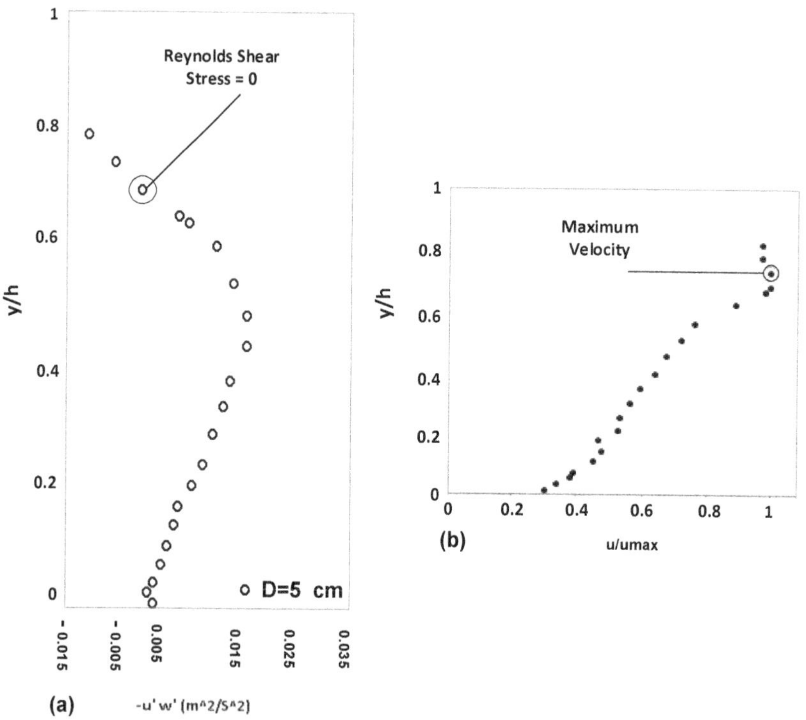

Figure 6. (**a**) Location of zero Reynolds stress over submerged vegetation cover. (**b**) Location of maximum point velocity over submerged vegetation cover.

Since the zero Reynolds shear always corresponds to the zero-velocity gradient or maximum velocity, Figure 6 demonstrates that the location of zero RS in the Reynolds stress profiles is superposed with the location of maximum velocity for the flow over submerged vegetation cover. However, for the flow over gravel with vegetation banks, such a superposition is not considered, but it is possible to observe the change of location of zero Reynolds stress with the maximum velocity location with approaching the vegetated banks.

Comparison of the results of experimental set up to those of Parvizi et al. over an artificial pool [2] shows that the vegetation arrangement (submerged and emergent), and the sediment size and bed geometry influence the flow velocity distribution and Reynolds stress distribution, but not the log law validation and convex form of Reynolds stress distributions (Figures 7 and 8). Interestingly, no irregular velocity profile is observed for submerged vegetation over an artificial pool.

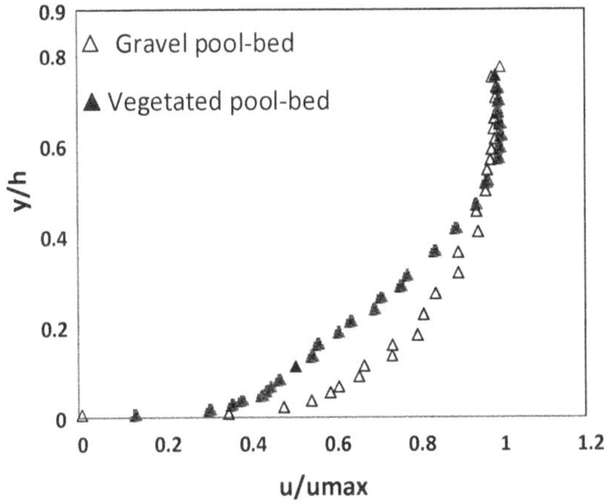

Figure 7. Velocity distributions in pool bed by Parvizi et al. [2].

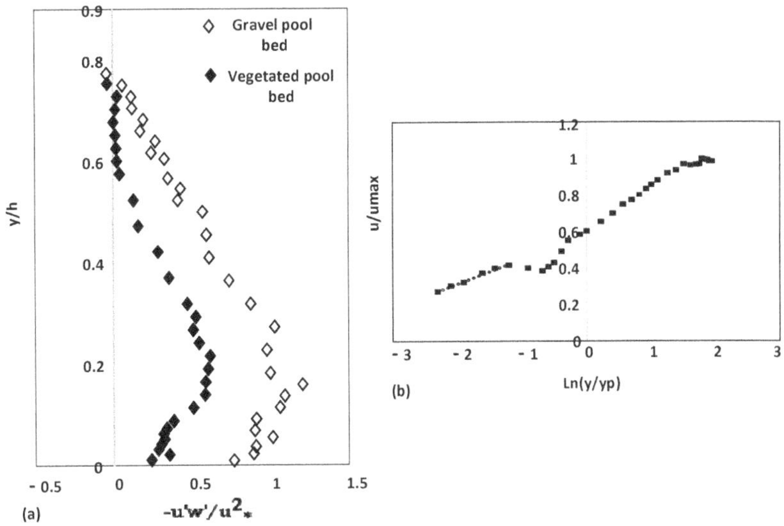

Figure 8. Flow over Pool with submerged vegetation by Parvizi et al. [2]: (a) Distribution of Reynolds shear stress over gravel bed and vegetated canopy; (b) Validation of logarithmic law in vegetated pool.

In the artificial pool examined by Parvizi et al. [2], the velocity, flow depth and shear stress change at different sections along the pool, generating various flow conditions, including decelerating flow in the entrance region, quasi-uniform flow at the middle section and accelerating flow at the exit region of the pool.

In an artificial pool with an aspect ratio of 2 covered by submerged vegetation, the main difference is the decrease of the turning point keeping the convex distribution of Reynolds shear stress. This difference directly affects estimation of shear velocity and drag coefficiency. The negative values of Reynolds stress are rarely observed for this submerged vegetation over bed form, while it is observed for emergent vegetation due to dip phenomenon (the location u_{max} from the water surface). The above considerations show that aspect ratio does not play a significant role on the estimation of Reynolds stress, as well as drag coefficient determination in submerged vegetation compared to the emergent vegetation, which plays a significant role. This will help engineers consider the role of distance from wall vegetation in estimation of hydraulic parameters only for emergent vegetation.

4. Discussion

Figure 3 shows that the log law is valid for both submerged and emergent vegetation but with different thickness near the bed. Deviation from the log law is significant in the outer layer for the submerged vegetation; this is due to vegetation roughness, which is different from gravel roughness. The significant deviation in the log law application for submerged vegetation causes the von Karman constant to reduce $\kappa = 0.16$ as reported by Afzalimehr et al. [38]. This value (κ) is obtained from the average of many different profiles. The small values of von Karman constant have been reported by Yalin [39] as $\kappa = 0.16$ for sand-bed streams and as $\kappa = 0.21$ by Vanoni [40] for suspended sediment. Therefore, submerged vegetation causes the value of von Karman constant decreases making significant difference with the reported value in literature for universal von Karman constant 0.4 in gravel-bed streams. Distance from the bank vegetation does not significantly affect the log law thickness, showing suitable performance of this law across a natural stream with gravel-bed and bank vegetation in emergent cases.

Figure 4 shows that when the aspect ratio decreases ($W/h < 5$), the velocity distribution is influenced more significantly by the emergent vegetation in banks than the submerged vegetation on the bed. However, this is not observed for the aspect ratio ($W/h > 5$). This is partly due to certain limitations of ADV to collect data in small aspect ratio. The reason for observing the dip phenomenon for emergent case with ($W/h < 5$) is a better performance of ADV to reveal the interaction of gravel-bed with the bank vegetation in generation of secondary currents, which forces the location of the maximum velocity reach near the bed. The change of u_{max} location is important in application of the boundary layer theory to estimate drag coefficient because the emergent vegetation decreases the thickness of the boundary layer zone whose data is used to determine shear velocity. In Figure 4 for h = 11 cm where ($W/h > 5$), the turning point is not observed in the velocity distribution, showing lack of measuring by ADV probe for shallow flow near the water surface. Experience with ADV shows that near the water surface, data keep the same trend that they have at last points measured by ADV towards the water surface. Using additional probes to collect data near the water surface region may provide potentially useful information but requires additional resources. However, comparison of data obtained near the bed with that near the water surface confirms a good agreement between bottom shear velocities, calculated using the Reynolds stress (using all data in Reynolds stress profile) and the logarithmic law methods, which uses only the near-bed data.

The density of vegetation covering the experimental set up of this study is almost the same as the pool set up built by Parvizi et al. [2], but with a very thin layer of vegetation in comparison to the present work. The authors did not exclude the influence of the aspect ratio (W/h) considering Figures 5b and 8a. In fact, for Figure 5b the aspect ratio is 4 or 5.4 but for Figure 8a, it is 2. This difference forces the location of maximum Reynolds stress approaching the bed for $W/h = 2$.

The different size of gravel in this study and that used in Parvizi et al. [2] have no significant impact on the shape of RS distribution, as both Figures 5b and 8a show a convex form, and for both cases a polynomial fit can be used to determine bed shear stress. All measurements were conducted over the vegetation patch in the region where flow reached the fully developed condition, and the submerged vegetation front area has no effect in the estimation of hydraulic parameters.

In rivers and natural channels, submerged and emergent vegetation covers are observed in banks frequently. However, no roughness coefficient or sediment transport parameter such as Shields parameter considers the simultaneous effect of submerged and emergent vegetation in their estimations. Many hydraulic models in river engineering are based on the assumptions to simplify the calculation process, which leads to over or under estimation of results, increasing the cost of river restoration projects and decreasing the accuracy of key parameter estimations. Accordingly, in future studies, researchers should apply a mixed estimation of key hydraulic parameters in rivers, which are affected by the presence of vegetation on beds and banks.

The limitation of ADV to detect flow characteristics very near the bed ($h < 4$ mm) calls to use better instruments such as PIV, which are very suitable for collecting data with more details in laboratory studies.

5. Conclusions

Understanding the effects of submerged and emergent vegetation covers on the velocity and Reynolds stress distribution is important for hydraulic designers and engineers who work in fluvial projects. Using two setups in the presence of submerged and emergent vegetation over flat bed and an artificial bedform, the following results are obtained in this study:

1. In emergent vegetation, the velocity distribution is influenced by a small aspect ratio ($W/h < 5$) forcing the maximum velocity to move towards the bed. However, in the submerged vegetation, the velocity distribution is less influenced by the aspect ratio, showing the location of maximum velocity far from the bed.
2. The log law is valid up to $y/h = 0.23$ for the flow over submerged vegetation in different aspect ratios. However, for emergent vegetation, the validation zone depends on the distance from the vegetation bank; at small aspect ratio and the central axis of the flume, this law is valid up to $y/h = 0.75$.
3. Reynolds stress (RS) distributions show a convex form for submerged and emergent vegetation covers at different aspect ratios. However, approaching the bank vegetation (emergent case), the power of secondary currents increases, forcing the maximum RS shifts towards the bed.
4. For the aspect ratio of $W/h = 4$ and flow over submerged vegetation, the location of zero shear stress superposes that of maximum velocity. However, in emergent vegetation, such a superposition is not considered. Further, approaching the bank vegetation and shifting the maximum velocity towards the bed, the location of zero shear stress approaches the bed.
5. The results of this research show that estimation of key parameters of fluvial projects, velocity and Reynolds stress is influenced by the vegetation arrangements, submerged or emergent. Therefore, a better estimation of these parameters influences drag coefficient and sediment determinations, improving bank stability with reasonable channel design, reducing the cost of project and modifying the input data for hydraulic models.

Author Contributions: Methodology, P.P., M.R.T.M. and H.A.; Validation, S.A. and M.R.T.M.; Formal analysis, M.R.T.M.; Investigation, M.R.T.M.; Writing—original draft, M.R.T.M.; Writing—review & editing, P.P., M.R.T.M., H.A. and S.A.; Visualization, M.R.T.M.; Supervision, H.A. and S.A.; Project administration, P.P. and M.R.T.M. All authors have read and agreed to the published version of the manuscript.

Funding: This research received no external funding.

Data Availability Statement: Not applicable.

Conflicts of Interest: The authors declare no conflict of interest.

Abbreviations

d_{50}	Median diameter of sediment particles
C	Constant of the log law
D	Distance from the wall
y_p	Plant height (for run 1 and 2)
h	Water depth above the vegetation (for run 1 and run 2) and water depth over gravel bed (for run 3 and run 4)
W	Flume width
ks	Roughness scale (equivalent sand roughness)
Q	Flow discharge
u	Mean point velocity
u_{ave}	Average velocity at a section;
u_{max}	Maximum velocity
u'	Turbulence intensity in longitudinal direction
w'	Turbulence intensity in vertical direction
u^*	Shear velocity
y	Distance from top of gravel or vegetation cover
Y	Distance from the bed
κ	von Karman constant

References

1. Shi, H.; Huai, W. Flow structure of open channel flow with partly submerged vegetation. In Proceedings of the 12th International Symposium on Ecohydraulics, Tokyo, Japan, 19–24 August 2018.
2. Parvizi, P.; Afzalimehr, H.; Sui, J.; Raeisifar, H.R.; Eftekhari, A.R. Characteristics of Shallow Flows in a Vegetated Pool—An Experimental Study. *Water* **2023**, *15*, 205. [CrossRef]
3. Barahimi, M.; Sui, J. Effects of Submerged Vegetation Arrangement Patterns and Density on Flow Structure. *Water* **2023**, *15*, 176. [CrossRef]
4. Soltani, N.; Afzalimehr, H.; Shahiri Tabarestani, E.; Eftekhari, A.; Khabari, M.; Nazari-Sharabian, M.; Karakouzian, M. Turbulence Characteristics in Mild and Steep Entrance Slopes of Pool-Riffle Sequences. *Water* **2023**, *15*, 720. [CrossRef]
5. Kazem, M.; Afzalimehr, H.; Sui, J. Characteristics of Turbulence in the Downstream Region of a Vegetation Patch. *Water* **2021**, *13*, 3468. [CrossRef]
6. Sohrabi, S.; Afzalimehr, H.; Singh, V.P. Estimation of drag coefficient of emergent and submerged vegetation patches with various densities and arrangements in open channel flow. *ISH J. Hydraul. Eng.* **2022**, *29*, 297–307. [CrossRef]
7. Hopkinson, L.; Wynn, T. Vegetation impacts on near bank flow. *Ecohydrol Ecosyss Land Water Process. Interact. Ecohydrogeomorpho.* **2009**, *2*, 404–418. [CrossRef]
8. Bernhardt, E.S.; Palmer, M.A.; Allan, J.D.; Alexander, G.; Barnas, K.; Brooks, S.; Carr, J.; Clayton, S.; Dahm, C.; Follstad-Shah, J.; et al. Synthesizing U.S. river restoration efforts. *Science* **2005**, *308*, 636–637. [CrossRef] [PubMed]
9. Aberle, J.; Järvelä, J. Flow resistance of emergent rigid and flexible floodplain vegetation. *J. Hydraul. Res.* **2013**, *51*, 33–45. [CrossRef]
10. King, A.T.; Tinoco, R.O.; Cowen, E.A. A k–ε turbulence model based on the scales of vertical shear and stem wakes valid for emergent and submerged vegetated flows. *J. Fluid Mech.* **2012**, *701*, 1–39. [CrossRef]
11. Yagci, O.; Kabdasli, M. The impact of single natural vegetation elements on flow characteristics. *Hydrol. Process.* **2008**, *22*, 4310–4321. [CrossRef]
12. Järvelä, J. Effect of submerged flexible vegetation on flow structure and resistance. *J. Hydrol.* **2005**, *307*, 233–241. [CrossRef]
13. Thorne., C.R. Effects of vegetation on riverbank erosion and stability. In *Vegetation and Erosion*; Wiley: Chichester, UK, 1990.
14. Huai, W.X.; Zeng, Y.H.; Xu, Z.G.; Yang, Z.H. Three-layer model for vertical velocity distribution in open channel flow with submerged rigid vegetation. *Adv. Water. Res.* **2009**, *32*, 487–492. [CrossRef]
15. Stephan, U.; Gutknecht, D. Hydraulic resistance of submerged flexible vegetation. *J. Hydrol.* **2002**, *269*, 27–43. [CrossRef]
16. Kummu, M. Roughness Characteristics and Velocity Profile in Vegetated and Nonvegetated Channel. Master's Thesis, Aalto University, Espoo, The Netherland, 2002.
17. Shucksmith, J.D.; Boxall, J.B.; Guymer, I. Effects of emergent and submerged natural vegetation on longitudinal mixing in open channel flow. *Water Resour. Res.* **2010**, *46*, W04504. [CrossRef]
18. Kumar, P.; Sharma, A. Experimental investigation of 3D flow properties around emergent rigid vegetation. *Ecohydrology* **2022**, *15*, e2474. [CrossRef]
19. Afzalimehr, H.; Dey, S. Influence of bank vegetation and gravel bed on velocity and Reynolds stress distributions. *Int. J. Sediment. Res.* **2009**, *24*, 236–246. [CrossRef]
20. Setayesh, P.; Afzalimehr, H. Effect of Reedy Emergent Side-Vegetation in Gravel-Bed Streams on Bed Shear Stress: Patch Scale Analysis. *Iran. J. Sci. Technol. Trans. Civ. Eng.* **2021**, *46*, 1375–1392. [CrossRef]

21. Afzalimehr, H.; Sui, J.; Moghbel, R. Hydraulic parameters in channels with wall vegetation and gravel bed. *Int. J. Sediment Res.* **2010**, *25*, 81–90. [CrossRef]
22. Nezu, I.; Nakagawa, H. Turbulent in open-channel flow. In *IAHR MonographSeries*; Balkema: Rotterdam, The Netherlands, 1993.
23. Graf, W.H.; Altinakar, M.S. *Fluvial Hydraulics: Flow and Transport Processes in Channels of Simple Geometry*; Willey: New York, NY, USA, 1998.
24. Hadian, S.; Afzalimehr, H.; Ahmad, S. Effects of Channel Width Variations on Turbulent Flow Structures in the Presence of Three-Dimensional Pool-Riffle. *Sustainability* **2023**, *15*, 7829. [CrossRef]
25. Carollo, F.G.; Ferro, V.; Termini, D. Flow Velocity Measurements in Vegetated Channels. *J. Hydraul. Eng.* **2002**, *128*, 664–673. [CrossRef]
26. Nepf, H.M.; Vivoni, E. Flow structure in depth-limited, vegetated flow. *J. Geophys. Res. Oceans.* **2000**, *105*, 28547–28557. [CrossRef]
27. Velasco, D.; Bateman, A.; Redondo, J.M.; Medina, V. An Open Channel Flow Experimental and Theoretical Study of Resistance and Turbulent Characterization over Flexible Vegetated Linings. *Flow Turbul. Combust.* **2003**, *70*, 69–88. [CrossRef]
28. Yang, S.-Q.; Xu, W.-L.; Yu, G.-L. Velocity distribution in a gradually accelerating free surface flow. *Adv. Water Resour.* **2006**, *29*, 1969–1980. [CrossRef]
29. Nepf, H.M. Hydrodynamics of vegetated channels. *J. Hydraul. Res.* **2012**, *50*, 262–279. [CrossRef]
30. Wang, J.; He, G.; Dey, S.; Fang, H. Influence of submerged flexible vegetation on turbulence in an open-channel flow. *J. Fluid Mech.* **2022**, *947*, A31. [CrossRef]
31. Julien, P.Y. *Erosion and Sedimentation*; Cambridge University Press: Cambridge, UK, 2010.
32. Afzalimehr, H. Effect of non-uniformity of flow on velocity and turbulence intensities over a cobble-bed. *Hydrol. Process. Int. J.* **2009**, *24*, 331–341. [CrossRef]
33. Simons, D.B.; Şentürk, F. *Sediment Transport Technology: Water and Sediment Dynamics*; Water Resources Publication: Littleton, CO, USA, 1992.
34. Song, T.; Chiew, Y. Turbulence measurement in nonuniform open-channel flow using acoustic Doppler velocimeter (ADV). *J. Eng. Mech.* **2001**, *127*, 219–232. [CrossRef]
35. Afzalimehr, H.; Rennie, C.D. Determination of bed shear stress in gravel-bed rivers using boundary-layer parameters. *Hydrologic. Sci. J.* **2009**, *54*, 147–159. [CrossRef]
36. Afzalimehr, H.; Anctil, F. Velocity distribution and shear velocity behavior of decelerating flow over a gravel- bed. *J. Can. Civ. Eng.* **1999**, *26*, 468–475. [CrossRef]
37. Wang, Y. Reynolds stress model for viscoelastic drag-reducing flow induced by polymer solution. *Polymers* **2019**, *11*, 1659. [CrossRef]
38. Afzalimehr, H.; Moghbel, R.; Ghalichand, J.; Sui, J. Investigation of Turbulence characteristics in channel with dense vegetation over bed. *Int. J. Sediment Res.* **2011**, *26*, 255–268. [CrossRef]
39. Yalin, M.S. *Mechanics of Sediment Transport*, 2nd ed.; Pergamon: New York, NY, USA, 1992.
40. Vanoni, V. *Sedimentation Engineering*; ASCE Publications: Reston, VA, USA, 1975.

water

MDPI

Article

Turbulence Kinetic Energy and High-Order Moments of Velocity Fluctuations of Flows in the Presence of Submerged Vegetation in Pools

Mohammad Reza Tabesh Mofrad [1], Parsa Parvizi [2], Hossein Afzalimehr [1,*] and Jueyi Sui [3,*]

1 School of Civil Engineering, Iran University of Science and Technology, Tehran 16846-13114, Iran; mohamadreza_tabesh@yahoo.com
2 Department of Hydraulic, Isfahan University of Technology, Isfahan 84156-83111, Iran; p.parvizi@alumni.iut.ac.ir
3 School of Engineering, University of Northern British Columbia, Prince George, BC V2N 4Z9, Canada
* Correspondence: hafzali@iust.ac.ir (H.A.); jueyi.sui@unbc.ca (J.S.); Tel.: +98-913-2175524 (H.A.); +1-250-9606399 (J.S.)

Abstract: The flow in arid and semi-arid regions changes significantly during seasons, letting many vegetation patches develop in different parts of rivers. In the presence of aquatic plants in streams, different flow structures have resulted. When the water level increases in these rivers, the presence of vegetation patches influences the turbulent flow structures, which may considerably change the estimation of key hydraulic parameters. The results of earlier investigations indicated that a wide range of submerged and non-submerged vegetation influences the hydrodynamic features of flows in rivers and streams. In the present investigation, two pools with various slopes of entry and exit sections were used to conduct eight independent experiment runs. In addition, a vegetation patch over the entire pool section has been set up to investigate the effects of the vegetation patch on flow structures in pools. The effect of two slopes of 5 and 10 degrees for both entrance and exit of the pools on flow structure has been investigated. Considering two aspect ratios of 2.0 and 2.7, the distributions of flow velocity, Reynolds normal and shear stresses, turbulence intensities, turbulent kinetic energy (TKE), quadrant analysis, and spectral analysis have been studied at the trailing edge of the vegetation patch along an artificial pool. Results show that, for large entrance and exit slopes (10 degrees), the TKE distribution profiles have no specific form. However, the TKE values have a convex-shaped distribution pattern with the maximum TKE value near the bed when the slopes of the entrance and exit sections of the pool are small (5 degrees). Results showed that both ejections and sweeps govern the turbulence structures and coherent motions at the trailing edge of the vegetation patch along the pool. The geometry, entrance, and exit slopes of the pool have no effect on the validation of power spectral function compared to the presence of a vegetation patch in a flatbed.

Keywords: vegetation canopy; pool; Reynolds shear stresses; turbulent kinetic energy; skewness coefficients

Citation: Tabesh Mofrad, M.R.; Parvizi, P.; Afzalimehr, H.; Sui, J. Turbulence Kinetic Energy and High-Order Moments of Velocity Fluctuations of Flows in the Presence of Submerged Vegetation in Pools. *Water* **2023**, *15*, 2170. https://doi.org/10.3390/w15122170

Academic Editor: Roberto Gaudio

Received: 6 May 2023
Revised: 1 June 2023
Accepted: 6 June 2023
Published: 8 June 2023

1. Introduction

In wetland environments, the channel beds of streams or rivers are commonly covered by either partially or fully submerged diverse aquatic plants, such as grasses, shrubs, and bushes. Both vegetated and non-vegetated zones in these streams significantly influence the hydrodynamic features of the flow, such as velocity distributions, turbulence intensity, and coherent structures, as well as mass and momentum exchange. To date, a variety of research works have been published with respect to the flow structures [1–4].

Due to the interaction of the vegetated zone and main channel, the vertical distribution profile of the streamwise velocity follows an S-shape, which is caused by the secondary currents in a narrow channel created by riparian vegetation. In fact, an S-shaped velocity profile indicates that there are two vertical mixing layers with coherent structures induced by the development of vertical shear stress, resulting in variations of flow velocities in both longitudinal and vertical directions. The characteristics of the horizontal coherent structures near the vegetation boundary and main flow interface are also influenced by secondary transversal velocity, and different inclination directions of these coherent structures are observed at different depths corresponding to the local transversal velocity [5]. Furthermore, zones of accelerating and decelerating flows in a river are generated due to the changes in its depth and width. Given the significant impact of these factors on turbulence generation and sediment transport, a variety of river topographies, particularly pools and riffles, would be developed [6–8]. To assess the effects caused by a vegetation canopy and a pool bed, turbulence structures and coherent motions of the bursting process have been investigated in the present study. Turbulent flow structures generated by the channel bed and vegetation cover are significantly more complex in conditions with bank vegetation than that with submerged vegetation [9,10].

To investigate turbulence structures above the vegetation canopy, the turbulent kinetic energy (TKE) should be examined, and then the shear and wake generation can be estimated [11]. It should be noted that the wake production is the horizontal average of the product of the local variations of Reynolds stresses and velocity gradients. As reported by the previous researchers, the shear velocity in the presence of vegetation yields more turbulent energy than that caused by the wake of vegetation, except for those zones around the edge of the vegetation patches [12,13]. Furthermore, the presence of vortexes leads to an increase in the TKE [3,14–16]. In 2D flows in channels with emergent vegetation elements, instabilities caused by horizontal vortexes at the downstream edge of a vegetation patch play a dominant role in flow structures [17,18]. In the presence of a vegetation canopy layer in a flow, an analogy between the canopy layer and the flow's mixing layer has been reported [19,20].

In the near-bed region, by means of the quadrant analysis, the occurrence probabilities are dominated by the presence of sweep motions and depend on the thickness density of the vegetation patch, which causes a larger wake zone in the emergent region over the canopy. The majority of reported studies focused on evaluating the structures of coherent flows above a vegetation patch [4,5]. In the present study, coherent flow structures have been investigated upstream of the vegetation patch, across the entire vegetation patch in a pool (compared to those in a pool without vegetation), and at the vegetation canopy boundary. It has been found that the motion of ejections appears behind the vegetation patch. The dominant bursting process in flows over vegetation canopy belongs to the periodic occurrence of sweep and ejection events. A variety of research works has been conducted to study the interactions between strong vortices above vegetation canopies and quadrant occurrences [11]. This study's results of high-order moments of velocity fluctuations u' and w', also known as skewness fluctuations sk_u and sk_w, can be used to derive factors that provide useful information about asymmetry in streams [18]. In other words, a non-zero skewness of velocity fluctuations in the streamwise and vertical directions indicates an asymmetric probability density function (PDF) of the considered variable, which is associated with quadrant bursting events [21].

In practice, engineers are keen on the estimation of low resistance along a pool in the presence of vegetation in a channel bed since the resistance changes along different sections of a pool, including both entrance and exit slopes as well as different locations inside the vegetation patch. To the authors' knowledge, no research work has been reported on the investigation of the details of flow structures (including 3D velocity components distributions, 3D turbulence intensities, and turbulence kinetic energy distributions) at the trailing edge of vegetation patch (where the flow reach to gravel along a stream) in the presence of bed forms. Moreover, the application of the power spectral density of velocity components over 3D bed forms in the presence of vegetation patches in the channel bed and the validation of Kolomogrove $-5/3$ law has been investigated in this study.

2. Materials and Methods

In the present study, the desired two pool bed forms were built in a laboratory flume, which is 8-m long, 0.4-m wide, and 0.6-m deep. One was built with a constant slope of 5 degrees for both the entrance and exit sections, while the other was built with a constant slope of 10 degrees for both the entrance and exit sections. All experiment runs have been conducted with vegetation in the pools and without vegetation in the pools (Figure 1). For this experimental study, each desired pool bed form was set up in a section of the flume to ensure that the flow was under fully developed conditions. To avoid the effects of the downstream tailgate, the 1.5-m long pool was built at a distance of 4.5 m from the entrance of the flume. The tailgate located downstream end of the flume was used to control the flow depth and maintain the desired flow depth of 20 cm and 15 cm, respectively (Table 1). A water gauge mounted on the channel walls is used to determine the water depth.

Figure 1. Experimental setup. (**a**) The hydraulic flume. (**b**) The gravel pool bed form. (**c**) The gravel grains. (**d**) The vegetated pool bed form. (**e**) The artificial grass used in the experiments. (**f**) Measured velocity points over the pools.

Table 1. Experimental Conditions.

Pool Setup	Runs.	Entrance Slope	Exit Slope	H (cm)	Pool-Bed Material	U (m/s)	Q (lit/s)	F_r	*Re* (10^4)	w/h	H/d_{50}
Setup 1	Run 1	5°	5°	20	Gravel Bed	0.125	10.5 ± 0.1	0.09	2.5	2	20
	Run 2	5°	5°	20	Gravel Bed	0.5	40.5 ± 0.1	0.13	10	2	20
	Run 3	5°	5°	20	Vegetated Canopy	0.125	10.5 ± 0.1	0.09	2.5	2	-
	Run 4	5°	5°	20	Vegetated Canopy	0.5	40.5 ± 0.1	0.13	10	2	-
Setup 2	Run 5	10°	10°	20	Gravel Bed	0.125	10.5 ± 0.1	0.09	2.5	2	20
	Run 6	10°	10°	20	Vegetated Canopy	0.125	10.5 ± 0.1	0.09	2.5	2	-
	Run 7	10°	10°	15	Vegetated Canopy	0.125	10.5 ± 0.1	0.09	2.5	2.67	-
	Run 8	10°	10°	20	Vegetated Canopy	0.125	10.5 ± 0.1	0.09	2.5	2	-

The median diameter of bed material (d_{50}) in the pool bed for all experiments is d_{50} = 10.4 mm, which is determined by using one hundred random samples of the grains. The results of the grain size distribution show that the bed material is relatively non-uniform according to the geometric standard deviation $\sigma_g = (d_{84}/d_{16})^{0.5}$, where d_{16} and d_{84} are the 16th and 84th percentile of the particle-size distribution, respectively. In this study, the geometric standard deviation is smaller than 1.4, indicating that the bed material is non-monotonous. Figure 2 shows the grain size distribution curve, and the particle characteristics for this experiment are determined through the following method (Table 2). The granulation coefficient ($\mathbf{G_r}$) is determined by $(-0.5)(d_{84}/d_{50} + d_{50}/d_{16})$; the average particle size ($\mathbf{D_g}$) depicts as $(d_{84}/d_{16})^{0.5}$; $\mathbf{C_c}$ indicates the curvature coefficient calculated by $(d_{30})^2/(d_{10} \times d_{60})$, and $\mathbf{C_u}$ represents the uniformity coefficient calculated by (d_{60}/d_{10}).

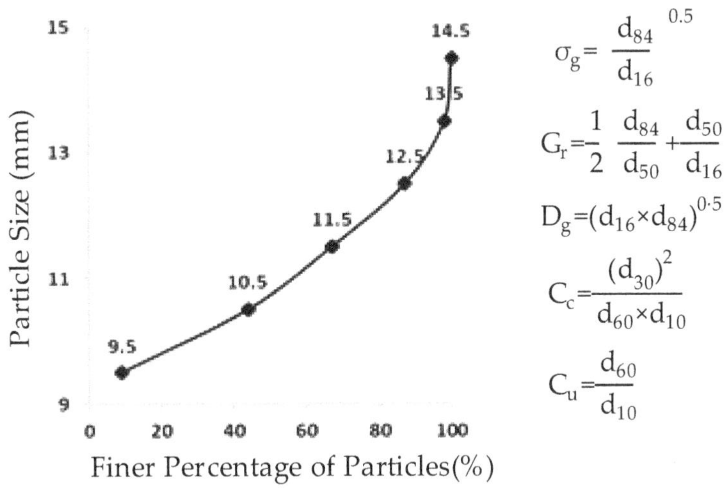

Figure 2. The grain size distribution curve and characteristics of bed material.

Table 2. Characteristics of bed material.

Parameters	σ_g	G_r	D_g	C_c	C_u	d_{10}	d_{16}	d_{30}	d_{50}	d_{60}	d_{84}
Sum.	1.17	1.17	4.6	0.95	1.22	8.8	9	9.5	10.4	10.8	12.3

The vegetation selected for this experimental study was a 2-cm tall artificial grass patch (measured under dry conditions) that covers the entire pool section of the flume. Table 2 summarizes data collected from all experimental runs. The experiments conducted in the pool with the 5-degree slope for both entrance and exit sections have been carried out for two different discharges of 10.5 and 40.5 lit/s. The experiments conducted in the pool with the 10-degree slope were carried out under two different aspect ratios (w/h; channel width to water depth) of w/h = 2 and 2.7. Along with the aforementioned information, Table 2 also presents the flow Froude number; $Fr = \left[U/(gH)^{0.5}\right]$, and the flow Reynolds number; $Re[= UH/v_m]$, where H is the water depth; U is mean velocity, and v_m is the flow kinematic viscosity. An electromagnetic flowmeter installed at the water pipe entering the flume was used to measure flow discharge.

A Nortek Vectrino Acoustic Doppler Velocimeter (ADV) was used to measure the instantaneous three-dimensional velocity components. For each experimental run, the ADV (Acoustic Doppler velocimeter) was used to measure the flow velocity at the distance of z = 2 mm from the bed to 50 mm below the water surface. Even though some data were filtered to remove outliers, most of the velocity data collected in this study was examined with high SNR (signal-to-noise ratio) and correlation. To investigate secondary currents, data for Run 8 were collected along the channel's second axis, which was located 10 cm away from the flume side wall, whereas data collected for all other experimental runs were obtained along the flume's center line (20 cm from the flume sidewalls).

3. Results

3.1. Velocity Distribution

As shown in Figure 3, the stream-wise velocity (x direction) profiles of 3D flows are displayed for all experimental runs. At each measurement point, the mean point velocity is divided by the maximum flow velocity (u_c) of the profile, and each depth from the bed (z) is divided by the flow depth (H); in this way, the dimensionless velocity (u/u_c) and flow depth (z/H) are obtained, respectively (Figure 3). As the water depth rises on the entrance slope, the flow velocity decreases, creating a decelerating flow section (CDF) over the pool's entrance section. The accelerating flow (CAF) develops at the end of the middle pool section, where the water depth decreases along the exit slope of the bed. For the flow with a low aspect ratio (flume width/flow depth H/B < 5), the location of the maximum velocity may fluctuate due to the influence of secondary currents. However, with a greater aspect ratio (H/B > 5), this effect is hardly noticeable. If the aspect ratio is less than 5, the dip phenomenon occurs, and the flow will be 3D [22,23]. Neither negative velocities nor flow separation at the bed of both entrance and exit sections of pools could be recorded because of the limitations of the ADV. As can be seen in Figure 3, there are different velocity profiles for flow in the pools with the change in the bed slope. However, flow velocities in the pool with a slope of 10 degrees are slower than those with a slope of 5 degrees. In a decelerating flow, the near-bed velocities are lower than those at the water surface are, and this discrepancy increases in the streamwise direction.

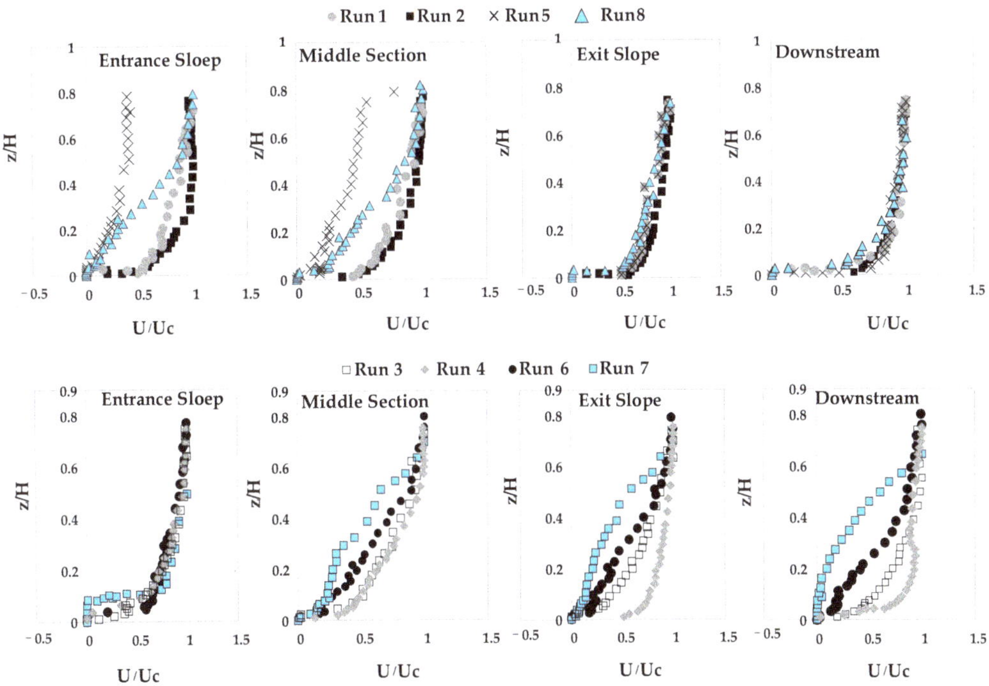

Figure 3. Distribution profiles of stream-wise velocity (u_c is the maximum flow velocity).

The effect of decelerating flow is still present in the middle pool section, preventing the development of a uniform flow. In the accelerating flow, velocity increases near the bed and decreases near the water surface as the flow continues along the exit section of a pool.

Despite the aspect ratios being less than 5, no evidence for the velocity dip phenomenon is observed within the range of flow depth of $z/h < 0.6$. Nevertheless, the highest flow velocity has been observed below the water surface at a depth of $z/H > 0.6$ along the decelerating flow section when the entrance slope is 5 degrees (Figure 3). As the flow enters the pool region, the maximum flow velocity moves toward the water surface in the zone of flow depth of $z/h > 0.8$. This finding agrees well with the reported result that the dip phenomenon occurs at a distance up to 0.67 H (or dip parameter) from the bed for uniform flow with a hydraulic rough bed [21]. According to some earlier studies, the value of the dip parameter for non-uniform flows is 0.22 H [22]. In a gravel bed river with flexible submerged vegetation patches and 3D bed forms, this value is up to 0.8 H [24]. However, the findings of the theoretical investigations based on the Navier–Stokes equations demonstrate that the value of the dip parameter in an accelerating flow is higher than that in both uniform and decelerating flows [6].

According to this research, as depicted in Figure 3, the velocity fluctuation is primarily dependent on the drag force caused by the vegetation in the zone up to a depth of $z/H = 0.1$ in the pool, and the viscous shear stress has only a minor impact on the velocity profile, leading to minimal changes in the velocity profile. The velocity profile in the upper layer of the flow ($z/H > 0.1$) increases gradually with the flow depth (z/H). With increasing the aspect ratio, the presence of vegetation canopy inside the pool can result in an S-shaped distribution profile of velocity. This is true for both pool entry and exit sections with a slope of 10 degrees (decreasing the water depth from 20 cm to 15 cm). In fact, the velocity gradient in the upper layer of the flow approaches null; the velocity gradient reaches the maximum in the middle and close to the bed, and the velocity distribution follows the logarithmic function. This outcome supports the findings of prior research [22]. Moreover, contrary to what was found in earlier research, the dimensionless flow velocities in the outer zone of the decelerating flow section (z/h > 0.2) are not necessarily higher than those in the accelerating flow section, and some discrepancies have been observed in the aforementioned trends. [25–27]. To further assess the impacts of secondary currents on the characteristics of flow in a pool with the presence of vegetation, flow velocity components in the other two directions (span-wise v and vertical direction w) are shown for four experimental runs in Figure 4.

The presence of both positive and negative values of velocity profiles in both span-wise and vertical directions indicates the pattern of the secondary currents [5,23]. While the average vertical velocity is negative in the decelerating and middle sections of the pool, revealing the occurrence of a downward flow, the positive values of vertical velocities in the accelerating section indicate the inclination of flow toward the water surface. The opposing effect of accelerating and decelerating stream portions on secondary currents along the pool bed has been noted by a number of researchers, which led to the flow convergence along the entrance slope and flow divergence along the exit slope section of the flow [7], despite the fact that in a balanced flow, it is not always the case that the vertical velocity is downward in an accelerating flow and upward in a decelerating flow [26]. The convergence and divergence patterns of the flow in the deceleration and acceleration sections are in agreement with field investigations and experimental studies [28,29].

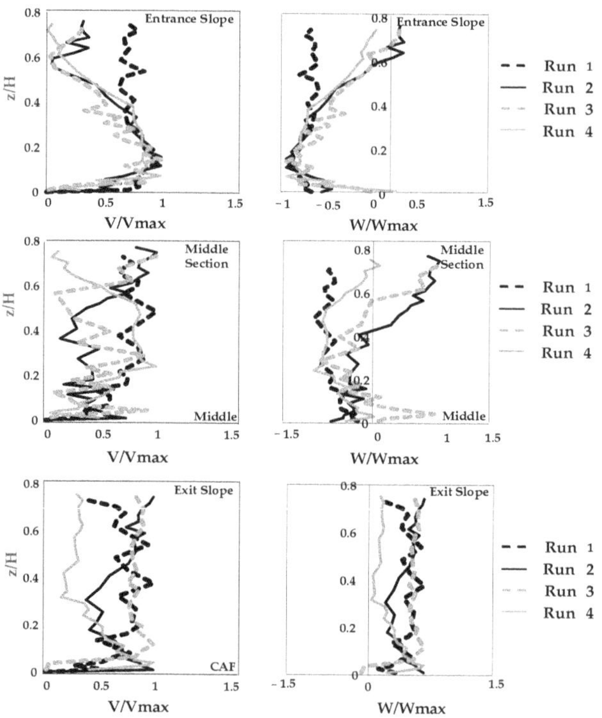

Figure 4. Span-wise and vertical velocity profiles.

3.2. *Reynolds Normal and Shear Stress Distributions*

In this study, using the following equation, the dimensional Reynolds normal stresses have been estimated at three locations (entrance slope, middle pool section, and exit slope) along the pool bed, which is depicted in Figure 5.

$$\sigma_x = -\rho\overline{u'^2},\ \sigma_y = -\rho\overline{v'^2},\ \sigma_z = -\rho\overline{w'^2} \tag{1}$$

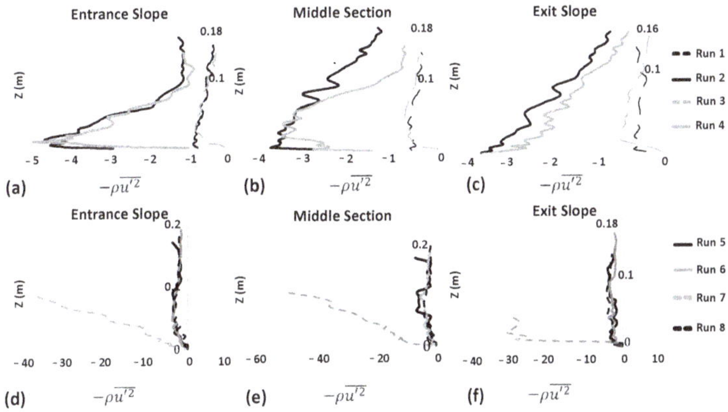

Figure 5. Dimensional Reynolds Normal Stress at different sections along the pool. (**a**) Entrance Slope Run 1 to 4, (**b**) Middle section Run 1 to 4, (**c**) Exit slope Run 1 to 4, (**d**) Entrance slope Run 5 to 8, (**e**) Middle section Run 5 to 8, (**f**) Exit slope Run 5 to 8.

Under the condition of a constant flow discharge, the normal stresses on the pool bed with vegetation patch are clearly greater than those on the gravel surface without vegetation (Runs 1 and 3). The magnitude of the normal stress decreases as the flow rate rises (Runs 2 and 4). Moreover, the magnitude of the normal stress decreases as the bed slope decreases (Runs 3 and 6). Additionally, when the aspect ratio rises, the magnitude of the normal stress decreases at the water surface (Run 7). This phenomenon occurs due to the stronger secondary currents in shallow flows (W/h = 2.7). For all experimental runs, the magnitudes of the normal stresses in three directions of the flow are shown in Figure 6A for the entrance slope, for the middle pool section in Figure 6B, and for the exit slope in Figure 6C. The results indicate that the values of the normal stress in most parts of the stream are higher in the stream-wise direction than those in both lateral and vertical directions. However, the magnitude of the normal stress is affected by the presence of secondary currents, resulting in shallow channels with different roughness from the bed (Figure 5). In addition, the Reynolds normal stress distributions are also affected by the bed form slope (Figure 5). This makes it difficult to provide a general pattern for the distribution of the Reynolds normal stress in the bed forms with a vegetation cover.

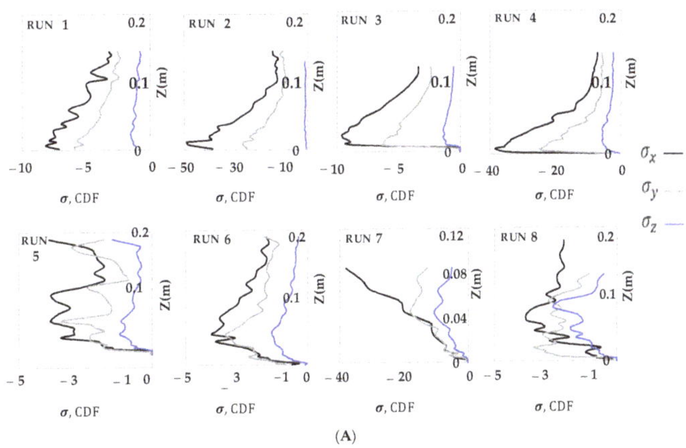

Figure 6. *Cont.*

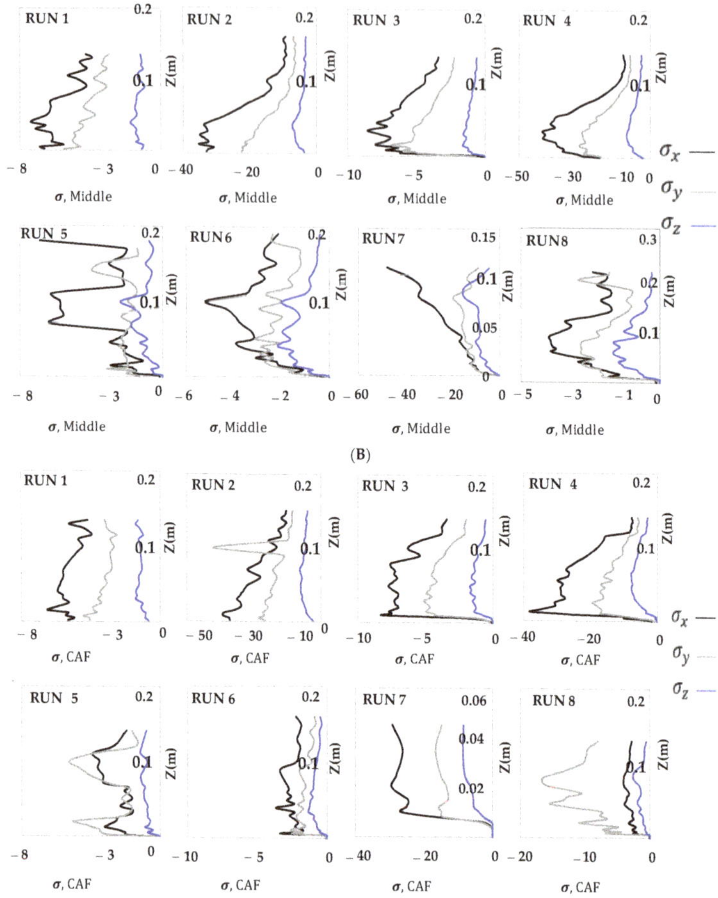

(C)

Figure 6. (**A**). Dimensional Reynolds Normal Stress in entrance slope section (CDF). (**B**) Dimensional Reynolds Normal Stress in middle pool section (Middle). (**C**) Dimensional Reynolds Normal Stress in exit slope section (CAF).

In Figure 7, Reynolds shear tresses ($-\rho\overline{u'w'}$) calculated using Equation (2) are normalized with the squared friction velocity u_*^2, in which u_* is the shear velocity calculated from the velocity profiles near the bed, which follows the logarithmic law as presented in Equation (3). Wherein k is the von Karman coefficient, which is equal to 0.41; z_0 is the reference bed level that varies with the roughness height of the bed material. In both uniform and non-uniform flows, it has been claimed that the inner zone of each velocity can be extended to a relative depth of z/z_0 = 0.2 [26].

$$\overline{u'w'} = \frac{1}{N}\sum_{i=1}^{N}(u-\overline{u})(w-\overline{w}) \tag{2}$$

$$\frac{u}{u_*} = \frac{1}{k}\ln\frac{z}{z_0} \tag{3}$$

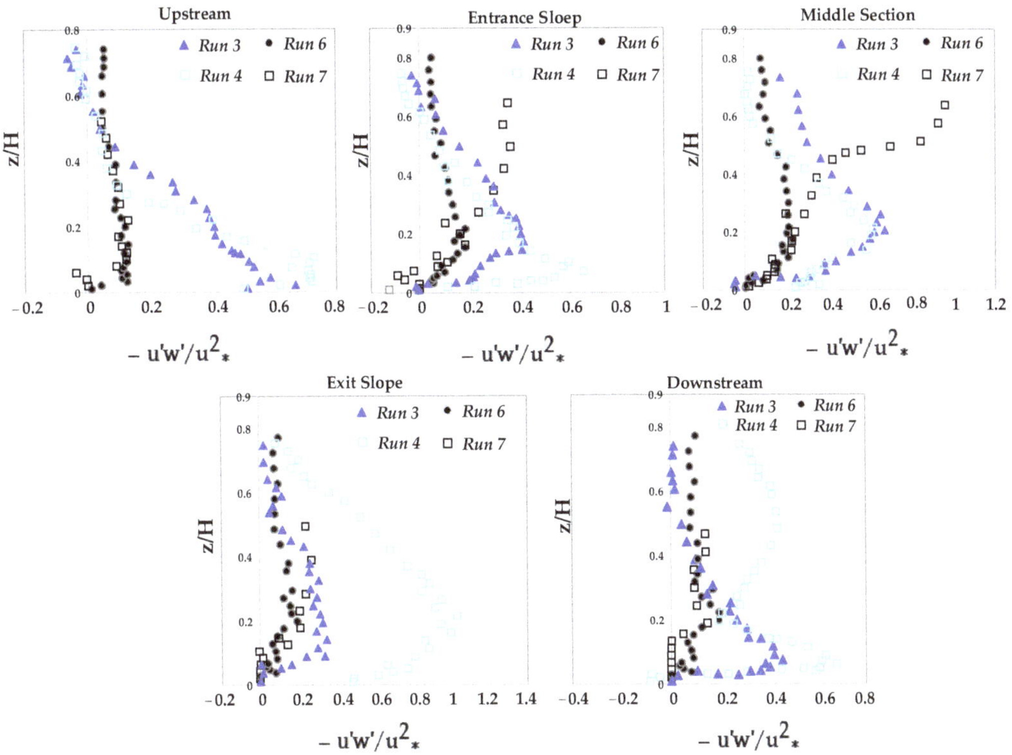

Figure 7. Reynolds Shear Stress distribution profiles in the presence of vegetation in pools.

In the presence of vegetation canopy in the pool bed, four experimental runs have been conducted. The dimensionless Reynolds shear stress (RSS) profiles are displayed in Figure 7 along the flow direction from upstream to downstream. One can also observe various trends of Reynolds stress for different experimental runs from Figure 7. If the channel bed is smooth, the RSS distribution often shows a linear profile. The rough sub-layer near the bed may cause the Reynolds stress distribution to grow up to 20% of the flow depth before decreasing [22,30–32]. Results of the present study indicate that the secondary currents are intensified due to the presence of the vegetation canopy, the non-zero vertical velocities, and the smaller aspect ratio of less than 5. Several researchers have demonstrated that the secondary currents in shallow flows cause the shear stress distribution to deviate from the expected linear shape [6,33].

Depending on the variation in flow velocities, the maximum shear stress can be observed in the upstream portion of the channel, which is located in the zone close to the channel bed within the distance of $z/H < 0.15$, revealing that the RSS values are completely affected by the roughness elements and relative submergence (H/d_{50}) [34]. The positive RSS values close to the water surface support the existence of the dip phenomenon. The greater the slope of the pool entrance section (Runs 6 and 7), the higher the Reynolds stress values are induced along the flow. Furthermore, the shear stress reduced toward the water surface as the flow depth increased, and its distribution profile appeared to have a convex shape. Moreover, the positive pressure gradient (also known as the reverse pressure gradient) and the negative pressure gradient (also known as the favorable pressure gradient) have an impact on the Reynolds stress distribution in the decelerating and accelerating flows, respectively [35].

The shear stress profiles in the middle pool section do not exhibit a tendency toward a linear shape, suggesting that the flow is not homogeneous there. In the presence of a vegetation patch in the pool, the increased flow velocity leads to an increase in the shear stress values in the zone of $z/H < 0.2$ due to higher turbulence intensities. For all experimental runs, the maximum shear stress inside the middle pool section occurred away from the channel bed depending on the flow velocity and slopes of entrance and exit sections, specifically the maximum RSS is observed at a depth of $z/H = 0.25$ in the middle part of the flow (Run 3, 4, and 6) [24,36]. However, the impact of secondary currents resulting from the increase in the aspect ratio has the opposite effect on the location of the maximum shear stress and is shifted toward the water surface (Run 7). For the same flow discharge, as the results showed from Runs 3 and 6, with the increase in the slope of the entrance section (namely, with the decrease in flow velocities), the RSS values increased in the zone of $z/H < 0.5$ (except for the middle pool section).

Due to significant disturbances resulting from bed forms and roughness, the shape of shear stress distribution profiles in the downstream section of the pool differs from that in the upstream section of the pool. Negative Reynolds stresses are also observed sometimes, as illustrated in Figure 7. Most likely, this results from the flow being transmitted to each component's wake zone by vegetation-induced drag. Nonetheless, the previous investigations in the presence of vegetation and bed morphologies also reported the existence of negative RSS values. These results demonstrated that, even if the sweep and ejection events result in positive RSS values in the flow field, the outward and inward events in the bursting process might be responsible for the negative stresses.

In some places, particularly in the zones where the flow was accelerating, the RSS value increased dramatically. It may be explained by the fact that when the flow is accelerated, or even when the channel bed has a larger slope, the turbulent velocity components undergo some changes, as indicated in Figures 7 and 8. Moreover, the distribution of the flow responds more strongly and quickly to the changes in the bed slope near the channel's side walls than it does near the channel's centerline (Figure 8). It has been observed from Figure 9 that, over the gravel pool bed, the highest Reynolds stress region grows to the end of the middle pool section. Figure 8 also demonstrates this result for flow over a gravel bed form, as the maximum RSS is highest along the middle and exit sections of the pool. However, with a lower flow discharge, the presence of vegetation resulted in an obvious decrease in the shear stress (Run 3). While with the higher flow discharge, the vegetated canopy minimizes shear stresses on the entrance slope, the zone with the greatest stress is only visible in the middle section of the pool bed (Run 4).

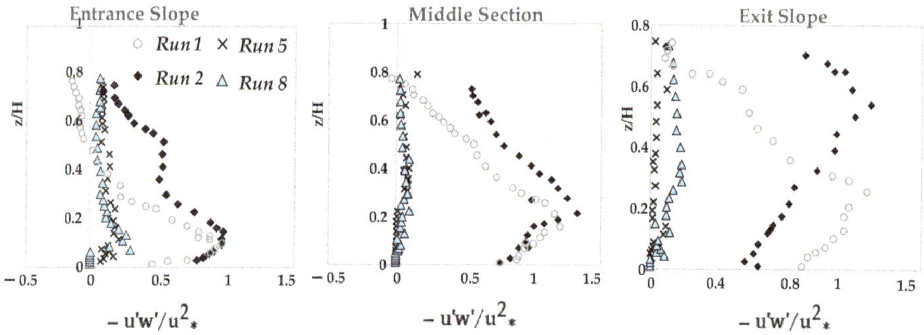

Figure 8. Reynolds Shear Stress distribution profiles at three points of the pools.

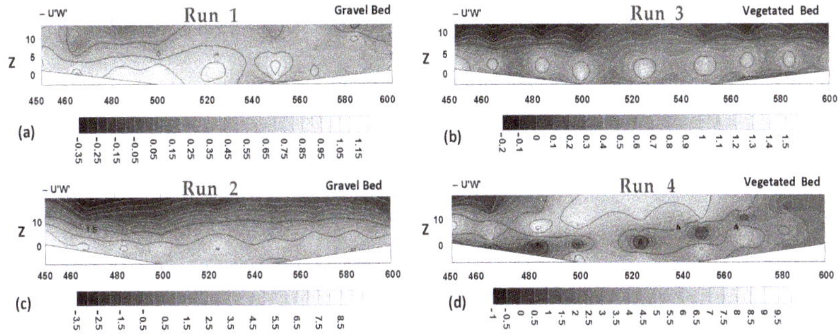

Figure 9. Reynolds Shear Stress contour distributions: (**a**) Run 1; (**b**) Run 3; (**c**) Run 2; (**d**) Run 4.

3.3. *Turbulence Kinetic Energy (TKE)*

As seen in Figure 3, the velocity profile in the inner layer of the flow is almost constant due to the significant wake effects resulting from the vegetation canopy; this zone exhibits negligibly little vertical momentum transmission, although it may occasionally appear with a slight contour gradient. Similar to the results reported by the previous researchers regarding open-channel flow in the presence of either emergent or submerged vegetation, this inner layer zone refers to the "longitudinal exchange zone", which has a negligibly limited vertical momentum conveyance [37–39]. The logarithmic law is used to describe the turbulence structure in this zone. Figure 10 represents the variation of the dissipation rate of turbulent kinetic energy ($\text{TKE} = 0.5c_2\rho\left[\overline{u'^2} + \overline{v'^2} + \overline{w'^2}\right]$), normalized with the squared friction velocity $\left(u_*^2\right)$, with respect to flow depth in central pool section where vegetation patch presents. In the mixing layer (as shown in Figure 11), due to the vertical transfer of momentum, the unstable inflection point of velocity is induced by secondary currents. Moreover, in the mixing layer of flow at the canopy part, the turbulent diffusion has an obvious impact on the turbulence kinetic energy (TKE) budget.

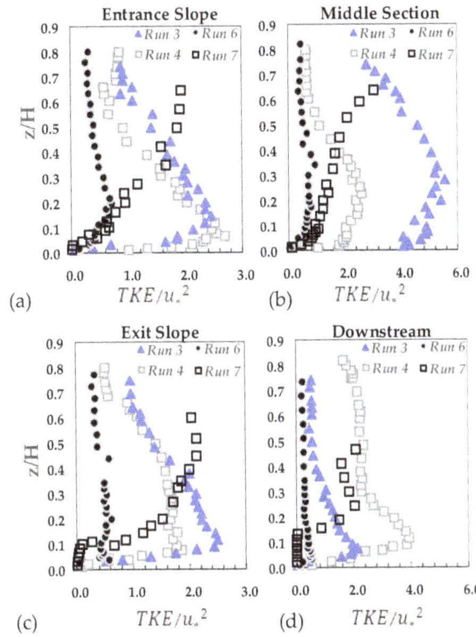

Figure 10. Turbulence Kinetic Energy for various flows (Runs 3, 4, 6, and 7) along the pool, (**a**) Entrance Slope, (**b**) Middle Section, (**c**) Exit Slope, (**d**) Downstream.

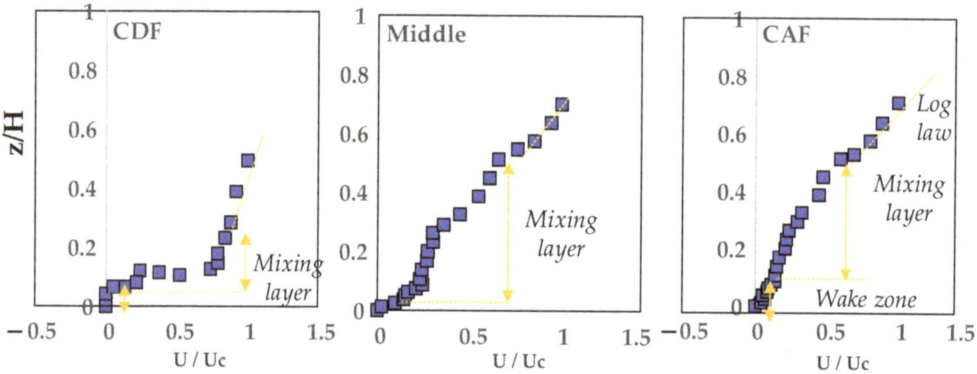

Figure 11. Stream-wise velocity profiles (Run 7) at different sections along the pool; Entrance slope, middle section, and exit slope.

Figure 10 shows that the great TKE values are observed at different locations in several flow sections. The energy losses are reduced; i.e., as the entrance slope increases, the pressure losses per unit length of the bed are reduced, leading to lower RSS values [17]. In fact, the mixing eddies resulting from the velocity gradient over the vegetation canopy decreased, and the reduction in the entrance slope of a pool led to the turbulence becoming weaker. The profiles of TKE for Run 7 values do not follow a convex shape, as they indicate the higher entrance and exit slopes, justifying the significant role of bed-form geometry on the TKE distribution. In addition, due to the stronger secondary currents and higher turbulence velocity, an increase in flow velocity by reducing flow depth resulted in larger TKE values in Run 7.

The maximum TKE value is predicted to occur generally in a zone that is relatively close to the bed ($z/h < 0.2$) since the production of Von Karman vortexes is typically associated with the occurrence of TKE values that peak a severe distortion in the velocity profile [9]. Moreover, other studies have revealed lower ranges for the wake zone in streams with broad and heavily submerged vegetation as well as in streams with broad, partially submerged vegetation in the bed. Results of the present study support the existence of a small wake zone over the vegetation canopy [2,13] only for the small bed-form slopes (Runs 3 and 4).

3.4. Turbulence Intensities

By calculating the root mean square (RMS) of flow velocities, turbulence intensity is displayed in Figure 12 to show the values of intensities for pool setup 1. It has been found that the vertical turbulence intensity is highest in the upper layer above the canopy of vegetation, about in the zone of $z/H < 0.2$, while stream-wise and lateral turbulence intensities reach their maximum slightly above the canopy of vegetation.

Research results showed that the maximum intensity of the streamwise turbulence (u′) occurs right above the canopy, while the maximum intensity of the vertical turbulence occurs right below the emergent canopy. Due to the non-uniformity of flow in experiments, the values of u′, in particular, exhibit a convex shape rather than a concave one, which deviates from the exponential law proposed by Nezu [38]. This suggests that vegetation considerably modifies the equilibrium state of turbulent kinetic energy and, thus, the redistribution of turbulence intensities. In fact, it should be noted that the isotropic turbulence inside vegetation (including the vegetation stems and canopy) occurs when there is a strong wake of turbulence over those objects. As a result, the wake caused by vegetation elements may cause convex patterns of turbulence intensities in a cascade process [31]. Similar convex-shaped distribution of turbulence intensity has been reported in studies of the aquatic canopies [39].

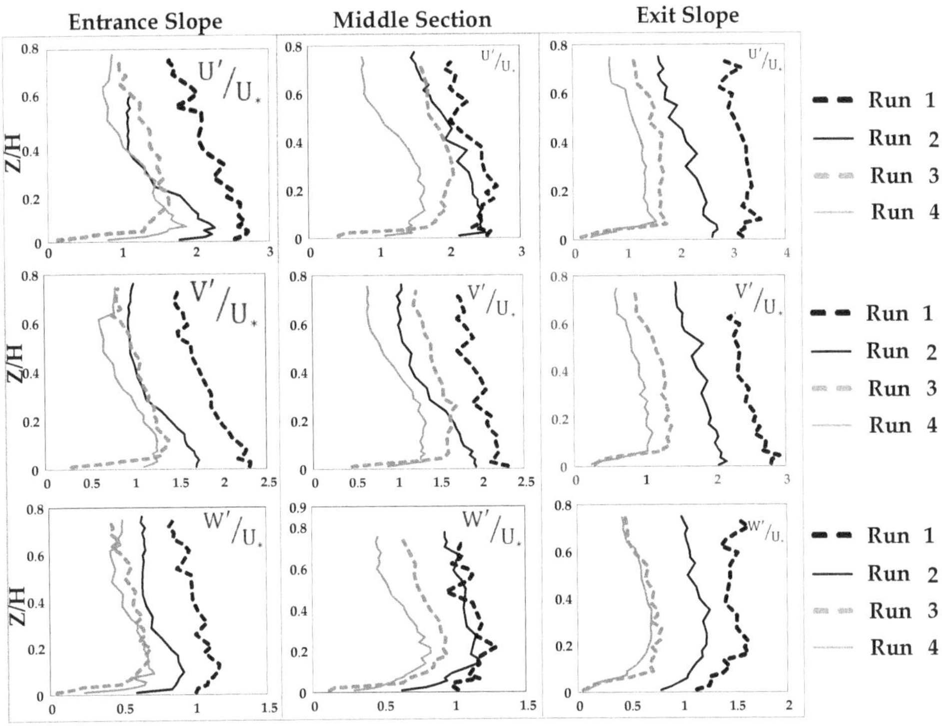

Figure 12. Turbulence intensities for the case of pool setup-1.

It appears that the presence of vegetation has a considerable impact on the kinetic energy and the distribution of turbulence intensities [17,40]. Figure 12 displays the possible relationship between the highest disturbance intensity and the bed slope of the flume. Along the accelerating part of the flow, the value of disturbance over the gravel bed form is larger than the values recorded in the decelerating region. Streams with vegetation patches, on the other hand, did not exhibit the same pattern.

The maximum amount of disturbance over the gravel bed form along the accelerating section of the flow has greater values than that along the decelerating section. However, streams with vegetation patches did not show a similar trend. The turbulence intensities over the decelerating and accelerating sections did not have a significant difference.

The zone of the inner layer with the largest turbulence intensities can be seen around the zone of $z/H < 0.2$, while in the central region of the flow, it appeared to be in the zone of $z/H < 0.4$. These zones (or flow depths) are also thought to be the spots where shear stress is at its highest. Instead of the accelerating flow, the decelerating flow intensifies the turbulence and increases Reynolds shear stress, indicating that flow structures are affected by the non-uniformity of the flow [6,32,41].

3.5. Quadrant Analysis

A quadrant analysis has been applied to predict detailed statistical characteristics of turbulent flow structures [42,43]. However, to the authors' knowledge, no study has been reported on the quadrant analysis of flow structure at the trailing edge of a vegetation patch in a gravel-bed stream. This study focuses on the exit border of flow from the vegetation patch in a gravel-bed channel. The important question is, what happens for turbulent flow structures at the trailing edge of a vegetation patch over a pool, and how does it affect the estimations of hydraulic parameters in this study?

As mentioned by Parvizi et al. [35], the outward motion of the flow is a result of the positive values of u′ and v′ in bursting process occurrences; the sweep motion is recognized with the positive value of u′, and the negative value w′ generally corresponds to the flow stream over the vegetation canopy (often denotes the flow stream above the vegetation canopy). However, the ejection motion of the bursting phenomenon, which has a positive value for w′ and a negative value for u′, along with the sweep motion, have been observed over smooth walls in the boundary layer [42,43]. Lastly, the flow has been recognized to be moving inward if both u′ and w′ have negative values. The presence of vegetation patches over the bed form plays a role in the flow structure. However, this aspect of fluvial hydraulics has been considered less often by researchers. This part of the study especially concentrates on the border between vegetation and grave. The important question is, what happens in the trailing edge of a vegetation patch over a pool, and how does this affect the turbulence features and their distributions? It appears that the sweep motion is the primary event of the flow downstream of the pool (trailing edge of a vegetation patch), where the flow reaches the gravel bed again. Nonetheless, according to the result of experiment Run 7, the isotropic event often occurs near the boundary between the vegetation patch and gravel bed (trailing edge of the vegetation patch). The contribution of each quadrant event is presented in Figure 13, indicating the dominant role of the ejection motion in the bursting process of the flow and the sweep motion as the second dominant event in the stream.

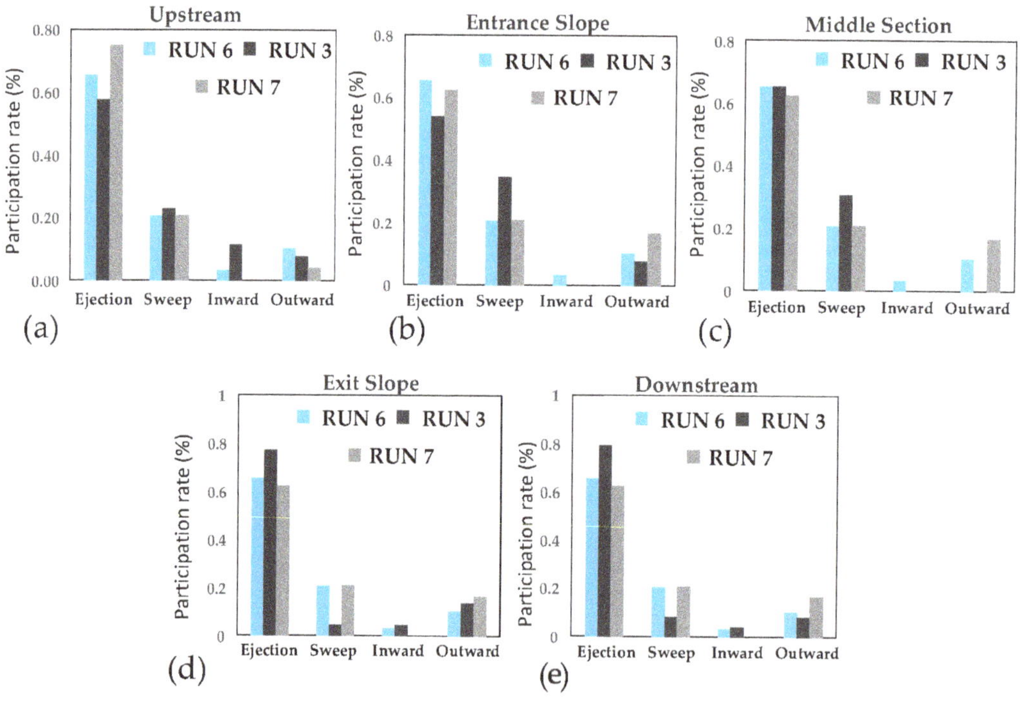

Figure 13. Bursting process for different sections along the pool, (**a**) Upstream of the pool section, (**b**) Entrance slope, (**c**) Middle section, (**d**) Exit slope, (**e**) Downstream of the pool section.

Along the decelerating part of the flow (CDF), sweep motion appears above the vegetation canopy in the inner layer of the flow in the zone of $z/H < 0.2$, while the modification of the entrance slope doesn't lead to the change of the zone for the occurrence of the sweep motion in the stream (Runs 3 and 6). However, by adjusting the aspect ratio in the flow (Run 7, from $w/H = 2.0$ to 2.7; $w/H < 5.0$), the sweep phenomenon has been seen in a higher layer of the stream in the zone of $z/H < 0.5$, indicating that the change of flow velocity results in the occurrences of bursting events. By increasing the slope of the entrance section of the pool, a larger region for the sweep phenomenon resulted, while a higher sweep zone ($z/H < 0.4$) was observed by altering the aspect ratio. The sweep motion has been noticed right above the vegetation canopy in the acceleration section of the flow. However, a larger zone of sweep motion has been generated because of the modification of the slope of the exit section of the pool. In addition, Figure 13 shows that the ejection motion is the primary event of the flow at the trailing edge of vegetation.

3.6. *Skewness Coefficients*

Skewness coefficients (sk_u and sk_w) of velocity fluctuations are the third central moment of the factors u′ and w′, which include positive, negative, or zero values in turbulent flows. Asymmetry in a flow field can be better understood by looking at the skewness factors [18]. Positive skewness means that the variable u′ is more likely to take on a large positive value than a large negative one. In fact, a non-zero skewness of velocity fluctuations in the stream-wise and vertical directions indicates an asymmetric probability density function (PDF) of the considered variable, namely, the dominated processes in one direction are more possible than in the other one depending on the symptom of the statistics.

The positive skewness indicates that the PDF has a longer tail for $u' > 0$ than that for $u' < 0$. A zero-skewness value shows an isotropic or homogenous turbulence. The possible explanation is that instability of the inflection point of velocity profiles is responsible for the generation of coherent eddies in the presence of flexible vegetation inflows. This implies that coherent eddies, such as ejections and sweeps, will resonate with the flexible vegetation [35].

Figure 14 displays the skewness for measured velocity fluctuations (sk_u and sk_w) from experiments based on data collected from experimental Runs 3, 6, and 7 along the upstream, entrance slope with a decelerating flow (CDF), middle-pool section, and exit sections which leads to an accelerating flow (CAF) of the pool while maintaining the same flow rate and vegetation canopy. The sweep motion is the major event right above the gravel grains in the zone of $z/H < 0.1$ in the upstream region of the pool, where the fluctuations of flow velocity mostly depend on two parameters, flow depth (H), and median grain size of bed material (d_{50}) (Runs 3 and 6). To precisely study flow dynamics, data was collected at the boundary of the gravel bed and vegetation patch at the upstream portion of the pool, as shown in Figure 13c.

One can see from Figure 14 that sk_u has slightly positive values at point c together with a positive sk_w values, where vegetation patch begins at the gravel border, implying that the outward motion has occurred closely above the gravel barrier in the zone of $z/H < 0.1$.

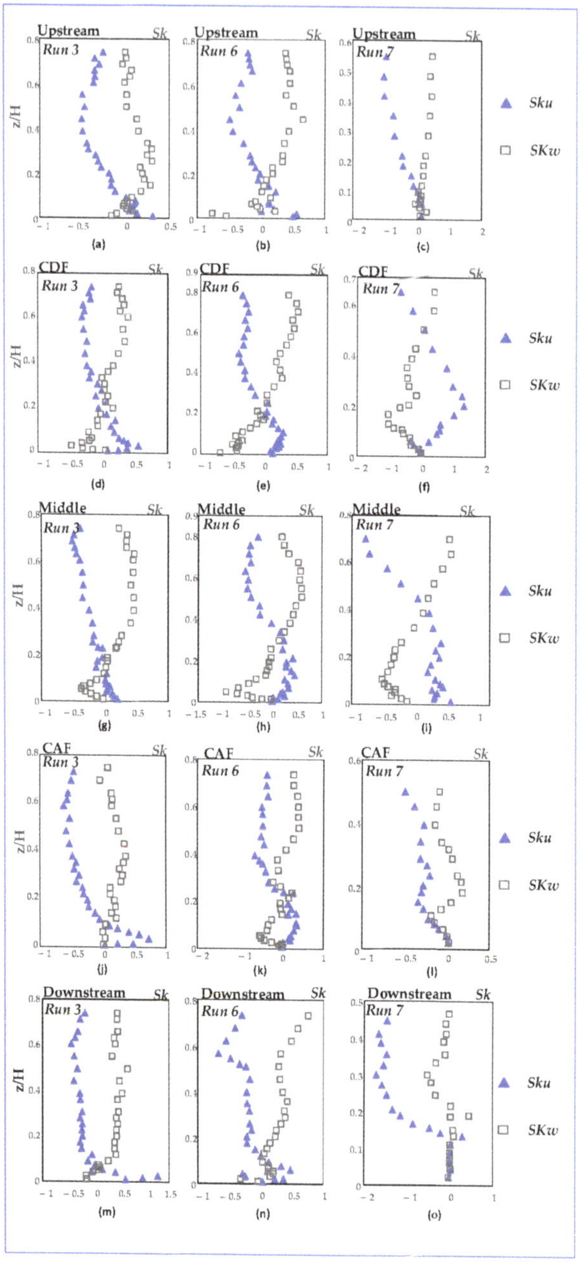

Figure 14. Skewness coefficients. (**a–c**) Upstream of the pool Section. (**d–f**) Decelerating flow along the entrance slope (CDF). (**g–i**) Middle pool section. (**j–l**) Accelerating flow along the exit slope (CAF). (**m–o**) Downstream of the pool Section.

3.7. Spectral Analysis

There are three distinct scaling regimes observed in turbulent flows across flat surfaces. At low frequencies, there is a scaling sub-range, which is referred to as the production subrange, characterized by a −1 spectral slope. At intermediate frequencies, Kolmogorov claimed that an inertial subrange with a spectral slope of −5/3 is seen. When the Reynolds number is high enough, Kolmogorov claims [40] that there is no loss of energy during the shift from large to tiny eddies scale. The third scaling subrange is the viscous subrange, where spectra decay significantly faster than in the inertial subrange [41]. It is expected that the presence of topography and bed forms in the regimes both affect the transition of eddies between the production and inertial subranges [42].

Spectral analysis has been performed for velocity components along the bed form at the same points located at 6 mm from the vegetation cover along the entrance slope, middle pool section, and exit slope of the pool. The "6 mm" was selected because this is almost the closest point to the bed where the ADV can collect data without any difficulty. Moreover, the distance of 6 mm from the vegetation cover is selected in this study to compare the results of this study to those of other studies, such as Najafabadi et al. [43].

The power spectral densities of velocity components in Figure 15 are presented for three velocity components; stream-wise velocity is presented as blue curves; spanwise velocity components are presented as green curves, and vertical velocity components as red curves. Figure 15 shows a slight deviation in Kolmogorov's −5/3 power law for the vertical velocity component from the inertial subrange with the presence of a smaller bed slope, but it is generally valid (Figure 15a–c). The slight deviation resulted from the vegetation on the bed limiting its effect to near the vegetation. In addition, the Kolmogorov scale may mainly represent a balance between the kinetic and viscous energy in the flow field and not necessarily an eddy scale, which could explain this discrepancy in the velocity deviation of the Kolmogorov law.

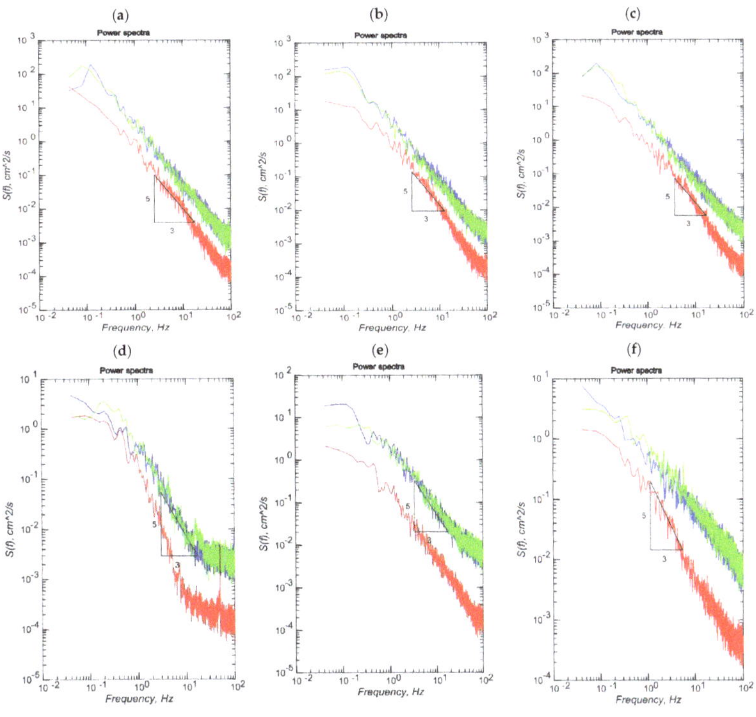

Figure 15. Spectral analysis. for two slopes at different sections along the pool, (**a**) Entrance slope vegetated pool 5°, (**b**) Middle section vegetated pool 5°, (**c**) Exit slope vegetated pool 5°, (**d**) Entrance slope vegetated pool 10°, (**e**) Middle section vegetated pool 10°, (**f**) Exit slope vegetated pool 10°.

Najafabadi et al. (2018) reported that Kolmogorov's −5/3 power law is prominent only at the deepest point of the bed form [43]. However, the result of the present study shows that the law rests universally for the trailing edge of the vegetation patch in the 3D bed form, as it is reported for the 2D pool with the 3D flow in a laboratory experiment over the gravel bed [44,45]. Figure 15 shows that the shedding frequency falls between 1.8 and 4.0 Hz. However, Nepf (1999) reported that this range is (1.8–3.6 Hz) for a random cylinder as emergent vegetation [46], but (1.0–1.8 Hz), as reported by Lacy and Roy (2007) [47]. Over the gravel bed between vegetation patches, Afzalimehr et al. (2021) pointed out that this range is from 3.0 to 7.7 Hz [48]. This comparison shows that the shedding frequency is affected by the changes in the bed-form slope as well as the presence of 3D bed forms and vegetation patches, resulting in higher values than those reported in the literature [49].

4. Discussions

In this section, a comparison of the results of this study to those of Nepf and Ghisalberti (2008) [4] and Wang et al. (2022) [42] has been presented in order to clarify the contribution of the present research. In the study of Nepf and Ghisalberti (2008) [4], they did not consider the effect of the 3D bed forms and a dense vegetation patch without flexibility, but they considered plant stems as rigid circular cylinders with a specific distance between the rigid circular cylinders. Thus, the turbulent structure in their study is completely different from that of the present study. Nepf and Ghisalberti (2008) [4] could investigate the flow structures inside the vegetation patch because the setup of vegetation stems was spaced apart, and this type of setting enabled them to collect data inside the vegetation patch using an ADV. A comparison of Reynolds stress (RS) distribution reported by Nepf and Ghisalberti (2008) [4] and that of the present study shows that the location of the bed form (entrance, middle, and exit sections) plays a significant role in the RS distribution and, thus, leads to different results. For example, the Reynolds stress distribution (Figure 5 in the study of Nepf and Ghisalberti (2008) [4]) displays a clear convex distribution. In the present study, however, the RS distribution is affected by the entrance slope of the pool and vegetation cover. As shown in some profiles, the negative RS values are due to the different contributions of bursting events and the role of the 3D bed form in the generation of secondary currents. In some runs (Run 7 in the middle pool section of Figure 7), the RS has an increasing trend rather than a decreasing trend toward the water surface. Their objective was to investigate the RS distribution and to identify the effect of vegetation deflection on creating a smaller vegetation height under stronger flow conditions [4]. In this study, it is emphasized that data are unable to collect inside the vegetation patch due to a very thin vegetation layer (the height of vegetation is only 2 cm) and no change in vegetation flexibility. However, Nepf and Ghisalberti (2008) [4] investigated flexible canopies, where the passage of the Kelvin–Helmholtz vortices generated a wave called Monami, showing a progressive wave along the canopy interface. Furthermore, they reported that the stem geometry played a significant role in turbulent flow structures. In the present study, we investigated the interaction of vegetation cover without rigid stems and with no flexibility at some important locations, such as the trailing edge of the vegetation patch. The results of this study indicate that the application of the specific (constant) values of velocity and Reynolds stress for all parts of a bed form will lead to incorrect estimation of hydraulic parameters (e.g., resistance).

Wang et al. (2022) [42] compared the influences of flexible vegetation on the flow velocity to those of rigid vegetation and found that the vegetation tilt has little effect on the difference between flow velocity inside the vegetation and that of vegetation canopy as well as the turbulence structure in the flow. However, the present study does not consider velocity structure inside the thin vegetation layer. The results of this study indicate that the bed-form slopes (entrance and exit sections), vegetation patches, and the location of the pool (entrance, middle, and exit sections) affect the velocity distribution. Results of the spectral analysis show that Kolomogrove −5/3 law over the 3D vegetation patch rests valid as it does for the 2D pools and 3D flows. Wang et al. (2022) [42] used a flatbed with vegetation with the height of 0.2 and 0.1 m and arranged vegetation elements in five rows in the streamwise direction and nine rows in the spanwise direction (there is a distance between pellets), where the flow passed through the vegetation patch. By comparing the TKE values in the flow with inclined rigid vegetation to those with vertical rigid vegetation, they found that the TKE in the former case was obviously smaller than that with vertical rigid vegetation. Wang et al. (2022) [42] investigated the energy spectra at the height of the vegetation canopy directly above a row of plants and in the middle of the two rows of plants. They reported that Kolmogorov's $-5/3$ scaling law is satisfied in the inertial subrange of flow with rigid vegetation. However, they haven't investigated the validation of this law at the trailing edge of the 3D vegetation patch.

5. Conclusions

The flow in any rivers in arid and semi-arid regions decreases dramatically during summer. As a consequence, vegetation patches develop in different parts of the rivers. When the water level increases in these rivers during spring, fall, and winter, the hydraulic conditions in rivers with the presence of vegetation patches are completely different from those without vegetation. Thus, some hydraulic parameters, including flow velocity and Reynolds stress, can be over- or underestimated due to the presence of the vegetation patches in channels. To better understand the changes in hydraulic parameters along 3D vegetation patches, experiments have been carried out in a laboratory flume with an artificial bed form. Two slopes of 5 and 10 degrees for both the entrance and exit sections of the pools have been utilized. The following results have been drawn with respect to the distribution of velocity profiles, Reynolds normal and shear stresses, TKE, turbulence intensities, skewness coefficients, and bursting process:

(1) In general, the bed-form slopes of both entrance and exit sections, the vegetation patches, and the location of the pool (entrance, middle pool, and exit section) affect the velocity profile, Reynolds shear stresses, TKE distribution, and the contribution of each bursting events on the turbulent flow structures. For a pool with entrance and exit slopes of 10 degrees, the TKE distribution has no specific form (Run 7). However, for a pool with a small entrance and exit slope of 5 degrees (Runs 3 and 4), the TKE distribution has a convex shape with the maximum value near the bed. Results of the quadrant analysis reveal that the bursting events at the trailing edge of the vegetation patch, where the flow exits from the vegetation patch to the gravel bed, display different distributions compared to other locations along the 3D vegetation patch. This difference plays a significant role in estimating flow resistance in open channels. The validation of the spectral analysis of the Kolmogorov power law for different bed-form slopes has been conducted;

(2) The Reynolds normal stress in the stream-wise direction is greater than those in both lateral and vertical directions. There is a disruption of normal stress values in the stream-wise direction due to the presence of secondary currents generated due to the different roughness of the channel bed and sidewalls of the flume. Therefore, the difference in the roughness and bed slope influences the normal Reynolds stress distributions. In the stream-wise direction, the region with the highest Reynolds shear stress (RSS) moves away from the bed. The RSS values in the zone of $z/H > 0.5$ decrease as the bed slope increases. A decrease in the reverse pressure gradient and the favorable pressure gradient affects the distribution of Reynolds stresses in both decelerating flow zone (entrance section of a pool) and the accelerating flow zone (exit section of a pool). When the entrance slope of the pool is smaller, the distribution of Reynolds stress is more regular toward the water surface;

(3) The quadrant analysis in this study focuses on the role of different bursting events at the trailing edge of the vegetation patch where the flow leaves the vegetation patch to the gravel bed. The results of this study clarified that both ejections and sweeps govern the turbulence structures and coherent motions at the trailing edge of the vegetation patch. The presence of the vegetation patch, as well as the changes of both

the entrance and exit slopes of a pool, generate the non-uniformity of the flow and increase the turbulence intensity and TKE in the downstream section of the vegetation patch. The sweep motion occurs in a narrower zone above the vegetation canopy. The sweep motions of bursting events are the dominant processes directly above the vegetation canopy, while the outward motion with slightly positive values has been observed at the front edge of the vegetation patch;

(4) The Kolmogorov $-5/3$ power law rests valid for the 3D vegetation patch. The geometry of the bed form, including both entrance and exit slopes of a pool, does not influence this law compared to the presence of a vegetation patch in a flat bed;

(5) The shedding frequency is affected by the changes in the bed-form slopes and the presence of vegetation, resulting in higher values than those reported in the literature.

Author Contributions: Conceptualization, M.R.T.M., P.P., H.A. and J.S.; Methodology, P.P., H.A. and J.S.; Validation, M.R.T.M.; Formal analysis, P.P. and H.A.; Investigation, M.R.T.M., P.P., H.A. and J.S.; Data curation, H.A., M.R.T.M. and P.P.; Writing—original draft, P.P.; Writing—review & editing, H.A., J.S. and M.R.T.M.; Supervision, H.A.; Project administration, H.A. All authors have read and agreed to the published version of the manuscript.

Funding: This research received no external funding.

Data Availability Statement: Data are available under request.

Conflicts of Interest: The authors declare no conflict of interest.

References

1. Graf, W.H.; Altinakar, M.S. *Fluvial Hydraulics: Flow and Transport Processes in Channels of Simple Geometry*; Wiley & Sons: New York, NY, USA, 1998.
2. Nepf, H.M. Hydrodynamics of vegetated channels. *J. Hydraul. Res.* **2012**, *50*, 262–279. [CrossRef]
3. Dey, S. *Fluvial Hydrodynamics, Hydrodynamic and Sediment Transport Phenomena*; Springer: Berlin/Heidelberg, Germany, 2014.
4. Nepf, H.; Ghisalberti, M. Flow and transport in channels with submerged vegetation. *Acta Geophys.* **2008**, *56*, 753–777. [CrossRef]
5. Shi, H.; Zhang, J.; Huai, W. Experimental study on velocity distributions, secondary currents, and coherent structures in open channel flow with submerged riparian vegetation. *Adv. Water Resour.* **2023**, *173*, 104406. [CrossRef]
6. MacVicar, B.J.; Rennie, C.D. Flow and turbulence redistribution in a straight artificial pool. *Water Resour. Res.* **2012**, *48*, W02503. [CrossRef]
7. Sawyer, A.M.; Pasternack, G.B.; Moir, H.J.; Fulton, A.A. Riffle-pool maintenance and flow convergence routing observed on a large gravel-bed river. *Geomorphology* **2010**, *114*, 143–160. [CrossRef]
8. Thompson, D.M. The role of vortex shedding in the scour of pools. *Adv. Water Resour.* **2006**, *29*, 121–129. [CrossRef]
9. Huai, W.-X.; Zhang, J.; Wang, W.-J.; Katul, G.G. Turbulence structure in open channel flow with partially covered artificial emergent vegetation. *J. Hydrol.* **2019**, *573*, 180–193. [CrossRef]
10. Tang, C.; Yi, Y.; Jia, W.; Zhang, S. Velocity and turbulence evolution in a flexible vegetation canopy in open channel flows. *J. Clean. Prod.* **2020**, *270*, 122543. [CrossRef]
11. Ortiz, A.C.; Ashton, A.; Nepf, H. Mean and turbulent velocity fields near rigid and flexible plants and the implications for deposition. *J. Geophys. Res. Earth Surf.* **2013**, *118*, 2585–2599. [CrossRef]
12. Zong, L.; Nepf, H. Vortex development behind a finite porous obstruction in a channel. *J. Fluid Mech.* **2012**, *691*, 368–391. [CrossRef]
13. Liu, C.; Hu, Z.; Lei, J.; Nepf, H. Vortex Structure and Sediment Deposition in the Wake behind a Finite Patch of Model Submerged Vegetation. *J. Hydraul. Eng.* **2018**, *144*, 04017065. [CrossRef]
14. Raupach, M.R.; Finnigan, J.J.; Brunei, Y. Coherent eddies and turbulence in vegetation canopies: The mixing-layer analogy. *Boundary Layer Meteorol.* **1996**, *78*, 351–382. [CrossRef]
15. Okamoto, T.-A.; Nezu, I. Spatial evolution of coherent motions in finite-length vegetation patch flow. *Environ. Fluid Mech.* **2013**, *13*, 417–434. [CrossRef]
16. Houra, T.; Tsuji, T.; Nagano, Y. Effects of adverse pressure gradient on quasi-coherent structures in turbulent boundary layer. *Int. J. Heat Fluid Flow* **2000**, *21*, 304–311. [CrossRef]
17. Ghisalberti, M.; Nepf, H.M. Mixing layers and coherent structures in vegetated aquatic flows. *J. Geophys. Res.* **2002**, *107*, 3-1–3-11.
18. Kumar, P.; Sharma, A. Experimental investigation of 3D flow properties around emergent rigid vegetation. *Ecohydrology* **2022**, *15*, e2474. [CrossRef]
19. Kazem, M.; Afzalimehr, H.; Sui, J. Characteristics of Turbulence in the Downstream Region of a Vegetation Patch. *Water* **2021**, *13*, 3468. [CrossRef]

20. Liu, M.; Huai, W.; Ji, B. Characteristics of the flow structures through and around a submerged canopy patch. *Phys. Fluids* **2021**, *33*, 035144. [CrossRef]
21. Bassey, O.B.; Agunwamba, J. Derived Models for the Prediction of Cole's and Dip Parameters for Velocity Gradients Determination in Open Natural Channels. *J. Civ. Environ. Res.* **1998**, *8*, 1–19.
22. Kironoto, B.; Graf, W.H.; Reynolds. Turbulence characteristics in rough uniform open-channel flow. *Proc. ICE Civ. Eng. Mar. Energy* **1994**, *106*, 333–344.
23. Dey, S.; Nath, T.K. Turbulence Characteristics in Flows Subjected to Boundary Injection and Suction. *J. Eng. Mech.* **2010**, *136*, 877–888. [CrossRef]
24. Afzalimehr, H.; Nosrati, K.; Kazem, M. Resistance to Flow in a Cobble-Gravel Bed River with Irregular Vegetation Patches and Pool-Riffle Bedforms (Case study: Padena Marbor River). *Ferdowsi Civ. Eng. JFCEI* **2021**, *2*, 35–50.
25. Song, T.; Chiew, Y.M. Turbulence Measurement in Nonuniform Open-Channel Flow Using Acoustic Doppler Velocimeter (ADV). *J. Eng. Mech.* **2001**, *127*, 219–232. [CrossRef]
26. Coles, D. The law of the wake in the turbulent boundary layer. *J. Fluid Mech.* **1956**, *1*, 191–226. [CrossRef]
27. MacWilliams, M.L.; Wheaton, J.M.; Pasternack, G.B.; Street, R.L.; Kitanidis, P.K. Flow convergence routing hypothesis for pool-riffle maintenance in alluvial rivers. *Water Resour. Res.* **2006**, *42*, W10427. [CrossRef]
28. MacVicar, B.; Roy, A.G. Hydrodynamics of a forced riffle pool in a gravel bed river: 1. Mean velocity and turbulence intensity. *Water Resour. Res.* **2007**, *43*, W12401.
29. Finnigan, J. Turbulence in Plant Canopies. *Annu. Rev. Fluid Mech.* **2000**, *32*, 519–571. [CrossRef]
30. Thornton, C.I.; Abt, S.R.; Morris, C.E.; Fischenich, J.C. Calculating shear stress at channel-overbank interfaces in straight channels with vegetated floodplains. *J. Hydraul. Eng.* **2000**, *126*, 929–936. [CrossRef]
31. Nepf, H.M.; Vivoni, E. Flow structure in depth-limited, vegetated flow. *J. Geophys. Res. Oceans* **2000**, *105*, 28547–28557. [CrossRef]
32. Yang, S.-Q.; Chow, A.T. Turbulence structures in non-uniform flows. *Adv. Water Resour.* **2008**, *31*, 1344–1351. [CrossRef]
33. McLean, S.R.; Nikora, V.I. Characteristics of turbulent unidirectional flow over rough beds: Double-averaging perspective with particular focus on sand dunes and gravel beds. *Water Resour. Res.* **2006**, *42*, W10409. [CrossRef]
34. Nezu, I.; Nakagawa, H. *Turbulence in Open-Channel Flows*; Routledge: Abingdon, UK, 2017.
35. Parvizi, P.; Afzalimehr, H.; Sui, J.; Raeisifar, H.R.; Eftekhari, A.R. Characteristics of Shallow Flows in a Vegetated Pool—An Experimental Study. *Water* **2023**, *15*, 205. [CrossRef]
36. Nezu, I.; Sanjou, M. Turburence structure and coherent motion in vegetated canopy open-channel flows. *J. Hydro-Environ. Res.* **2008**, *2*, 62–90. [CrossRef]
37. Shivpure, V.; Devi, T.B.; Kumar, B. Turbulent characteristics of densely flexible submerged vegetated channel. *ISH J. Hydraul. Eng.* **2016**, *22*, 220–226. [CrossRef]
38. Nezu, I. Turbulence intensities in open channel flows. in Proceedings of the Japan Society of Civil Engineers. *J. JSCE* **1977**, *261*, 61–76. (In Japanese)
39. Wilson, C.A.M.E.; Stoesser, T.; Bates, P.D.; Batemann Pinzen, A. Open channel flow through different forms of submerged flexible vegetation. *J. Hydraul. Eng.* **2003**, *129*, 847–853. [CrossRef]
40. Wilkinson, S.N.; Keller, R.J.; Rutherfurd, I.D. Phase-shifts in shear stress as an explanation for the maintenance of pool–riffle sequences. *Earth Surf. Process. Landf.* **2004**, *29*, 737–753. [CrossRef]
41. Nikora, V. 3 Hydrodynamics of gravel-bed rivers: Scale issues. *Dev. Earth Surf. Proc.* **2007**, *11*, 61–81.
42. Wang, J.; He, G.; Dey, S.; Fang, H. Influence of submerged flexible vegetation on turbulence in an open-channel flow. *J. Fluid Mech.* **2022**, *947*, A31. [CrossRef]
43. Najafabadi, E.F.; Afzalimehr, H.; Rowiński, P.M. Flow structure through a fluvial pool-riffle sequence—Case study. *J. Hydro-Environ. Res.* **2018**, *19*, 1–15. [CrossRef]
44. Nikora, V.; Goring, D. Flow Turbulence over Fixed and Weakly Mobile Gravel Beds. *J. Hydraul. Eng.* **2000**, *126*, 679–690. [CrossRef]
45. Singh, A.; Porté-Agel, F.; Foufoula-Georgiou, E. On the influence of gravel bed dynamics on velocity power spectra. *Water Resour. Res.* **2010**, *46*, W04509. [CrossRef]
46. Nepf, H.M. Drag, turbulence, and diffusion in flow through emergent vegetation. *Water Resour. Res.* **1999**, *35*, 479–489.
47. Lacey, R.W.J.; Roy, A.G. A comparative study of the turbulent flow field with and without a pebble cluster in a gravel bed river. *Water Resour. Res.* **2007**, *43*, W05502. [CrossRef]
48. Afzalimehr, H.; Riazi, P.; Jahadi, M.; Singh, V.P. Effect of vegetation patches on flow structures and the estimation of friction factor. *ISH J. Hydraul. Eng.* **2021**, *27*, 390–400. [CrossRef]
49. Yue, W.; Meneveau, C.; Parlange, M.B.; Zhu, W.; Van Hout, R.; Katz, J. A comparative quadrant analysis of turbulence in a plant canopy. *Water Resour. Res.* **2007**, *43*, W05422. [CrossRef]

water
MDPI

Article

Double Parameters Generalization of Water-Blocking Effect of Submerged Vegetation

Chunlin Qiu [1], Jiesheng Huang [1], Shihe Liu [1,*] and Wenhao Pan [2]

[1] State Key Laboratory of Water Resources and Hydropower Engineering Science, Wuhan University, Wuhan 430072, China
[2] Changjiang Institute of Survey, Planning, Design and Research, Wuhan 430010, China
* Correspondence: shiheliu@126.com

Abstract: Submerged vegetation has strong vitality, and the root system is highly developed. Because this vegetation has a good bank-solidifying-and-beautifying effect, it is widely used in ecological river construction. However, the open channel flow field and water-blocking mechanism of submerged vegetation are complicated. It is not convenient to use this kind of original model directly in engineering calculation, but it can be much more convenient if the original model is generalized into a simple model. However, there are not many generalization models, so it is necessary to propose a simple generalization model of the water-blocking effect of submerged vegetation to facilitate engineering calculation. Upon theoretical analysis, numerical calculation and experiment data analysis, the following conclusions are obtained: As the basis of generalization, in order to make up for the deficiency of experimental results, a new numerical simulation model for the flow field of submerged vegetation open channel flow was firstly proposed. For the purpose of this research, a simple generalization model of the water-blocking effect of submerged vegetation was proposed. Finally, two parameters of generalized roughness coefficient and virtual channel elevation were obtained to reflect the water-blocking effect. They can be substituted directly into a planar two-dimensional model in engineering. It achieves the ultimate goal of convenient engineering calculation.

Keywords: submerged vegetation; open channel flow; numerical simulation; generalization model; water-blocking effect

Citation: Qiu, C.; Huang, J.; Liu, S.; Pan, W. Double Parameters Generalization of Water-Blocking Effect of Submerged Vegetation. *Water* **2023**, *15*, 764. https://doi.org/10.3390/w15040764

Academic Editors: Jueyi Sui and Hossein Afzalimehr

Received: 26 December 2022
Revised: 6 February 2023
Accepted: 8 February 2023
Published: 15 February 2023

1. Introduction

Submerged vegetation refers to herbs that are submerged in water. Such plants are vigorous and have strong roots. They do well in solidifying banks and beaches, greening and beautification. Therefore, they are widely used in slope protection and bank protection engineering and river ecological transformation. However, the flow field and water-blocking mechanism of submerged vegetation are complicated. In particular, the flow field of submerged vegetation open channel flow is a three-dimensional model with complex descriptions. No matter wether the models are constructed by commercial software or experiments, they have this feature. It is not convenient to use this kind of original model directly in engineering calculation, but it can be much more convenient if the original model is generalized into a simple model. From the perspective of engineering applications, planar two-dimensional mathematical model calculation hasbecome an important means to study the characteristics of river flow. Therefore, the simple generalized model also needs to be connected with the planar two-dimensional model in the engineering calculation. How to build a simple and applicable generalization model of water-blocking effect of submerged vegetation is necessary and vital. It is directly related to the accuracy of eco-river flow simulation results.

At present, there are not many simple generalization models. However, results of original models are abundant. The characteristics of open channel flow with submerged vegetation are generally studied experimentally and by mathematical model calculation. Such experiments include those by Dunn et al. [1], who used wooden sticks to simulate submerged vegetation. These sticks were arranged in uniform flow in an open channel. The authors measured the flow velocity and Reynolds stress. The data are complete and detailed and can reflect the characteristics of the water flow with submerged vegetation. Therefore, the present paper uses that experimental data for analysis. The stratification phenomenon of flow with submerged vegetation was explained by Raupach et al. [2] based on experiments. An empirical formula for inflection depth (i.e., the height from the top of the vegetation down to the boundary between the upper layer and lower layer) was given by Nepf et al. [3] based on experiments. Both groups concluded that there is a transition zone in the submerged vegetation flow layer, which inspired the definition of influenced depth and related research in this article. Plastic sheets were used to simulate submerged vegetation by Nezu et al. [4], who used a laser Doppler anemometer (LDA) and a particle imaging velocimeter (PIV) to measure the flow field between sheets. The data are accurate and abundant and can reflect the characteristics of submerged vegetation flow. The experimental data were also used to verify numerical flow field in this article. Juez et al. [5] used synthetic grass and wood stick to simulate vegetation. PIV was used to measure data, then, cross-sectional distribution of the streamwise velocity and Reynolds shear stress were obtained to study the characteristics of flow field. The total sediment mass settled and the surface occupied by the sediments were also assessed. Vargas-Luna et al. [6] used the grass-type plastic vegetation, provided with sticks mimicking roots, to represent the vegetation. Vertical velocity and flow discharge were measured, and sediment transport rates were also estimated. The abundant experimental data can also be used for in-depth analysis. The flow structure in vegetated flow can also be affected by vegetation morphology, distribution, water depth and discharge, etc. [7–9]. Structures and phenomena such as mixing layers, vegetated shear layers, vortex, "monami", coherent motions can be found in the vegetated flow throughexperiments [10–15]. For numerical simulation, the direct numerical simulation method (DNS) was used to solve the N-S equation. Related results were obtained by Coceal et al. [16], Ji et al. [17], etc. However, due to the limitation of computing ability, it is difficult to apply these results to the study of the characteristics of vegetation group flow. Therefore, the double averaging method is always used to simulate the pore flow in the vegetation layer and the free flow above the vegetation layer. The related results can be seen in articles of Choi et al. [18], Kang et al. [19], etc. In addition, numerical analysis models such as turbulence models, shallow water models, three-zone model and longitudinal diffusion model are proposed to describe different vegetation flows [20–27]. From the point of engineering application, people often focus on the influence of submerged vegetation on channel discharge capacity. For this, boundary treatment is needed to generalize the water-blocking effect of submerged vegetation, which is also one of the research directions of this article.

Many experimental studies have been carried out on open channel flow with submerged vegetation. Therefore, abundant experimental data have been accumulated. The flow field characteristics and roughness coefficient value are also studied. In addition, the flow field is numerically simulated by researchers using the double-average method. The existing research results show the following feature: the flow field characteristics and associated scientific problems are given more attention than general characteristics and engineering problems. In order to facilitate engineering calculation, a simple generalization model of water-blocking effect of submerged vegetation is needed. However, there are few such simple models at present, so the research objective of this article is to propose a simple generalization model of the water-blocking effect of submerged vegetation. The research of this article is mainly divided into two steps: (1) As the basis of generalization, in order to make up for the deficiency of experimental results, a new numerical simulation model for the flow field of submerged vegetation open channel flow was firstly proposed. (2) For the purpose of this research, a simple generalization model of the water-blocking effect of submerged vegetation was proposed. In the past, researchers focused only on changes in the roughness coefficient with or without vegetation. In fact, except that the roughness coefficient of open channel flow with submerged vegetation is different from that without vegetation, the position of the lower boundary (i.e., theoretical zero point's position; see Section 2.2.1 for definitions) should be adjusted accordingly.

2. Materials and Methods

2.1. Numerical Simulation of Open Channel Flow with Submerged Vegetation

2.1.1. Mathematical Model

The double-average method is used to study the characteristics of open channel flow with submerged vegetation. The so-called double average refers to the Reynolds average (i.e., time average for the constant flow studied in this paper) and the local spatial average for the governing equations (N-S equations) of the flow. By means of the Reynolds average, the instantaneous variable ψ is decomposed into the time mean variable $\overline{\psi}$ and fluctuation variable ψ'. The spatial local average is averaged over a sufficiently large plane area. The time mean variable $\overline{\psi}$ can be further decomposed into the local spatial mean $\langle\overline{\psi}\rangle$ and local spatial pulsation ψ'' to eliminate the uneven distribution of flow field statistics caused by the uneven distribution of vegetation. For the constant uniform open channel flow with submerged vegetation, we assume that (1) the stem of vegetation is rigid, and the surface meets the condition of no slippage; (2) vegetation is uniformly distributed in space or consists of a number of evenly distributed patches of plants. The two-dimensional governing equation of the section after double averaging and sealing is:

$$gi_b - \frac{2(1-\theta)}{\pi D} C_D \langle\overline{u_1}\rangle^2 + \frac{d}{dx_3}\left((\nu+\nu_t)\frac{d\langle\overline{u_1}\rangle}{dx_3}\right) = 0 \quad (1)$$

where

$\langle\overline{u_1}\rangle$: longitudinal flow velocity after double averaging (m/s);
i_b: bottom slope of the open channel;
D: diameter of submerged vegetation (m);
θ: porosity of the flora;
C_D: drag coefficient;
ν: molecular viscosity coefficient (m^2/s);
ν_t: turbulence viscosity coefficient (m^2/s).

There are two different turbulent generating mechanisms in open channel flow with submerged vegetation. That is, the large-scale shear turbulence and the smaller scale stem turbulence, of which the scale is comparable to the diameter of submerged vegetation stem. In this paper, the subscripts s and w are used to represent the corresponding turbulence generating mechanism. The turbulent kinetic energy k and the turbulent kinetic energy dissipation rate ε are decomposed into two parts:

$$k = k_s + k_w \quad (2)$$

$$\varepsilon = \varepsilon_s + \varepsilon_w \quad (3)$$

The governing equations of k_s, k_w and ε_s are as follows.

$$\frac{d}{dx_3}\left(\left(\frac{\nu_t}{\sigma_k}+\nu\right)\frac{dk_s}{dx_3}\right) + G_s - W_D - \varepsilon_s = 0 \quad (4)$$

$$\frac{d}{dx_3}\left(\left(\frac{\nu_t}{\sigma_k}+\nu\right)\frac{dk_w}{dx_3}\right) + G_w + W_D - \varepsilon_w = 0 \quad (5)$$

$$\frac{d}{dx_3}\left(\frac{\nu_t}{\sigma_\varepsilon}\frac{d\varepsilon_s}{dx_3}\right) + C_{\varepsilon 1}\frac{\varepsilon_s}{k_s}G_s - C_{\varepsilon 2}\frac{{\varepsilon_s}^2}{k_s} = 0 \quad (6)$$

For submerged vegetation, in the vegetation layer, the wake turbulence kinetic energy generation mechanism and dissipation mechanism are similar to those of emergent vegetation. Therefore, the equation of ε_w is directly taken to be:

$$\varepsilon_w = \beta_3 \frac{{k_w}^{3/2}}{D} \tag{7}$$

The values of σ_k, σ_ε, $C_{\varepsilon 1}$, and $C_{\varepsilon 2}$ in Equations (4)–(6) are consistent with those in the standard k-ε model. W_D is a term reflecting the transformation of shear turbulent kinetic energy to stem-scale wake kinetic energy. The formula is

$$W_D = 4\beta_1 a \langle \overline{u_1} \rangle C_D k_s \tag{8}$$

The turbulent viscosity coefficient can be calculated by

$$\nu_t = C_\mu \frac{k_s}{\varepsilon_s} \tag{9}$$

where $C_\mu = 0.09$.

2.1.2. Verification

The data of Nezu et al. [4] and Dunn et al. [1] were used to verify the mathematical model.

1. Verification 1: data from Nezu et al. [4];
2. Verification 2: data from Dunn et al. [1].

The flume in the experiment of Nezu et al. [4] was 10 m long and 0.4 m wide. Thin sheets of plastic (5 cm high, 0.8 cm wide and 0.1 cm thick) were used to simulate vegetation. Plastic sheets were arranged in rows in the 9 m long experimental section. An LDA and PIV were used to measure the flow velocity under different vegetation densities. The measurement vertical lines were arranged 7 m behind the entrance of the experimental area to ensure that the flow fully developed to be two-dimensional turbulence. Three data series were selected for verification calculation, which were measured in high Reynolds number conditions (A-10, B-10 and C-10) by the LDA. Vertical distribution of velocity and Reynolds stress data were used to verify. The main flow features of the experiments used to validate the mathematical model are shown in Table 1.

Table 1. Main flow features of the experiments in Nezu et al. [4].

Case	Water Depth H (cm)	Height of Vegetation Element h (cm)	Mean Bulk Velocity U_m (cm/s)	U_h [1] (cm/s)	Friction Velocity U* [2] (cm/s)	Fr	Reynolds Number Re
A-10	15.0	5.0	12.0	5.83	2.76	0.10	1.8×10^4
B-10	15.0	5.0	12.0	5.25	2.53	0.10	1.8×10^4
C-10	15.0	5.0	12.0	5.77	2.31	0.10	1.8×10^4

[1] U_h is the space-averaged (horizontally averaged) mean velocity at the vegetation top edge. [2] U* here is defined as the value of Reynolds stress at the vegetation top edge.

The experiments in Dunn et al. [1] were carried out in an experimental flume with a length of 19.50 m, a width of 0.91 m and a depth of 0.61 m. The flume was long enough to form uniform flow in the middle section. Wooden sticks with a diameter of 0.64 cm and a height of 12 cm were staggered in the water flume to simulate rigid vegetation. The flow velocity was measured by ADV. Ten groups of data under different experimental conditions were verified, including vertical distribution of velocity and Reynolds stress data. The main flow features of the experiments used to validate the mathematical model are shown in Table 2.

Table 2. Main flow features of the experiments in Dunn et al. [1].

Experiment	Q (L/s)	Bed Slop S (%)	Bulk FlowVelocity (m/s)	Water Depth (m)	Reynolds Number Re	Manning's n ($m^{1/6}$)	Fr
1	179	0.36	0.587	0.335	2.24×10^5	0.034	0.33
2	88	0.36	0.422	0.229	1.13×10^5	0.041	0.29
3	46	0.36	0.308	0.164	0.57×10^5	0.048	0.24
4	178	0.76	0.709	0.276	1.91×10^5	0.038	0.36
5	98	0.76	0.531	0.203	1.25×10^5	0.045	0.37
6	178	0.36	0.733	0.267	1.96×10^5	0.025	0.39
7	95	0.36	0.570	0.183	1.20×10^5	0.027	0.42
8	180	0.36	0.506	0.391	2.58×10^5	0.042	0.29
9	58	0.36	0.298	0.214	0.70×10^5	0.056	0.19
10	180	1.61	0.746	0.265	2.03×10^5	0.052	0.40
11	177	0.36	0.625	0.311	2.22×10^5	0.031	0.35
12	181	1.08	0.854	0.233	2.38×10^5	0.036	0.58

2.2. Study of the Water-Blocking Effect Generalization of Submerged Vegetation

For uniformly distributed submerged vegetation, the vegetation density, a, is an important parameter to generalize its influence (Dunn et al. [1]). The definition is formulated as follows:

$$a = \frac{A}{V} = \frac{D \cdot h}{(\Delta x_1 \cdot \Delta x_2)h} \tag{10}$$

where

a: vegetation density (1/m);

A: inflow area of a single cylinder (m^2);

V: volume affected by a single cylinder (m^3);

D: diameter of the cylinder (m);

h: height of submerged vegetation (m);

Δx_1: length of the volume V in direction x_1 when averaged locally (m), x_1 and x_2 can be streamwise direction and spanwise direction, respectively;

Δx_2: length of the volume V in direction x_2 when averaged locally (m).

To describe the effect of submerged vegetation, we multiply vegetation density, a, by the height of submerged vegetation, h, and then divide it by water depth, H, creating a comprehensive parameter ah/H to describe the water-blocking effect of submerged vegetation on water. The larger ah/H is, the stronger resistance vegetation act on water, i.e., the denser the vegetation groups are. Otherwise, the sparser the vegetation groups are.

2.2.1. Generalization Model and Principle

Based on the above numerical simulation results of the double-average velocity and Reynolds stress of open channel flow with submerged vegetation, the water-blocking effect of submerged vegetation is generalized by adjusting the position of the lower boundary and modifying the roughness coefficient, as shown in Figure 1. Here, the meaning of "generalization" can be explained as: the original complex flow field model of submerged vegetation open channel flow is simplified into a simple flow field model to ensure that before and after simplification, the water-blocking effect remains equivalent. In this way, it is convenient to use the generalized simple flow field model parameters to do engineering calculation. In Figure 1, Δ_s is the height of the theoretical zero point, and the flow velocity here is zero after generalization; Δ_p is the height of the equivalent virtual obstacle after generalization, and it's called virtual channel height; Δ_h is the thickness of the affected region in which submerged vegetation changes the longitudinal velocity above the top of vegetation; and $\Delta_s + \Delta_h$ is defined as the influence depth. Before generalization, the real flow is the pore flow in the vegetation layer and the free flow above the vegetation layer, both of which are described uniformly by means of double averaging. However, flow after generalization (called generalized flow) is constant uniform flow in the open channel described by the Reynolds average method. The latter's effective water depth H_m can be defined as $H_m = H - h + \Delta_s$.

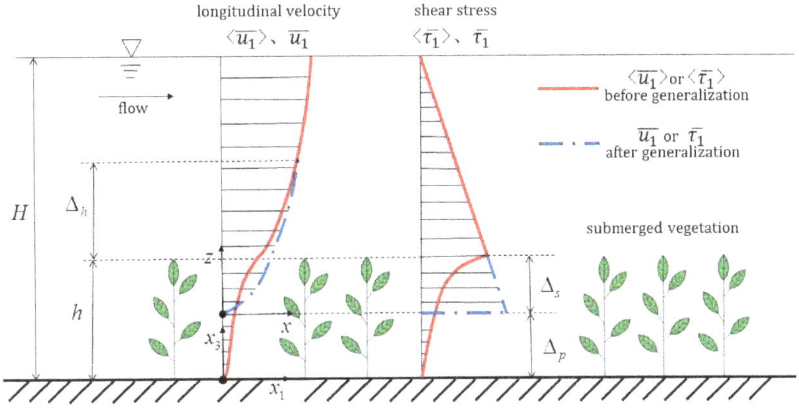

Figure 1. Generalization of flow longitudinal velocity and shear stress.

The principles of generalizing the water-blocking effect of submerged vegetation are as follows:

(1) The quantity of flow before generalization equals that after generalization. This is required by the conservation of mass. As shown in Figure 1, the equation applies:

$$\int_0^{h+\Delta_h} \langle \overline{u_1} \rangle dx_3 = \int_0^{\Delta_h+\Delta_s} \overline{u_1} dz \tag{11}$$

In Equation (11), Δ_s is the height discrepancy from the theoretical zero point of velocity (i.e., the top of virtual channel after generalization) to the top of the submerged vegetation layer.

(2) At the top of the influence depth (i.e., $z = \Delta_h + \Delta_s$), the double-average velocity before generalization equals the velocity after generalization.

$$\langle \overline{u_1} \rangle |_{x_3=h+\Delta_h} = \overline{u_1}|_{z=\Delta_h+\Delta_s} \tag{12}$$

2.2.2. Handling of Key Issues

In the process of solving Equations (11) and (12) to calculate Δ_s and Δ_h, the following problems need to be solved. One is the vertical distribution of the longitudinal velocity of the constant uniform flow, which is generated after generalization. In this paper, we choose the logarithmic velocity distribution formula (the constant shear stress hypothesis is also used, i.e., $\tau_b = \rho u_{*b}^2$). The other is that many flow characteristic variables after generalization are interrelated and need a coupling solution. This paper uses the following methods to obtain simultaneous equations.

(1) At $x_3 = \Delta_p = h - \Delta_s$, $\overline{u_1} = 0$, i.e., the theoretical zero point of velocity is at the place of $x_3 = \Delta_p$. Let $z = x_3 - \Delta_p$, and take the theoretical zero point as the origin to establish a new coordinate axis z. Then, the flow velocity distribution after generalization is:

$$\overline{u_1} = \frac{u_{*b}}{\kappa} ln \frac{(z+\alpha)}{\Delta_s} + 8.5 u_{*b} \tag{13}$$

Because at $z = 0$, $\overline{u_1} = 0$, we further obtain

$$\alpha = e^{-8.5\kappa} \cdot \Delta_s \tag{14}$$

We substitute Equation (13) into Equation (11) to obtain

$$\int_0^{h+\Delta_h} \langle \overline{u_1} \rangle dx_3 = \int_0^{\Delta_h+\Delta_s} \left[\frac{u_{*b}}{\kappa} ln \frac{(z+\alpha)}{\Delta_s} + 8.5u_{*b} \right] dz \tag{15}$$

(2) Above the top of influence depth, i.e., $z > \Delta_h + \Delta_s$, the flow velocity after generalization equals that before generalization. In the calculation process, the absolute value of the difference between the two is set to be less than a minimal value ε_m. That is, 0

$$\frac{\sum_{i=1}^{m} |\langle \overline{u_1} \rangle - \overline{u_1}|_i}{m} < \varepsilon_m \tag{16}$$

where i represents the velocity data i above the top of the influence depth. $i = 1, 2, \ldots, m$. Here, $|\langle \overline{u_1} \rangle - \overline{u_1}|_i$ is the absolute value of the difference between velocity data i before generalization and that after generalization. m is the total number of velocity data points above the top of the influence depth.

(3) According to the generalization of the shear stress distribution in Figure 1, we take

$$\tau_h = \beta \tau_b \tag{17}$$

In Equation (17), τ_h and τ_b, respectively, represent the shear stress at the top of the vegetation layer before generalization and the shear stress at the top of the virtual channel after generalization. We take $\beta = 0.75$ (determined after trial). By combining Equations (15)–(17), Δ_s, Δ_h and u_{*b} can be solved. The roughness coefficient n_p is determined by the Chezy formula:

$$C = \frac{U}{\sqrt{H_m S_f}} = \frac{U\sqrt{g}}{u_{*b}} = \frac{1}{n_p} H_m^{\frac{1}{6}} \tag{18}$$

Then,

$$n_p = \frac{u_{*b} H_m^{\frac{1}{6}}}{\sqrt{g} U} \tag{19}$$

In Equation (19), H_m is the effective water depth (the definition is in Section 2.2.1), $U = \frac{Q}{BH_m}$, Q is the flow discharge quantity, B is the channel width, and g is the acceleration of gravity.

2.2.3. Data for Generalization

Among the published experimental data, Dunn et al. [1] is relatively systematic and complete. The 9 series of experimental velocity data from Dunn et al. [1] in different working conditions were selected for double-average treatment and numerical calculation, and then, the water-blocking generalization mentioned above was carried out using these data.

3. Results

3.1. Numerical Simulation of Open Channel Flow with Submerged Vegetation

3.1.1. Verification 1

As shown in Figure 2, the calculated results were in good agreement with the experimental results. Figure 2 presents the verification calculation results of the vertical distribution of velocity and Reynolds stress. The dimensionless results are obtained after original results being divided by the frictional velocity u_* at the top of the vegetation layer, and are plotted in the figure. The frictional velocity is defined as $u_*^2 = -\left\langle \overline{u_1' u_3'} \right\rangle \Big|_{x_3=h}$.

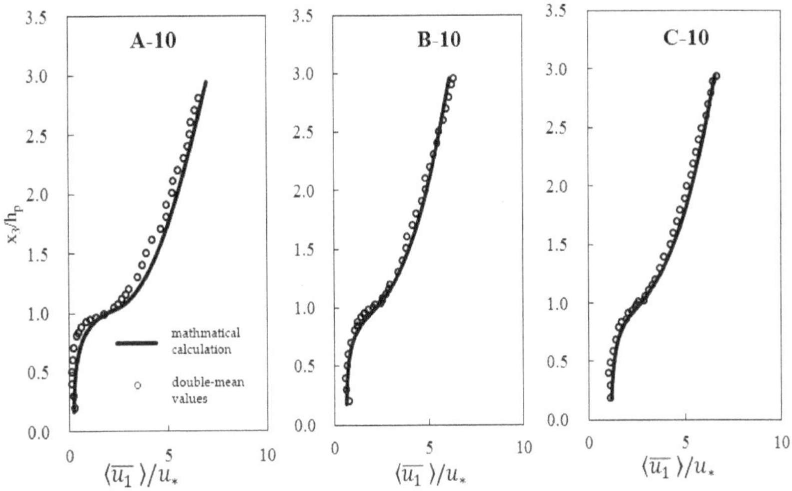

(a)

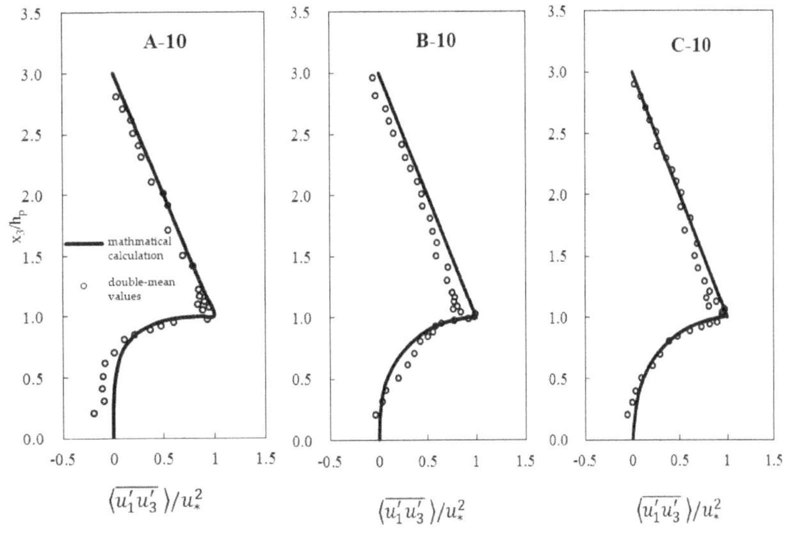

(b)

Figure 2. Verification calculation results of the double-mean longitudinal velocity and Reynolds stress (experimental data source: Nezu et al. [4]): (**a**) verification calculation results of the double mean longitudinal velocity; (**b**) verification calculation results of the double-mean Reynolds stress.

3.1.2. Verification 2

As shown in Figure 3, the verified results were basically consistent with the experimental results. Due to space limitations, Figure 3 only presents several of the verification results (including vertical distribution of velocity and Reynolds stress).

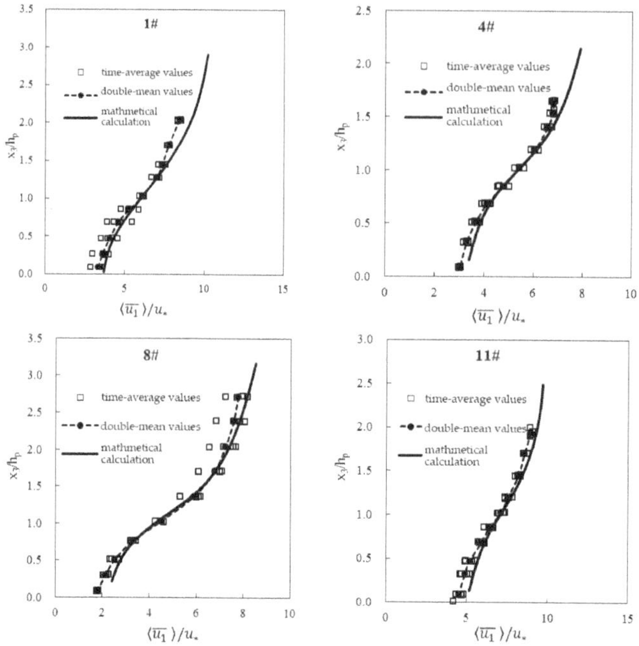

(a)

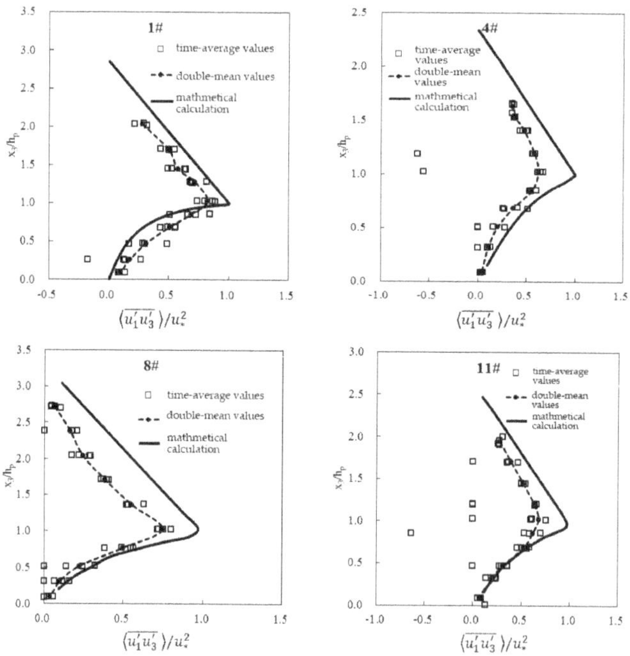

(b)

Figure 3. Verification calculation results of the double-mean longitudinal velocity and Reynolds stress (experimental data source: Dunn et al. [1]): (**a**) verification calculation results of the double mean longitudinal velocity; (**b**) verification calculation results of the double-mean Reynolds stress.

3.2. *Water-Blocking Effect Generalization of Submerged Vegetation*

3.2.1. Preliminary Result

Figure 4 shows the comparison between the double-average velocity and generalized velocity in different working conditions. Figure 4 verifies the accuracy of the generalization mode shown in Figure 1. Illustrations in Figure 4: Δ_p is the height of the generalized obstacle—the virtual channel after generalization. At this height, the generalized flow velocity is zero. From here up, the generalized flow velocity increases from zero. When it rises to the height of vegetation, h, the generalized flow velocity still does not recover to the original flow velocity due to the effect of vegetation obstruction. The height difference in this section is the theoretical zero point's height of the velocity, Δ_s. Then continue to rise to the highest dotted line, generalized flow velocity adjustment is completed, and it returns to the original flow velocity. That is, these two kinds of velocity overlap. The height difference of this section is the influence thickness, Δ_h. Then these two kinds of flow velocity remain consistently overlap, basically restore the original flow.

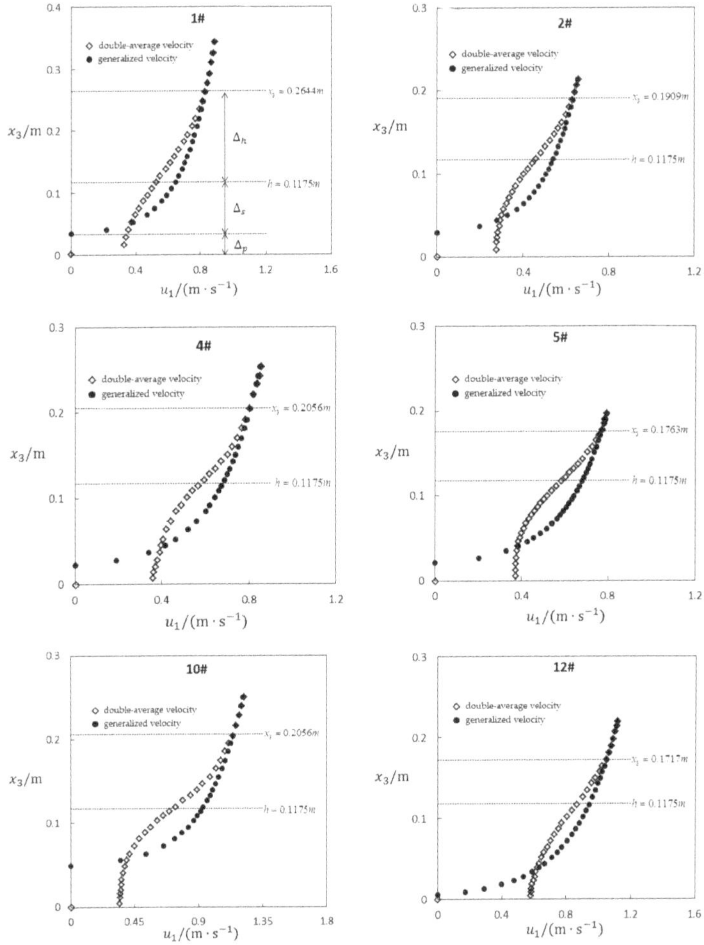

Figure 4. Comparison of the double-mean velocity $\langle \overline{u_1} \rangle$ and generalized velocity $\overline{u_1}$. (The upper dotted line is the depth of influence, the middle dotted line is the height of vegetation and the lower dotted line is the top of the virtual canal).

The results for the theoretical zero point's position and roughness coefficient are introduced below.

3.2.2. Height of the Theoretical Zero Point Δ_s

The height of the theoretical zero point Δ_s is nondimensionalized by vegetation height h, and a dimensionless parameter Δ_s/h is obtained. Figure 5 shows the relationship between Δ_s/h and ah/H. The fitting formula is

$$\frac{\Delta_s}{h} = \exp\left(-0.4774\frac{ah}{H} - 0.0159\right) \quad (20)$$

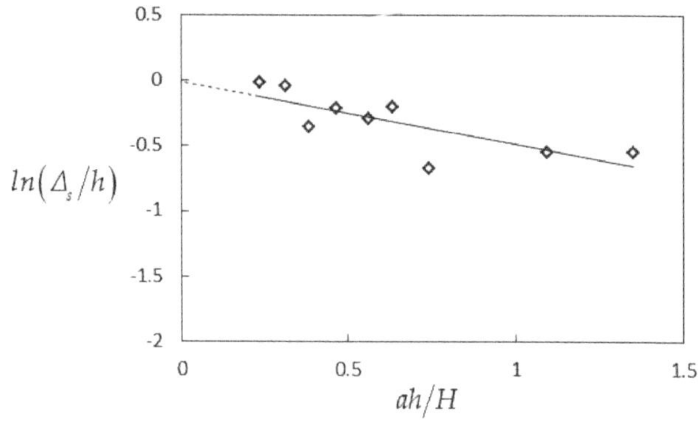

Figure 5. Height of the theoretical zero point Δ_s varies with the comprehensive parameter *ah/H*.

Figure 5 and Equation (20) show a trend of gradually decreasing Δ_s/h with increasing *ah/H*. That is, when the water depth H and vegetation height h remain unchanged, the denser the vegetation arrangement is, the higher the theoretical zero point (in other words, the top of virtual channel rises higher). Analysis for the reason: increasing *ah/H* means a denser vegetation arrangement. Then, there is more resistance to the flow, and the water-blocking effect of vegetation is strengthened. After the generalization, the relative height of virtual channel (Δ_p/h) will increase. Accordingly, the relative height of the theoretical zero point (Δ_s/h) goes down. When *ah/H* is zero, there is no vegetation holding back water, so there is no virtual channel after generalization. That is, the virtual channel height Δ_p is zero. Then, the theoretical zero point's height Δ_s is equal to the height of vegetation, h. Therefore, the y-coordinate value is zero. The extension prediction of fitting line accords with the actual law. The fitting line is suitable to express the relation between vegetation density and generalization results.

Figure 5 shows the variation trend of Δ_s. Here, $\Delta_p = h - \Delta_s$, so we can obtain the variation trend of Δ_p. Virtual channel elevation Δ_p is an important parameter obtained after generalization of water-blocking effect. Because it can be directly put into the plane two-dimensional model for engineering calculation.

3.2.3. Influence Depth $\Delta_s + \Delta_h$

The influence depth $\Delta_s + \Delta_h$ is nondimensionalized by vegetation height h. Figure 6 shows the $(\Delta_s + \Delta_h)/h$~ah/H graph. The fitting formula is

$$\frac{\Delta_s + \Delta_h}{h} = exp\left(-0.1924\frac{ah}{H} + 0.2525\right) \quad (21)$$

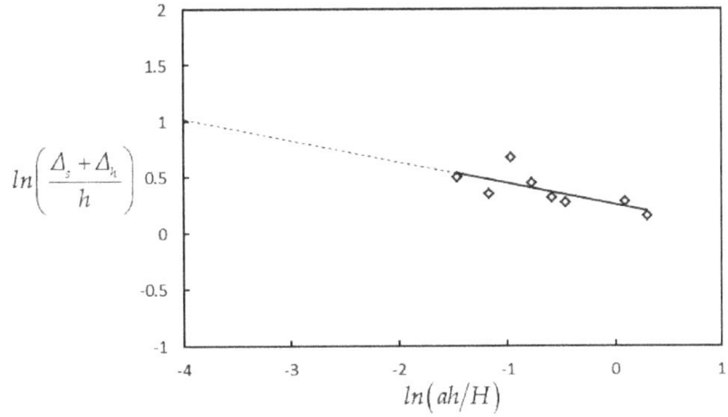

Figure 6. Influence depth $\Delta_s + \Delta_h$ varies with the comprehensive parameter *ah/H*.

As shown in Figure 6 and Equation (21), in the case of the same vegetation height h, the denser the vegetation arrangement is, the smaller the influence depth $\Delta_s + \Delta_h$ is. Further analysis: similarly, the greater *ah/H*, the stronger the water-blocking effect of vegetation. Then, the water needs a higher depth in the vertical direction to restore its original flow. Therefore, the greater the range of influence thickness zone is, i.e., the greater Δ_h is. However, the actual increase amplitude of influence thickness, Δ_h, is small. This is related to the change in the influence depth, $\Delta_h + \Delta_s$, subsequently. According to Section 3.2.2, the smaller the theoretical zero point's height Δ_s is, and the reduction is greater than the increase in Δ_h. Therefore, the influence depth $\Delta_s + \Delta_h$ shows a decreasing trend as a whole. The height of vegetation h is fixed, so the ordinate value in Figure 6 shows a decreasing trend. When the abscissa value in the Figure 6 is -4, the extension value of the fitting line is exactly about 1. In addition, the relative value of the influence depth $(\Delta_s + \Delta_h)/h$ is about e. In other words, in the condition of this vegetation arrangement, the influence depth $\Delta_s + \Delta_h$ is about e times of the height of vegetation h (about 2.72 times).

3.2.4. Generalized Roughness Coefficient n_p

Figure 7 shows the calculation results of generalized roughness coefficient n_p. Here, n_p calculated from the experimental data in Dunn et al. [1] basically remains at approximately 0.029. However, there is also a trend of a slight increase in n_p with increasing *ah/H*. The denser the vegetation arrangement is, the higher generalized roughness coefficient n_p is. The fitting formula is

$$n_p = 0.003ln(\frac{ah}{H}) + 0.029 \tag{22}$$

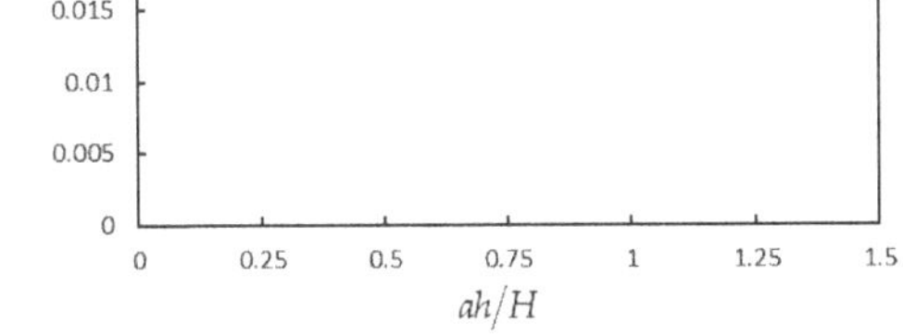

Figure 7. Generalized roughness coefficient n_p varies with the comprehensive parameter *ah/H*.

As shown in Figure 7 and Equation (22), the denser the vegetation arrangement is, the greater the generalized roughness coefficient n_p is. Because of the dense vegetation arrangement, the water-blocking effect of vegetation will be enhanced. Then the equivalent roughness coefficient after generalization, n_p, will increase.

The generalized roughness coefficient n_p is another important parameter obtained after generalization of water-blocking effect. Because it can also be directly put into plane two-dimensional model for engineering calculation.

3.2.5. Ratio of Single-Width-Discharges q/q_0

The influence of submerged vegetation on discharge capacity is further analyzed by using the generalization results of the water-blocking effect mentioned above. Here, q and q_0 represent the single-width discharge of open channel flow with or without submerged vegetation, respectively. The following formula can be obtained.

$$q_0 = \frac{1}{n_0} H^{5/3} i_b^{1/2} \tag{23}$$

$$q = \frac{1}{n_p} (H_m)^{5/3} i_b^{1/2} \tag{24}$$

We further obtain:

$$\frac{q}{q_0} = \frac{n_0}{n_p} \left(\frac{H_m}{H} \right)^{5/3} \tag{25}$$

We substitute n_p and Δ_s (from Figures 5 and 7) into Equation (25) to obtain the calculated values of q/q_0. Figure 8 shows the variation in q/q_0 with the changes in ah/H. The corresponding experimental values of q/q_0 (data from the experiment performed by Dunn et al. [1]) are also given in Figure 8. We also extend the calculated values of q/q_0. Figure 8 shows the following: (1) The calculated values are basically consistent with the experimental values of Dunn et al. [1]. This shows that the calculation method of the water-blocking effect proposed in this paper has high precision; (2) q/q_0 decreases monotonically with increasing ah/H. This result shows that the denser the submerged vegetation is or the higher the submerged vegetation is, the greater the discharge capacity decreases. This is in line with the objective law. Because of the dense vegetation arrangement, the water-blocking effect of vegetation will be enhanced. That is, the flow of water becomes more obstructed. With the decrease in flow velocity and flow space, the single-width-discharge, q, will become smaller. Then, divided by the single-width-discharge without vegetation obstruction, i.e., q_0, the single-width-discharges ratio q/q_0 will decrease. That is, the flow loss will increase, so Figure 8 shows a downward trend. It can be seen from Figure 8 that the cut-off point is the single-width-discharges ratio q/q_0 equals to 0.4 (ah/H equals to about 0.75). Before and after this cut-off point, the single-width-discharges ratio q/q_0 changes from fast to slow. When ah/H is zero, that is, there is no vegetation obstruction, the single-width-discharges ratio q/q_0 is equal to 1. There is no flow loss, in line with the objective law. When ah/H is between 0 and 0.75, the single-width-discharge loss is rapid. (q/q_0 rapidly decreases from 1 to 0.4. ah/H as short as 0.75 range, the single-width-discharge loses 60%.) It can be seen that in the early stage of vegetation density increase, a little change in vegetation density will cause a large loss of single-width-discharge. When ah/H is greater than 0.75, the single-width-discharge loss slows down. Over a long ah/H interval, the single-width-discharge ratio q/q_0 only decreases from 0.4 to 0.2. That is, the single-width-discharge loss only increases by 20%. This is a stable interval of high loss rate of single-width-discharge. It can be seen that after the early stage of vegetation density increase, even if vegetation density increases again, it has little impact on the loss of single-width-discharge. Figure 8 can be used as a reference for estimating single-width-discharge variation and loss under the action of vegetation arrangement parameter, ah/H.

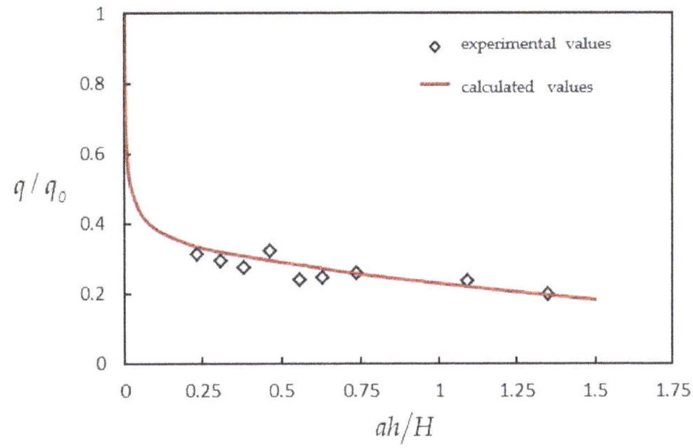

Figure 8. Ratio of single-width-discharges q/q_0 varies with the comprehensive parameter ah/H.

3.2.6. Comparison Results Related to Velocity Integration

The above is the water-blocking-effect generalization calculation for numerically simulated velocity. Then, the original velocity data were also generalized by the same way. Dunn et al. [1] conducted 18 groups of experiments, each of which measured data in four random cross sections. The original velocity data of these random sections were generalized with the same method mentioned above. Then we calculated four parameters: Δ_s, $\Delta_s + \Delta_h$, n_p, q/q_0. The velocity of four sections is different, so the single-width discharge of each section is calculated (i.e., the integral area of the velocity distribution diagram). These four sections are then sorted by the size of the single-width discharge in every group, labeled in descending order as Max1, Max2, Max3, and Max4. Then the influence of velocity integral areas in four sections on the four parameters is explored.

(1) Height of the Theoretical Zero Point Δ_s

Figure 9 shows the result of theoretical zero point height Δ_s, which is calculated after generalization using original flow velocity data. Then, Δ_s is also nondimensionalized in the same way. Figure 9 shows that: (1) When using original velocity to be generalized, the overall trend of the parameter Δ_s is consistent with that of the numerical calculation velocity in Section 3.2.2, i.e., both show a decreasing trend of Δ_s/h with the increase in ah/H. That is, when water depth H and vegetation height h remain unchanged, the denser vegetation layout is, the higher the theoretical zero point is (the higher virtual channel upper boundary is lifted). (2) For the same ah/H, the larger the integrated area of the flow velocity is, or say the single width flow discharge increases, (Max1 blue line is the largest, Max4 purple line is the smallest), the larger parameter Δ_s is, and the smaller it is otherwise. Further analysis for (2): for the same ah/H (i.e., the same group), although water depth H is equal, the greater integrated area of flow velocity in four cross sections is, the larger velocity value is, meaning flow moves faster. Therefore, the generalized virtual channel will be lower, i.e., Δ_p will be smaller, and the corresponding proportion Δ_p/h will be smaller too. Thus, faster flow can be allowed. Accordingly, the proportion of theoretical zero point height, Δ_s/h, is larger.

As seen in Figure 9, the relationship between velocity integration and theoretical zero point height is positive. Because $\Delta_p = h - \Delta_s$, the relationship between velocity integration and virtual channel elevation is negative.

It can be seen that even with the same *ah/H*, the difference in the integrated area of flow velocity (i.e., single-width discharge) can also affect the water-blocking-effect generalization results.

4. Discussion

The influence of numerical calculation and average treatment on the water-blocking-effect generalization is discussed below. Three cases are discussed, which are: original velocity is numerically calculated firstly, and then water-blocking-effect generalization is performed to obtain parameters (i.e., Sections 3.2.2–3.2.5); original velocity is generalized firstly to obtain original parameters, and then these original parameters are averaged (i.e., Section 3.2.6); original velocity is firstly treated by double-average method and then water-blocking-effect generalized to obtain parameters. Finally, we discuss four parameters (i.e., Δ_s, $\Delta_s + \Delta_h$, n_p and q/q_0) from three cases mentioned above.

4.1. Height of the Theoretical Zero Point Δ_s in Three Cases

The theoretical zero height Δ_s processed in three cases, and the fitting line obtained in Section 3.2.2 (here called the central line) are drawn into the coordinates together, as shown in Figure 13 below.

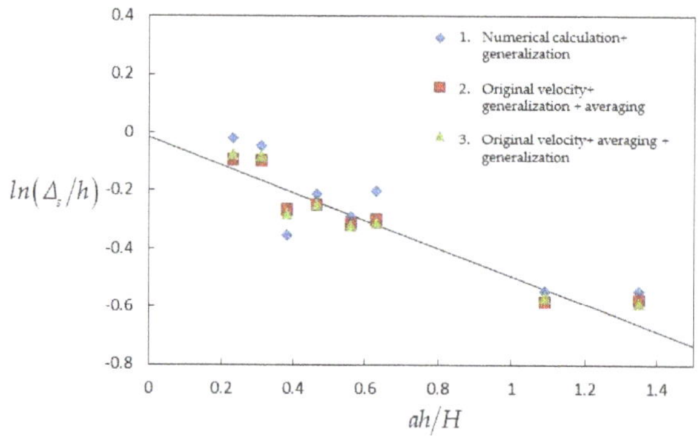

Figure 13. Height of the theoretical zero point Δ_s varies with the comprehensive parameter *ah/H* (3 cases).

It can be seen from Figure 13 that: (1) The results of three cases have the same overall trend. The larger the *ah/H* is, the smaller the Δ_s/h is. That is, with the dense vegetation layout, the theoretical zero height Δ_s is smaller, and the virtual channel after generalization is elevated (i.e., Δ_p is larger). (2) Comparing three cases' data, those of numerical calculation (case 1) are scattered and deviate most from the central line. Data of other two cases (case 2 and case 3) are concentrated relatively. It shows that the two mean-methods can better reflect the central trend. (3) There is little difference between data of case 2 and case 3, indicating that there is little difference between cases of first-generalization and first-averaging for this parameter Δ_s/h.

4.2. Influence Depth $\Delta_s + \Delta_h$ in Three Cases

The influence depth $\Delta_s + \Delta_h$ processed in three cases, and the fitting line obtained in Section 3.2.3 (here called the central line) are drawn into the coordinates together, as shown in Figure 14 below.

It can be seen from Figure 14 that: (1) the results of the three cases have the same general trend, the larger the *ah/H* is, the smaller the $(\Delta_s + \Delta_h)/h$ is. The *h* in the figure is 0.1175 m, so $(\Delta_s + \Delta_h)$ decreases with the increase in *ah/H*. (2) Comparing data in three cases, the difference is small, and they all concentrate around the central line. It shows that there is little difference among the three cases for reflecting the central trend.

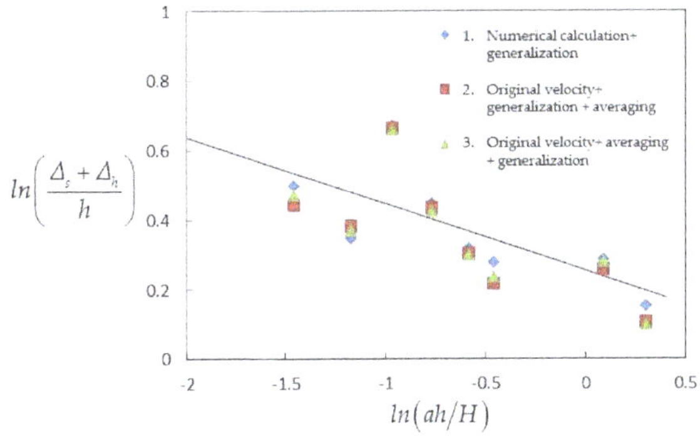

Figure 14. Influence depth $\Delta_s + \Delta_h$ varies with the comprehensive parameter *ah/H*. (3 cases).

4.3. Generalized Roughness Coefficient n_p in Three Cases

The generalized roughness coefficient n_p processed in three cases, and the fitting lines obtained in Section 3.2.4 (here called the central line) are drawn into the coordinates together, as shown in Figure 15 below.

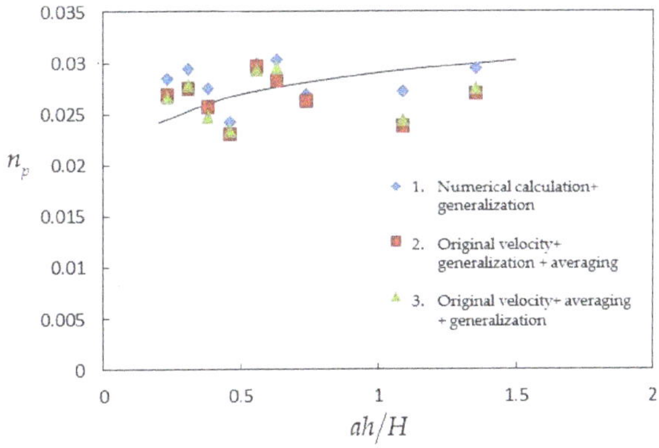

Figure 15. Generalized roughness coefficient n_p varies with the comprehensive parameter *ah/H*. (3 cases).

It can be seen from Figure 15 that: (1) The results of the three cases have the same general trend, the larger *ah/H* is, the larger n_p is. That is, with the dense vegetation layout, the generalized roughness coefficient n_p is larger, and all data generally distribute around the center line. (2) Comparing data in three cases, data in case 1 (i.e., numerical calculation data) are relatively larger. Data in case 2 and case 3 are relatively smaller and there is little difference between them. It shows that data processed by numerical calculation are larger than those processed by mean calculation, and there is little difference between the two latter cases.

4.4. Ratio of Single-Width-Discharges q/q_0 in Three Cases

The ratio of single-width-discharges q/q_0 processed in three cases, and the calculated line obtained in Section 3.2.5 (here called the central line) are drawn into the coordinates together, as shown in Figure 16 below.

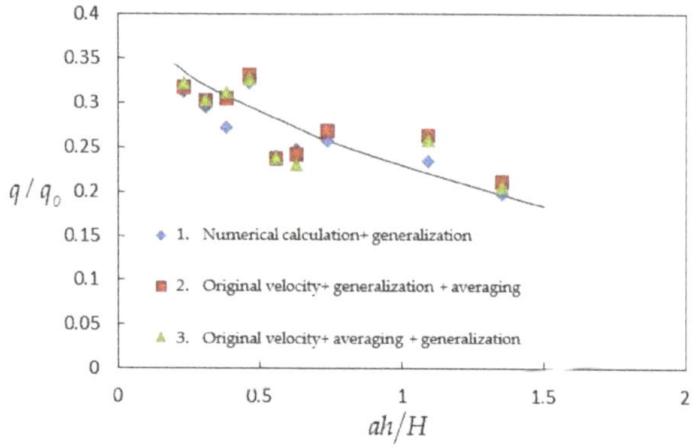

Figure 16. Ratio of single-width-discharges q/q_0 varies with the comprehensive parameter *ah/H*. (3 cases).

It can be seen from Figure 16 that: (1) the results of the three cases have the same general trend, the smaller *ah/H* is, the larger q/q_0 is. That is, with the sparse vegetation layout, smaller flow discharge loss is, and all data generally distribute around the center line. (2) Comparing data in three cases, data in case 2 and case 3 (i.e., mean calculation cases) are generally equal. Data in case 1 are also generally equal to that of case 2 and case 3, except two are relatively smaller. Therefore, we can basically say that there is little difference in q/q_0 of three cases, and they can all reflect the same center trend.

Brief summary of Section 4.1 to Section 4.4: average-treatment and numerical simulation treatment have effects on the comprehensive roughness coefficient and virtual channel elevation value, that is, they are different. In addition, average-treatment before or after water-blocking effect generalization makes little difference in the comprehensive roughness coefficient and virtual channel elevation.

4.5. Influences of Arrangement of Vegetation

A new mode for the description of field characteristics and general characteristics of submerged vegetation open channel flow is proposed in this article. In this work, space average method is used. This method can eliminate the effect of the arrangement of vegetation, whether it is regular or random. After space average method, specific arrangement forms can be further researched. That is, fine models of regular or random arrangements can be studied later. Therefore, a special research for this topic can be carried out. Moreover, submerged vegetation is complex in nature. To describe this complex morphology, porosity and water-blocking elevation are also considered in this article. In addition, flow with simple boundary conditions is studied. Then, other arrangement patterns will be studied in the next step. However, the generalized model is not affected by the arrangement mode, whether it is a regular or random arrangement. Therefore, in this work, regular arrangement is used to represent a common type of arrangement. Arrangement mode can be changed in the next step of more detailed research for fine models.

4.6. Discussion of Implications for Environmental Flows and Hydraulic Engineering Problems

The model in this manuscript can be used to calculate some environmental flows with submerged vegetation. For example, the numerical model can be used to determine sediment transport rates. The steps are as follows: Equations (1) and (4)–(6) are a set of equations. Four unknowns (k_s, k_w, ε_s, u_1) are solved together. The numerical method is to discretize the equations using an unsteady finite volume method. Each equation is discretized as the governing equation of unsteady flow. Add the time partial derivative of the variable to the left-hand side of each governing equation. The central difference scheme is used to differentiate the time term. The flow is iteratively calculated with the number of steps in a certain time. In addition, then we put in the boundary conditions. In the boundary conditions, k_s and ε_s are the same as the standard k-ε equation, and the gradient of the newly introduced k_w is zero at the bottom of the open channel and on the free surface. Finally, Gauss–Seidel iterative method is used to solve discrete algebraic equations. After that, the longitudinal velocity profile of open channel flow with submerged vegetation is obtained. These velocity values can then be used to calculate shear stress and determine the sediment transport rates. There are other implications as well, for example, a lot of aquatic vegetation is morphologically distinct from rigid cylinders (they have leaves, branches, etc.), so this would probably modify the velocity and shear stress profiles from the examples provided in this manuscript. The distribution pattern of velocity and shear stress is greatly affected by vegetation form and water flow condition. It is suggested that readers should first find experiments of submerged vegetation with leaves and branches for flow velocity distribution if they are interested. In addition, then you can think about whether to decompose k and ε into two or more parts. Accordingly, the governing Equations (4)–(6) may introduce new parts of k and ε. The variable expression in the equation may need to change some coefficients' values. Vegetation porosity is the main parameter.

This work has practical implications in real hydraulic engineering problems. The numerical simulation method proposed in this manuscript can be used to establish the corresponding model for the open channel flows with submerged vegetation, such as rivers, channels, and canals. Then the flow field can be predicted. The velocity and Reynolds stress distribution and sediment transport rates can be obtained. The generalization of the water-blocking effect can be used on basis of the numerical simulation to obtain generalized quantities. We can also consult the values of generalized quantities directly on the result charts with the specific value of comprehensive parameter ah/H, to obtain generalized quantities. These quantities can be calculated in the planar two-dimensional model or in the existing theory and general formulas of open channel flow. After these treatments, the overflow capacity and flood discharge capacity of vegetated open channels can be assessed and predicted, as well as the stabilization of sediment deposits inside riverbanks or in lateral embayments, and the shipping conditions of vegetated open channels. Submerged vegetation also has certain influences on channel morphology and stabilization of sediment deposits. The abundant vegetation in the riverbed can change the channel morphology into a wide and shallow form. For example, Southard et al. [28] have found that reaches with abundant bed vegetation are significantly wider (by an average of approximate to 50%), with shallower flows and lower velocities, than reaches with little bed vegetation. Dense submerged vegetation has the function of concentrating sediments, as found by Li et al. [29]: dense vegetation renders the vertical distribution profile uneven and captures sediment particles into the vegetation layer. Hyung et al. [30] also found that for certain flows, i.e., flows above the sediment motion threshold, sediment accumulated within and around the vegetation patch due to a reduction in bed shear stress. Vegetation-generated turbulence can cause an enhanced sediment transport, as pointed by Yang et al. [31]. In addition, sediment retention areas are formed by vegetation through baffling the flow and reducing the bed stress [32]. The generalization method proposed in this article can simplify original model and be applied to the calculation of channel morphology and sediments stability more conveniently.

5. Conclusions

Through theoretical analysis and numerical simulation, we discuss the flow field characteristics and generalization of the water-blocking effect of open channel flow with submerged vegetation, and draw the following conclusions:

(1) As the basis of generalization, in order to make up for the deficiency of experimental results, a new numerical simulation model for the flow field of submerged vegetation open channel flow was proposed before the main work. This is a simple model. The innovation is that the turbulence kinetic energy and the dissipation rate of turbulence

kinetic energy in the model are decomposed into two parts, large-scale shear turbulence part and stem-scale turbulence part, respectively. The interaction between the two is considered through a transformation term. The verification results show that the model has high simulation accuracy.

(2) For the purpose of the study, a simple generalization model of the water-blocking effect of submerged vegetation was proposed. Finally, two parameters of generalized roughness coefficient and virtual channel elevation were obtained to reflect the water-blocking effect. They can be put directly into a planar two-dimensional model in engineering. It achieves the ultimate goal of convenient engineering calculation.

(3) The influence of velocity integration and average-treatment on the water-blocking effect generalization model of submerged vegetation is further studied. The relationship between velocity integration and generalized roughness coefficient is positive, and the relationship between velocity integration and virtual channel elevation is negative. Average-treatment and numerical simulation treatment have effects on the generalized roughness coefficient and virtual channel elevation value, that is, they are different. In addition, average-treatment before or after water-blocking effect generalization makes little difference in generalized roughness coefficient and virtual channel elevation.

Author Contributions: Conceptualization, C.Q., S.L. and W.P.; methodology, C.Q., S.L. and W.P.; formal analysis, C.Q., S.L. and W.P.; investigation, C.Q. and W.P.; resources, C.Q. and W.P.; data curation, C.Q. and W.P.; writing—original draft preparation, C.Q., S.L. and W.P.; writing—review and editing, C.Q., S.L. and W.P.; visualization, C.Q. and W.P.; supervision, S.L. and J.H.; project administration, S.L.; funding acquisition, S.L. All authors have read and agreed to the published version of the manuscript.

Funding: This research was funded by Ministry of Science and Technology of the People's Republic of China, the National Key R&D Program of China, grant number 2018YFC0407600.

Institutional Review Board Statement: Not applicable.

Informed Consent Statement: Not applicable.

Data Availability Statement: The data presented in this study are available on request from authors.

Acknowledgments: The authors would like to express their gratitude to Changjiang River Scientific Research Institute.

Conflicts of Interest: The authors declare no conflict of interest. The funders had no role in the design of the study; in the collection, analyses, or interpretation of data; in the writing of the manuscript; or in the decision to publish the results.

Abbreviations

Symbol Key

ψ	instantaneous variable
$\overline{\psi}$	time mean variable
ψ'	fluctuation variable
$\langle\overline{\psi}\rangle$	local spatial mean
ψ''	local spatial pulsation
$\langle\overline{u_1}\rangle$	longitudinal flow velocity after double averaging (m/s)
$\overline{u_1}$	longitudinal flow velocity after generalization (m/s)
$\langle\overline{\tau_1}\rangle$	shear stress after double averaging (Pa)
$\overline{\tau_1}$	shear stress after generalization (Pa)
i_b	bottom slope of the open channel
D	diameter of submerged vegetation (m)
θ	porosity of the flora

Symbol Key

C_D	drag coefficient
ν	molecular viscosity coefficient (m^2/s)
ν_t	turbulence viscosity coefficient (m^2/s)
k	turbulent kinetic energy
k_s	turbulent kinetic energy of the large-scale shear turbulence
k_w	turbulent kinetic energy of the small-scale stem turbulence
ε	turbulent kinetic energy dissipation rate
ε_s	turbulent kinetic energy dissipation rate of the large-scale shear turbulence
ε_w	turbulent kinetic energy dissipation rate of the small-scale stem turbulence
W_D	a term reflecting the transformation of shear turbulent kinetic energy to stem-scale wake kinetic energy
ν_t	turbulent viscosity coefficient
a	vegetation density ($1/m$)
A	inflow area of a single cylinder (m^2)
V	volume affected by a single cylinder (m^3)
D	diameter of the cylinder (m)
h	height of submerged vegetation (m)
Δx_1	length of the volume V in direction x_1 when averaged locally (m), x_1 and x_2 can be streamwise direction and spanwise direction, respectively.
Δx_2	length of the volume V in direction x_2 when averaged locally (m).
ah/H	a comprehensive parameter to describe the water-blocking effect of submerged vegetation on water ($1/m$)
Δ_s	the height of the theoretical zero point, and the flow velocity here is zero after generalization (m)
Δ_p	the height of the equivalent virtual obstacle after generalization, and it's called virtual channel height (m)
Δ_h	the thickness of the affected region in which submerged vegetation changes the longitudinal velocity above the top of vegetation (m)
$\Delta_s + \Delta_h$	the influence depth (m)
H_m	effective water depth (m), defined as $H_m = H - h + \Delta_s$
τ_h	shear stress at the top of the vegetation layer before generalization (Pa)
τ_b	shear stress at the top of the virtual channel after generalization (Pa)
n_p	roughness coefficient
Q	flow discharge quantity (m^3/s)
B	channel width (m)
g	the acceleration of gravity (m^2/s)
u_*	frictional velocity (m/s), defined as $u_*^2 = -\left\langle \overline{u_1' u_3'} \right\rangle \Big|_{x_3=h}$
q	single-width discharge of open channel flow with submerged vegetation (m^3/s)
q_0	single-width discharge of open channel flow without submerged vegetation (m^3/s)

References

1. Chad, D.; Fabian, L.; Marcelo, G. Mean flow and turbulence in a laboratory channel with simulated vegetation. In *Civil Engineering Studies*; Hydrosystems Laboratory, Department of Civil Engineering, University of Illinois at Urbana–Champaign: Urbana, IL, USA, 1996; Hydraulic Engineering Series No. 51.
2. Raupach, M.R.; Finnigan, J.J.; Brunei, Y. Coherent eddies and turbulence in vegetation canopies: The mixing-layer analogy. *Bound-Layer Meteorol.* **1996**, *78*, 351–382. [CrossRef]
3. Nepf, H.; Ghisalberti, M.; White, B.; Murphy, E. Retention time and dispersion associated with submerged aquatic canopies. *Water Resour. Res.* **2007**, *43*, 436–451. [CrossRef]
4. Iehisa, N.; Michio, S. Turburence structure and coherent motion in vegetated canopy open-channel flows. *J. Hydro-Environ. Res.* **2008**, *2*, 62–90.
5. Carmelo, J.; Schärer, C.; Jenny, H.; Schleiss, A.J.; Franca, M.J. Floodplain land cover and flow hydrodynamic control of overbank sedimentation in compound channel flows. *Water Resour. Res.* **2019**, *55*, 9072–9091.
6. Vargas-Luna, A.; Duró, G.; Crosato, A.; Uijttewaal, W. Morphological adaptation of river channels to vegetation establishment: A laboratory study. *J. Geophys. Res.-Earth* **2019**, *124*, 1981–1995. [CrossRef]
7. Nepf, H.M.; Vivoni, E.R. Flow structure in depth-limited, vegetated flow. *J. Geophys. Res.-Oceans* **2000**, *105*, 28547–28557. [CrossRef]

8. Ghisalberti, M.; Nepf, H. Shallow flows over a permeable medium: The hydrodynamics of submerged aquatic canopies. *Transp. Porous Med.* **2009**, *78*, 309–326. [CrossRef]
9. Ortiz, A.C.; Ashton, A.; Nepf, H. Mean and turbulent velocity fields near rigid and flexible plants and the implications for deposition. *J. Geophys. Res.-Earth* **2013**, *118*, 2585–2599. [CrossRef]
10. Ghisalberti, M. Mixing layers and coherent structures in vegetated aquatic flows. *J. Geophys. Res.* **2002**, *107*, 3011. [CrossRef]
11. Ghisalberti, M.; Nepf, H.M. The limited growth of vegetated shear layers. *Water Resour. Res.* **2004**, *40*, 196–212. [CrossRef]
12. Ghisalberti, M.; Schlosser, T. Vortex generation in oscillatory canopy flow. *J. Geophys. Res.-Oceans* **2013**, *118*, 1534–1542. [CrossRef]
13. Takaaki, O.; Iehisa, N.; Michio, S. Flow-vegetation interactions: Length-scale of the "monami" phenomenon. *J. Hydraul. Res.* **2016**, *54*, 251–262.
14. Takaaki, O.; Iehisa, N. Spatial evolution of coherent motions in finite-length vegetation patch flow. *Environ. Fluid Mech.* **2013**, *13*, 417–434.
15. Zeng, Y.; Huai, W.; Zhao, M. Flow characteristics of rectangular open channels with compound vegetation roughness. *Appl. Math. Mech.* **2016**, *37*, 341–348. [CrossRef]
16. Coceal, O.; Dobre, A.; Thomas, T.G. Unsteady dynamics and organized structures from dns over an idealized building canopy. *Int. J. Climatol.* **2007**, *27*, 1943–1953. [CrossRef]
17. Ji, C.; Munjiza, A.; Williams, J.J.R. A novel iterative direct-forcing immersed boundary method and its finite volume applications. *J. Comput. Phys.* **2012**, *231*, 1797–1821. [CrossRef]
18. Choi, S.U.; Kang, H. Numerical investigations of mean flow and turbulence structures of partly-vegetated open-channel flows using the Reynolds stress model. *J. Hydraul. Res.* **2006**, *44*, 203–217. [CrossRef]
19. Kang, H.; Choi, S.U. Turbulence modeling of compound open-channel flows with and without vegetation on the floodplain using the Reynolds stress model. *Adv. Water Resour.* **2006**, *29*, 1650–1664. [CrossRef]
20. Coceal, O.; Thomas, T.G.; Castro, I.P.; Belcher, S.E. Mean Flow and Turbulence Statistics over Groups of Urban-like Cubical Obstacles. *Bound-Layer Meteorol.* **2006**, *121*, 491–519. [CrossRef]
21. Defina, A.; Bixio, A.C. Mean flow and turbulence in vegetated open channel flow. *Water Resour. Res.* **2005**, *41*, 372–380. [CrossRef]
22. Huai, W.; Song, S.; Han, J.; Zeng, Y. Prediction of velocity distribution in straight open-channel flow with partial vegetation by singular perturbation method. *Appl. Math. Mech.* **2016**, *37*, 1315–1324. [CrossRef]
23. Bai, F.; Yang, Z.; Huai, W.; Zheng, C. A depth-averaged two dimensional shallow water model to simulate flow-rigid vegetation interactions. In Proceedings of the 12th International Conference on Hydroinformatics (HIC)—Smart Water for the Future, Incheon, Republic of Korea, 21–26 August 2016.
24. Luhar, M.; Nepf, H.M. From the blade scale to the reach scale: A characterization of aquatic vegetative drag. *Adv. Water Resour.* **2013**, *51*, 305–316. [CrossRef]
25. Murphy, E.; Ghisalberti, M.; Nepf, H. Model and laboratory study of dispersion in flows with submerged vegetation. *Water Resour. Res.* **2007**, *43*, 1315–1324. [CrossRef]
26. Hu, Y.; Huai, W.; Han, J. Analytical solution for vertical profile of streamwise velocity in open-channel flow with submerged vegetation. *Environ. Fluid Mech.* **2013**, *13*, 389–402. [CrossRef]
27. Shi, H.; Liang, X.; Huai, W.; Wang, Y. Predicting the bulk average velocity of open-channel flow with submerged rigid vegetation. *J. Hydrol.* **2019**, *572*, 213–225. [CrossRef]
28. Southard, P.; Johnson, J.; Rempe, D.; Matheny, A. Impacts of Vegetation on Dryland River Morphology: Insights from Spring-Fed Channel Reaches, Henry Mountains, Utah. *Water Resour. Res.* **2022**, *58*, e2021WR031701. [CrossRef]
29. Li, D.; Yang, Z.; Sun, Z.; Huai, W.; Liu, J. Theoretical Model of Suspended Sediment Concentration in a Flow with Submerged Vegetation. *Water* **2018**, *10*, 1656. [CrossRef]
30. Kim, H.; Kimura, I.; Shimizu, Y. Bed morphological changes around a finite patch of vegetation. *Earth Surf. Process. Landf.* **2015**, *40*, 375–388. [CrossRef]
31. Yang, J.; Nepf, H. A Turbulence-Based Bed-Load Transport Model for Bare and Vegetated Channels. *Geophys. Res. Lett.* **2018**, *45*, 10428–10436. [CrossRef]
32. Cotton, J.; Wharton, G.; Bass, J.; Heppell, C.; Wotton, R. The effects of seasonal changes to in-stream vegetation cover on patterns of flow and accumulation of sediment. *Geomorphology* **2006**, *77*, 320–334. [CrossRef]

water
MDPI

Article

Characteristics of Shallow Flows in a Vegetated Pool—An Experimental Study

Parsa Parvizi [1], Hossein Afzalimehr [2,*], Jueyi Sui [3], Hamid Reza Raeisifar [2] and Ali Reza Eftekhari [2]

1 Department of Hydraulic, Isfahan University of Technology, Isfahan 8415683111, Iran
2 Department of Civil Engineering, Iran University of Science and Technology, Tehran 1311416846, Iran
3 School of Engineering, University of Northern British Columbia, Prince George, BC V2N 4Z9, Canada
* Correspondence: hafzali@iust.ac.ir; Tel.: +98-913-2175524

Abstract: Pools are often observed in gravel-bed rivers, together with the presence of vegetation patches. In the present study, a conceptual model of a gradual varied flow with both convective deceleration and acceleration flow sections has been constructed in a flume to study turbulent flow structures. Vegetation patches with extended canopies were planted in the pool sections in order to increase the thickness of the boundary layer inside the inner zone. The effects of different flows (namely decelerating, uniform and accelerating flows) along an artificial pool on flow velocity, shear stress and bursting events have been investigated. In addition, due to the occurrence of secondary currents in shallow streams, the characteristics of turbulent shallow flow have been investigated along two axes that are parallel to the sidewall of the flume. The results showed that the application of the log law should be used with care to estimate shear velocity along a pool with a vegetated bed. The presence of a vegetation patch causes an increase in Reynolds shear stress, especially along the entrance section of the pool where the flow decelerates. The results of the quadrant analysis reveal that the sweep and ejection events have the most dominant influence over the vegetation patch in the pool; however, the contributions of outward and inward events increase near the bed, especially in the entrance section of the pool where the flow is decelerating. The distribution of stream-wise RMS of turbulence intensity along the pool generally presents a convex shape.

Keywords: shallow currents; shear velocity; pool; quadrant analysis; vegetation patch

Citation: Parvizi, P.; Afzalimehr, H.; Sui, J.; Raeisifar, H.R.; Eftekhari, A.R. Characteristics of Shallow Flows in a Vegetated Pool—An Experimental Study. *Water* **2023**, *15*, 205. https://doi.org/10.3390/w15010205

Academic Editor: Giuseppe Oliveto

Received: 3 November 2022
Revised: 1 December 2022
Accepted: 13 December 2022
Published: 3 January 2023

1. Introduction

In natural rivers, bed morphology affects the flow patterns. Due to the process of the removal and deposition of sediments, and the combination of secondary currents with the sequence of convergent and divergent streams, regular patterns of sequential pool and riffle bed-forms appear very often in gravel-bed rivers. In other words, both the accelerating and decelerating flows often occur in a quasi-periodic manner along riffle and pool sections in gravel-bed rivers. For a flow along a river reach with both accelerating and decelerating sections, characteristics of the turbulent flow, sediment transportation, velocity and shear stress distributions are very different from those in a channel with a flat bed [1]. In recent years, researchers have investigated some of the important hydrodynamic processes of the flow in pools, including the generation of turbulent vortices and the lateral convergence and divergence of the flow associated with bed forms [2–9]. It has been reported that in narrow channels with an aspect ratio less than five, the secondary currents significantly affect the characteristics of the distribution of turbulent flow. For instance, the presence of negative vertical velocities indicates that the maximum velocity should occur under the water surface [10].

Conducted previous experiments to study flow characteristics in the artificial pools have been reported that the secondary currents along the central line of the channel indicate a convergent flow in the decelerating section and divergent flow in the accelerating sectio [11–14]. The presence of vegetation in a channel bed has significant impacts on the stability of hydrodynamic structures of the flow, and has been investigated through different previous studies [15–21]. The reduction in flow velocity due to the presence of vegetation patches in channel beds is another reason for determining flow resistance in the presence of vegetation patches [17,22]. On the other side, the presence of vegetation patches in a channel bed leads to the reduction in sediment erosion due to the decrease in shear stress, and increase in the thickness of the boundary layer. The increase in the thickness of the boundary layer leads to a decrease in the velocity in this layer [11,23–25] The momentum exchange between the top and bottom parts of vegetation makes it difficult to analyze vertical velocity distributions [25,26]. When an aspect ratio is less than five, the dip phenomenon occurs in open channels and can reach up to 35% of the flow depth. Under such a flow condition, the maximum flow velocity is clearly different from the velocity value at the water surface [23,27].

The results of the quadrant analysis used for assessing the structures of flow in both laboratory flumes and natural rivers showed that they are in good agreement, namely, ejection events appear near the bed, and sweep events occur near the water surface, which are dominant occurrences in the pool zone [12,28]. By studying simulated flow conditions in pool–riffle sequences, and comparing against their field observations have been investigated that the model simulations against field observations has little discrepancies [12]. To our knowledge, however, most experimental studies have been carried out in laboratory flumes with small aspect ratios ($W/h < 5$) in order to extend the results to natural rivers. In addition, most of the turbulent data available have been collected for artificial pools where the vegetation is not prevalent, which are different from the pools with vegetated cover [11,29].

Effective planning for the maintenance of natural bed-forms, such as pool–riffle sequences, is important for managers and requires extended studies of artificial pools and riffles in laboratories (e.g., [11]). In other words, to analyze the interaction between turbulent flow and vegetated pool beds, a better understanding of the hydraulic processes in a laboratory scale is required. This experimental study was carried out to investigate the effects of irregular bed shape and vegetation on hydraulic parameters such as shear velocity, friction coefficient and Reynolds stress to investigate their interaction with the desired laboratory conditions. Moreover, the impact of different aspect ratios on the characteristics of turbulent flow in a shallow stream over a vegetated pool bed has been considered for evaluation. Then, it is possible to obtain meaningful results through comparing vegetated pool data with the sandy pool data. In addition, the effect of vegetation patches on turbulent flow structures along a vegetated pool has been investigated by means of quadrant analysis. However, the limitations of the experiment included the formation of a balanced bed shape within a certain range of sediment particle size in a laboratory flume to create flow conditions comparable to those in natural rivers. The diameter of the sediment particles of the bed material in a flume should be selected in a way that their movement threshold is higher than the flow velocity. The limitation of the flume length forced us to build the desired pool in a fully developed flow condition. Since the boundary layer should develop along the entire water depth, the flow should travel a certain distance within the channel in order to achieve the same shape of velocity profiles. To ensure the creation of a developed flow, and to avoid the influence of the vortices caused by the downstream exit-gate of the flume, the longitudinal dimensions of all sections (entrance, pool and exit) were equal to 1.5 m. The 1.5 m pool section was constructed at the distance of 4.5 m to 6 m from the entrance of the flume. Moreover, studies in natural rivers have shown that the angle for pool-shape channels is normally less than 7.8 degrees. Thereby, considering an angle of 10 degrees in this research, the experimental conditions are close to the real conditions in natural rivers.

In addition, the limitations of the acoustic doppler velocity (ADV) device used for measuring the flow separation area have been noticed in this experiment. It happens because of the lack of recirculation in the flow rate, and the existence of the separation area on the bed surface. Furthermore, the study of shallow flow over rough beds shows that many methods and estimations do not have the necessary accuracy. Presently, to estimate the friction factor or roughness coefficient using the reported hydraulic models, one is unable to obtain results with the necessary accuracy. One of the main reasons for these poor predictions is the weakness of the estimation, since the existing methods do not use all data in a velocity profile to determine shear velocity, thus leading to poor predictions of the roughness coefficient, which is one of the key parameters of hydraulic models.

2. Materials and Methods

2.1. Theory

The flow velocities in the stream-wise (u), vertical (w) and lateral (v) directions were decomposed into time-averaged ($\overline{u}, \overline{v}, \overline{w}$), and turbulent components (u′, v′, w′). In addition, the velocity profiles near the bed follow the law of the wall, which is as follows:

$$\frac{u}{u_*} = \frac{1}{k}\ln\frac{z}{z_0} \tag{1}$$

where u_* is the shear velocity, k is the von Karman coefficient equivalent to 0.41, and z_0 is the reference bed level, which is dependent on the roughness height of the bed material. It has been found that the inner zone of each velocity profile can be extended to a relative depth of $z/z_0 = 0.2$ for uniform and non-uniform flows [13,30,31].

One of the methods used to determine the bed level involves solving the log law function so that it fits well with the measured data by using the trial-and-error method. However, in the present study, the equivalent roughness z_0 has been considered equal to d_{50} of the sandy bed material for the non-vegetated bed, and equal to vegetation height in the presence of vegetation in the channel bed [32]. The validation of the logarithmic law has been conducted by plotting (u) against $\ln(z/z_0)$, where the breakpoint between segments was allowed to vary within the range of $z/h \leq 0.2$ [33,34].

At present, there are several methods for determining the shear velocity, which is considered as the most fundamental scale for dimensioning the average velocity. Using the logarithmic method, one can obtain the shear velocity by calculating the regression fitting line for data in the inner layer with the flow depth of $z/h < 0.2$. Using data collected from laboratory experiments, the shear velocity can be obtained by the following equations, where u_c, δ_* and C are the maximum velocity, boundary layer displacement thickness, and constant coefficient, respectively. According to the results of other researchers, C is equal to 4.4 in sandy bed rivers, and θ is the momentum thickness of the boundary layer [31].

$$u_* = \frac{(\delta_* - \theta)u_c}{C\delta_*} \tag{2}$$

$$\delta_* = \int (1 - \frac{u}{u_c})dy \tag{3}$$

$$\theta = \int \frac{u}{u_c}\left(1 - \frac{u}{u_c}\right)dy, \tag{4}$$

Reynolds shear stress has been calculated from turbulent fluctuations in the stream-wise and vertical directions using the following equations, in which ρ is the water density. The Reynolds stress profiles are non-linear for non-uniform flows and follow a concave shape for the accelerated flows and convex shape for the decelerating flows.

$$\tau = -\rho\overline{u'w'} \tag{5}$$

$$\overline{u'w'} = \frac{1}{N}\sum_{i=1}^{N}(u - \overline{u})(w - \overline{w}) \tag{6}$$

To obtain the value of turbulence intensity, RMS parameters have been used, which can be expressed by the following equations:

$$\sqrt{\overline{u'^2}} = u_{RMS}, \sqrt{\overline{v'^2}} = v_{RMS}, \sqrt{\overline{w'^2}} = w_{RMS} \quad (7)$$

Quadrant analysis has been used to study coherent flow structures within or above a vegetation patch, and to investigate the relationship between time fluctuations of the velocity components (u', w'), especially their relationship with the trigonometric circle system in four quadrants of the coordinate system. In recent years, many studies have been conducted to assess the coherent flow structures in the downstream region of a vegetation patch. The results of previous studies reveal that there exists a small margin of ejection dominancy around the top edge behind a submerged vegetation patch. In addition, both the ejection and sweep events are dominant bursting events in the vicinity of the toe of the vegetation patch, which positively contribute to the downward momentum flux, and are involved in near-bed turbulence. Moreover, a notable feature of the flow has also been revealed, that is, the dominance of outward and inward interactions in the shear layer was induced by the flow above the vegetation patch [14,35,36].

To assess the contribution of each pseudo-oscillatory process event, a computer program has been formulated using MATLAB, and the results are presented in the following sections. According to the analysis of velocity fluctuations, the outward interaction occurs when $u' > 0$, $w' > 0$; the ejection event occurs when $u' < 0$, $w' > 0$; the inward interaction occurs when $u' < 0$, $w' < 0$; and the sweep interaction occurs when $u' > 0$, $w' < 0$ [32,37–39].

2.2. Experimental Setup

Laboratory experiments were conducted in a flume, which was 0.4 m wide, 0.6 m deep, and 8 m long with a maximum discharge of 50 L/s. To compare the results of the present study to the previous studies conducted using artificial pools, the flow rate of 10 L per second was chosen for this study. Due to the limitation of the flume length, and to ensure a fully developed flow in the pool, the desired pool bed form was built in the flume section located at a distance of 4.5 to 6 m from the flume entrance. Both the entrance and exit slopes of the pool were constructed at an angle of 10° over a stream-wise distance of 1.5 m to create a pool with a series of three sections, which were as follows: the entrance section, middle pool and exit section. Along the entrance section, the increase in the flow depth created the decelerating section. Along the second section (or middle pool section), the channel bed was flat, and was a transition section. The flow depth along the third section decreased and created an accelerating flow. The bed material had a smooth surface of the sand particles with a uniform distribution, and the median grain size (d_{50}) was 10.4 mm. The smooth surface of the channel bed was carefully prepared to eliminate large-scale variation in the bed topography. The artificial vegetation patch with a height of about 2.5 cm was placed over the middle pool section, which was 1.5 m long (Figure 1).

The laboratory experiments were carried out under two flow conditions, as given in Table 1, with a constant Reynolds numbers and different Froude numbers (condition 1: $R_e = 2.5 \times 10^4$, $Fr = 0.09$), and (condition 2: $R_e = 2.5 \times 10^4$, $Fr = 0.13$). In this study, two flow depths of 20 cm and 15 cm upstream of the pool section were used, and the corresponding aspect ratios were 2.0 and 2.7, respectively. Measurements of the flow velocity were conducted at 13 cross sections, with the first cross section located at 1.5 m upstream of the pool and the last was located at 1.0 m downstream of the pool. To assess the side effects of the flume wall on the turbulence, at each cross section, velocity profiles at two different locations were compared. One location was 20 cm from the sidewall, and the other was 10 cm from the sidewall. For each velocity profile, about 20 to 25 point velocities at different distances from the channel bed were measured.

The velocity data were collected using an ADV, and the minimum SNR was considered as 15. In addition, the data correlation coefficient value was observed to be more than

70 percent. Most of the velocity data collected in this study were considered with high SNR values and correlation, although some data were filtered to remove aberrant data. The objectives of this study were to investigate the velocity, Reynolds stress and turbulence intensity along a pool covered by sand and a submerged vegetation patch. The estimation of shear velocity was examined by different methods. In addition, the contribution of each bursting process along this pool was investigated using quadrant analysis.

(a) (b) (c)

Figure 1. Experimental setup. (**a**) Vegetated pool; (**b**) sandy pool; (**c**) vegetated boundary.

Table 1. Hydraulic conditions of laboratory experiments.

Q (L/s)	H (m)	W (m)	U_{avg} (cm/s)	W/H	Re × 10^4	Fr
10	0.20	0.4	12.5	2.0	2.5	0.09
10	0.15	0.4	16.7	2.7	2.5	0.13

3. Results and Discussions

3.1. Shear Velocity Distribution and Logarithmic Law

Figure 2 shows the distributions of shear velocity (u_*) under the conditions of different aspect ratios (W/H) and vegetation coverage (with or without the presence of vegetation). One can observe from Figure 2 that some irregularities exist in the value of shear velocity near the sidewall. Overall, the value of u_* decreased in the CDF (decelerating flow) section, and increased in the CAF (accelerating flow) section, indicating the impact of the secondary currents on the shear velocity. The presence of the vegetation patch in the pool bed resulted in an increase in the shear velocity, particularly in the CDF section. Under the condition of the same flow discharge, by increasing the aspect ratio (namely, the increase in flow velocity), the value of u_* obtained using the boundary layer method increased. By comparing results of shear velocity determined by the two methods, the estimated u_* using the boundary layer method was higher than that obtained using the logarithmic method, since the logarithmic method considers all point velocities of velocity profiles when calculating u_*. The results of this study confirm those of previous studies that the distribution of shear velocity over a pool bed and bedform is a function of the flow depth [11,12,40].

Figure 3 shows the results for assessing the validity of the logarithmic law over the pool in the presence of vegetation in the pool section along two axes of the channel at the distance of 10 and 20 cm from the sidewall, respectively. By using a trial-and-error method, it was found that the logarithmic law fits the data measured near the bed of the boundary layer well with a strong correlation, as the R^2 values show in Table 2. According to Figure 3, along the CAF (acceleration flow) section, the velocity data collected for the outer layer of the flow at a depth of $1.0 > z/h > 0.2$ (till water surface) are above the regression line, implying a dominant adverse pressure gradient ($\partial p/\partial x > 0$). However, a favorable pressure gradient ($\partial p/\partial x < 0$) is dominant in the CDF (deceleration flow) and middle pool sections of the pool, where the data points collected for the outer layer of flow ($1.0 > z/h > 0.2$) are below the regression line. Approaching the sidewall, the validity of the logarithmic law for the velocity data near the bed region has also been confirmed. For flow with a relatively low Reynolds number, the boundary layer tends to be a laminar boundary layer. As the Reynolds number increases, the boundary layer becomes unstable with small disturbances. Further growth of these disturbances results in the transition to a turbulent boundary layer. Most practical flows involve high Reynolds numbers and turbulent boundary layers. Because of the three-dimensional interchanges in momentum, a turbulent boundary layer is thicker and has a larger wall velocity gradient than that of a laminar layer with the same Reynolds number. The increased momentum near the wall allows a turbulent boundary layer to withstand a larger unfavorable pressure gradient than a laminar layer without separating, but results in higher wall shear stress and drag.

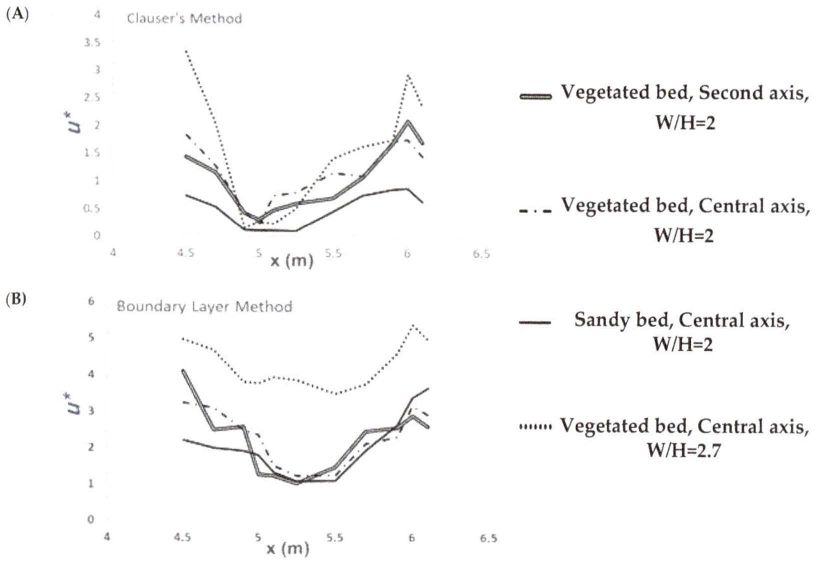

Figure 2. Distributions of shear velocity determined using two methods: (**A**) Clauser's method; (**B**) the boundary layer method.

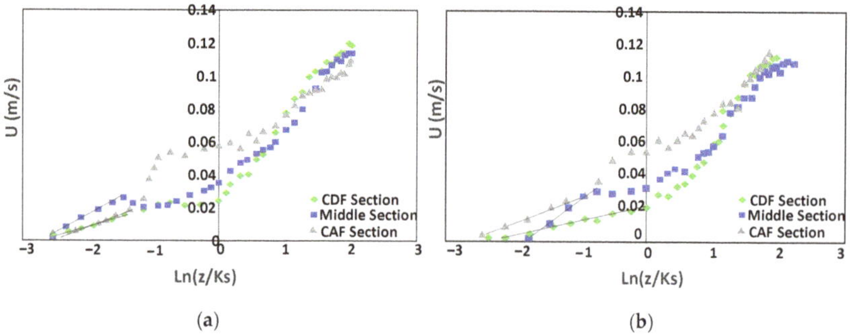

Figure 3. Validity of the logarithmic law method. (**a**) Central axis; (**b**) Second axis.

Table 2. Determinations of logarithmic law.

	x	Equation	R^2	U_*	f
Central Axis	CDF Section	$u/u_* = 0.0116\ln(z/k_s) + 0.0321$	0.98	0.00464	0.011
	Middle Sect.	$u/u_* = 0.02\ln(z/k_s) + 0.0538$	0.90	0.008	0.032
	CAF Section	$u/u_* = 0.0274\ln(z/k_s) + 0.0644$	0.96	0.0108	0.059
Second Axis	CDF Section	$u/u_* = 0.0106\ln(z/k_s) + 0.0204$	0.89	0.0042	0.009
	Middle Sect.	$u/u_* = 0.0151\ln(z/k_s) + 0.0333$	0.97	0.006	0.018
	CAF Section	$u/u_* = 0.0268\ln(z/k_s) + 0.054$	0.93	0.0105	0.058

The positive pressure gradient along the entrance section of the pool and the negative pressure gradient around the riffle zone play important roles in the gradient of the log law function. The positive pressure gradient helps the development of the flow separation zone where the turbulence intensity is high, thus, resulting in a relatively sharp pressure gradient in the middle pool section. In fact, the positive or unfavorable pressure gradient tends to decelerate the flow, and leads to an increased boundary layer thickness; thus, the velocity gradient at the wall will decrease. The unfavorable pressure gradient can cause the separation of the boundary layer, which often results in drastic changes in the flow patterns. Along the flow direction, the negative or favorable pressure gradient, which accelerates the flow, will result in a decrease in the boundary layer thickness and increase in the velocity gradient at the wall. However, there is no flow separation in the riffle zone due to a negative pressure gradient, which shows a negligible departure from the logarithmic law [31,41–43].

3.2. *Velocity Distribution along the Pool*

As shown in Figure 4, the dimensionless stream-wise velocity profiles are presented along the entire pool, where the mean point velocity (u) was converted to dimensionless velocity by dividing it by the maximum velocity (u_{max}) of each velocity profile. Also, the depth related to a velocity point measurement (z) became dimensionless depth by dividing it to the flow depth (h).

The results of the velocity distributions show that the maximum velocity occurs near the water surface. On the other hand, since it is difficult to measure velocity by ADV devices in the flow separation zone where the intensity of turbulence is high, negative stream-wise velocities were not observed in the present experimental study. Moreover, the difference between velocity profiles along the central axis and those along the second axis indicates the strength of the secondary currents in the presence of vegetation in the channel bed. The velocities near the bed in the decelerating flow section decreased with the increasing flow depth; and the velocities near the bed increased along the accelerating flow section with the decrease in the flow depth. These results are consistent with those of other studies [11,44].

Furthermore, dimensionless flow velocities in the outer zone ($z/h > 0.2$) in the decelerating flow section are higher than those in the accelerating flow section, which is consistent with the findings reported in the previous work [13,45]. In the presence of vegetation in the channel bed of the pool, by increasing the aspect ratio, the velocity profiles show an S-shaped distribution. For these velocity profiles, the velocity gradient in the upper part is very small; the velocity gradient is higher in the middle part; and in the region near the bed, the velocity follows a logarithmic distribution. This result confirms the findings of other studies [46]. The effect of the increased aspect ratio on the velocity profiles in the central axis of the pool channel resulted in an increase in the average flow velocity, and a larger velocity gradient along the entire channel bed of the pool. In addition, the region with the minimum velocity values was located at the end of the middle part of the pool section, leading to the presence of a non-uniform flow in the middle part of the pool section. Failure to create a uniform flow in the middle pool section was due to the residual effect of the decelerating flow. This observation is consistent with the results of previous study for a non-uniform open channel flow [47].

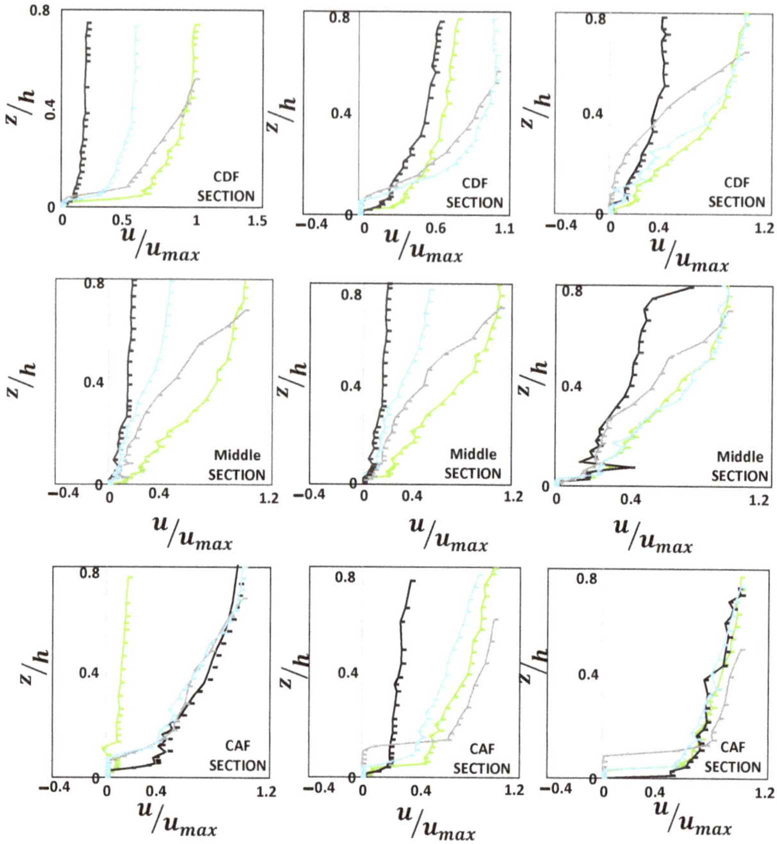

Figure 4. Distributions of stream-wise velocity profiles. Note: green line: central axis over vegetated bed with w/h = 2; black line: central axis over sandy bed with w/h = 2; gray line: central axis over vegetated bed with w/h= 2.7; blue line: second axis over vegetated bed with w/h = 2.

To further investigate the impacts of secondary currents on the characteristics of flow velocity, the velocity components in other two directions were examined, namely the spanwise velocity component (v), and the vertical velocity component (w). As shown in Figure 5, the velocity profiles (v, w) are presented at four cross sections along the pool, namely one cross section in the deceleration section, two cross sections in the middle pool section, and one cross section in the acceleration section. The results show that the occurrences of both positive and negative velocity values in the spanwise and the vertical directions are responsible for the difference in the pattern of secondary currents along the pool bed. The opposite effect of accelerating and decelerating flow on the secondary currents along the pool bed was also reported by other researchers [47].

3.3. *Quadrant Analysis of Bursting Events*

To determine the effects of the aspect ratio and vegetation cover on the distributions of bursting process events in the flow over the pool bed, quadrant analysis was investigated. One can observe from Figure 6 that in the inner zone over the vegetation patch, both the sweep and ejection interactions are dominant events along the central channel axis. This result is consistent with that reported by previous researchers for an open channel with dense vegetation [32]. However, for both the entrance and exit sections of the pool where the flow decelerates and accelerates, respectively, near the bed, the outward and inward events show more contribution to Reynolds shear stress at the central axis of the channel. Approaching the channel sidewall, the interaction of vegetation and decelerating flow reveals a complex pattern, indicating the contribution of all events to Reynolds shear stress. It has been also observed that, with a higher aspect ratio, the inward and outward interaction over the vegetation patch indicates that the flow velocity distribution is affected by the bursting process events. With the decrease in the flow depth along the pool with the vegetated bed, the sweep and ejection events were observed as the dominant events in the pool.

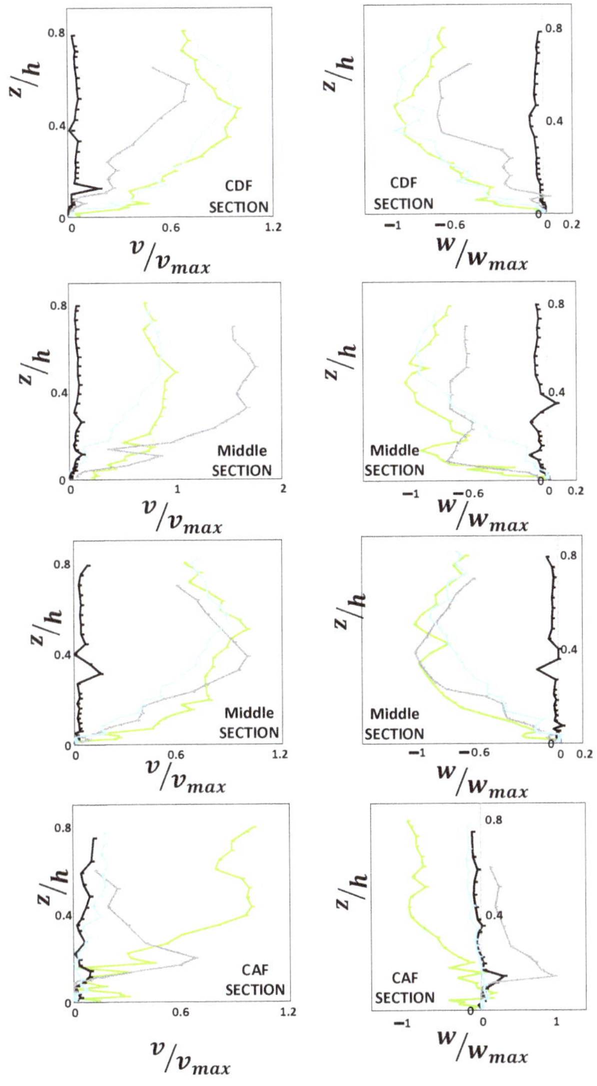

Figure 5. Distributions of velocity profiles. Note: green line: central axis over vegetated bed with w/h = 2; black line: central axis over sandy bed with w/h = 2; gray line: central axis over vegetated bed with w/h = 2.7; blue line: second axis over vegetated bed with w/h = 2.

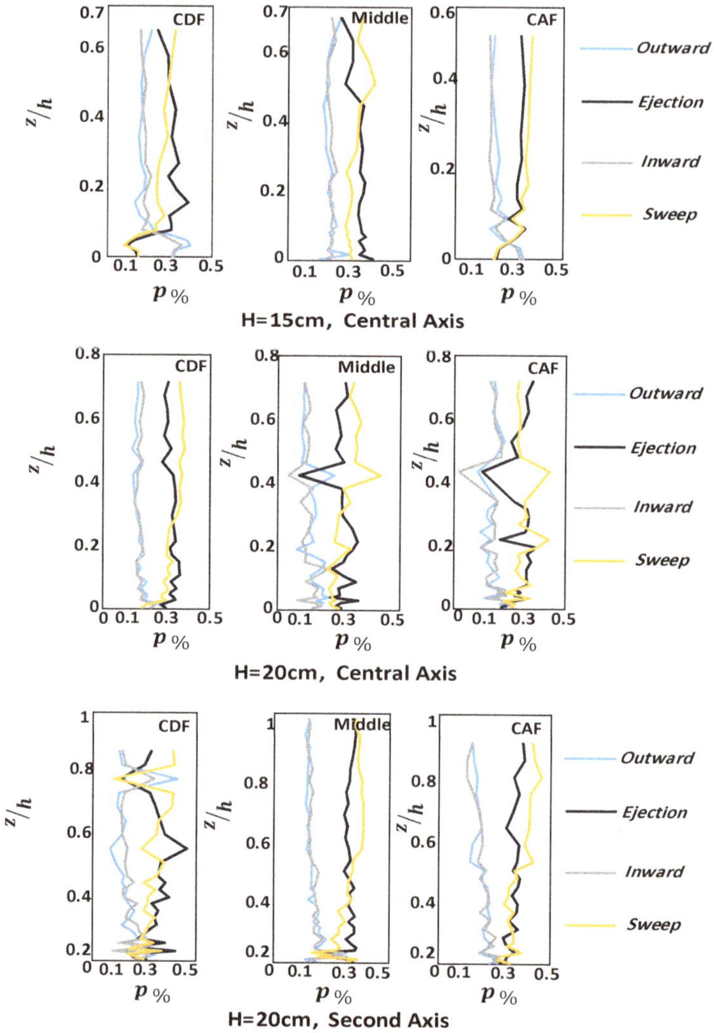

Figure 6. Bursting events in quadrant analysis.

Moreover, approaching the sidewall of the flume, the contribution of the sweep event was noted as dominant; however, the distribution of ejection varies in the CDF, middle and CAF section of the pool. Some considerable scattering in the distribution of the events is due to the presence of more pronounced (higher) roughness. The results of the quadrant analysis of the bursting process events along both the central and second axes indicate that, near the water surface, the dominant events are generally sweeps, followed by ejections, and then outward and inward interactions (Figure 6), thus resulting in positive Reynolds stress values, as shown in Figure 7.

3.4. Distribution of Shear Stress and Longitudinal Turbulence Intensities

The dimensionless Reynolds shear stress (RSS) profiles along the entire pool section are presented in Figure 7. The absence of negative values of shear stress near the water surface indicates that the maximum velocity occurs near the water surface. In addition, the less frequent outward interactions lead to negative values for $u'w'$, resulting in positive

Reynolds stress. In addition, in the middle pool section, the shear stress profiles do not follow the trend of a linear shape, indicating that there is no uniform flow in the middle pool section. By comparing the results of Reynolds shear stress along the central axis of the channel over the sand bed to those over the vegetated bed, it has been observed that the high values of RSS occur over the sand bed in the decelerating section of the pool. In the presence of vegetation in the bed, when approaching the sidewall, the distribution of Reynolds shear stresses shows smaller values. The shape and the location of the maximum value of the Reynolds stress distribution depend on the slopes of the entrance and the exit sections of the pool [18]. The RSS distribution in the decelerating flow section and middle pool section shows a convex shape. Previous researchers found that at the beginning of the accelerating flow section, the shear stress profiles have a convex shape [47]. However, in the present study, the Reynolds shear stress profiles along the accelerating flow section have a concave shape first and then convex shape at the final sections of the pool exit (close to the crest of the bed form).

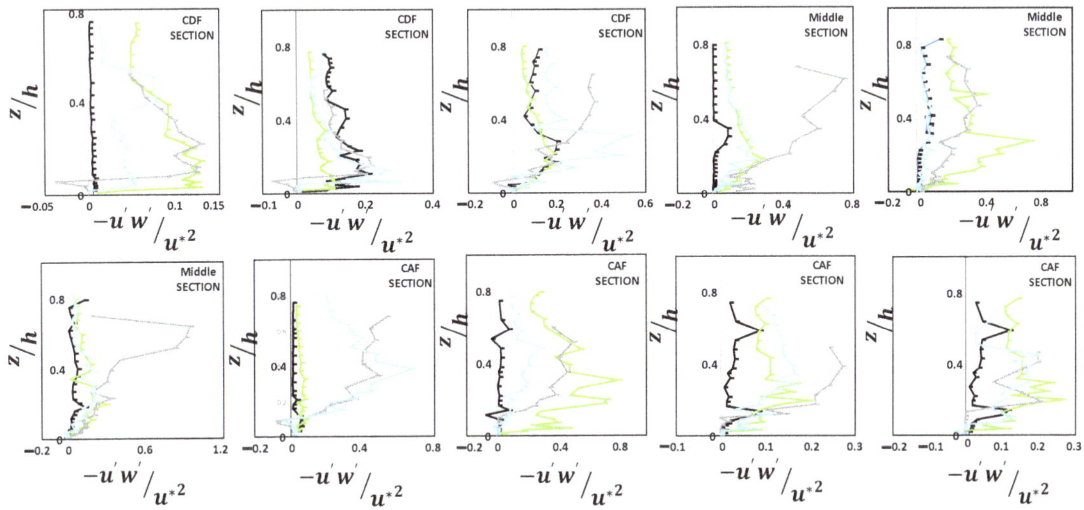

Figure 7. Distribution of Reynolds shear stress. Note: green line: central axis over vegetated bed with w/h = 2; black line: central axis over sandy bed with w/h = 2; gray line: central axis over vegetated bed with w/h = 2.7; blue line: second axis over vegetated bed with w/h = 2.

One can observe from Figure 7 that the distributions of turbulence intensities show different patterns. The non-uniform flow is generated due to the interaction of the vegetation and bed form, different pressure gradients, and the change in roughness from the sand bed to vegetated bed. Figure 8 shows the dimensionless turbulence intensities based on the mean square of the velocity fluctuations. The presence of vegetation patches in the bed leads to the higher turbulence intensities in the middle pool section and accelerating flow section. However, along the entrance section of the pool (decelerating flow section), the presence of vegetation in the bed resulted in a lower turbulence intensity. In the inner layer of the flow, the turbulence fluctuations are lower in the middle pool section, and higher in both the accelerating and decelerating sections of the pool. The effect of both lateral and vertical turbulence intensities was not clearly observed, and was not reported in this study.

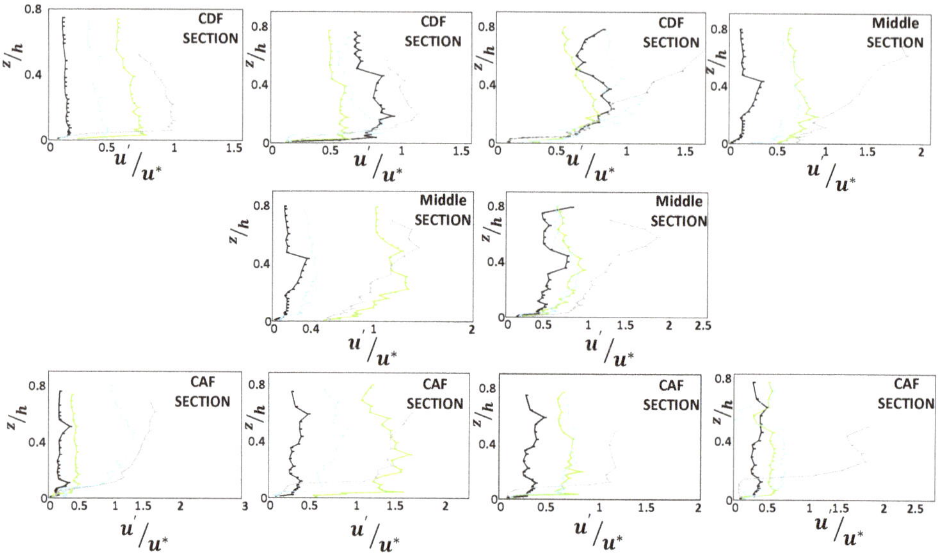

Figure 8. Distribution of stream-wise RMS of turbulence intensity. Note: green line: central axis over vegetated bed with w/h = 2; black line: central axis over sandy bed with w/h = 2; gray line: central axis over vegetated bed with w/h = 2.7; blue line: second axis over vegetated bed with w/h = 2.

4. Conclusions

The interaction between bed forms and vegetation patches in mountainous streams is a challenging subject for river engineers. Considering the importance of vegetation in stabilizing bed forms, both laboratory and analytical work has been conducted on bed forms in the presence of vegetation in the past several decades. However, our knowledge about the interaction of flow and bed forms with vegetation still needs to be clarified, including how bed forms and vegetation affect sediment transfer and production of turbulence, which is one of the reasons for this innovative study in the field of environmental hydraulics. To better understand this interaction, a laboratory study has been carried out in a flume with an artificial pool covered by vegetation patches. In addition, in fluvial hydraulics, it is essential to know the flow structures and estimate the related parameters for describing turbulence distributions. This study focused on the turbulent flow structures in an artificial pool in the presence and absence of vegetation patches over the pool. The results of our laboratory experiments revealed that the logarithmic law is valid for flow velocities within artificial pools in the presence of vegetation patches over the bed. However, towards the water surface, significant deviations of the velocity data from the logarithmic law have been observed. This suggests that, to calculate the fluvial parameters of a pool in the presence of vegetation patches in the bed, the logarithmic law should be used with care.

The shear velocities have been estimated by using both the boundary layer characteristics method and logarithmic method. The estimated shear velocities, obtained by the boundary layer characteristics method, are larger than those obtained via the logarithmic method, since the boundary layer characteristics method considers all point velocities of each profile for calculating shear velocity u*. The presence of vegetation patches in the pool bed resulted in an increase in the shear velocity, particularly in the entrance section of the pool where the flow decelerates. The results of the quadrant analysis reveal that sweep events play the most important role in different parts of the artificial pool, followed by ejection events. However, near the bed, especially along the pool entrance section where the flow is decelerating, the role of outward and inward events is considerable. Along the axis near the sidewall of the flume, the contributions of all events of the bursting process near the bed are complex. This is partly caused by the effect of vegetation patches on the turbulent flow structures. Near the water surface, in general, the sweep events play a dominant role, followed by ejection interactions.

The Reynolds shear stress and turbulence intensity distributions over the sand bed were compared to those in the pool with a vegetation patch. In the presence of vegetation in the channel bed of the pool, by increasing the aspect ratio, the velocity profiles show an S-shaped distribution. For these velocity profiles, the velocity gradient in the upper part is very small; the velocity gradient is higher in the middle part; and in the region near the bed, the velocity follows a logarithmic distribution. The presence of a submerged vegetation patch in the pool bed led to the occurrence of inward and outward events along the pool. The results of the quadrant analysis show that the occurrence of sweep and ejection events over the vegetation patch in the pool depends on the flow velocity distribution. With the decrease in the flow depth over the vegetated pool bed, sweep and ejection events were observed in the pool. In the presence of vegetation in the bed, by approaching the sidewall, the distribution of Reynolds shear stresses shows smaller values. Near the sidewall, the logarithmic law of the flow velocity in the inner region of the boundary layer near the bed is confirmed. The opposite effect of accelerating flow and decelerating flow on the secondary currents along the pool bed is due to the occurrence of positive and negative pressure gradients along the pool. Approaching the sidewall of the flume, the aspect ratio had no significant effects on the occurrence of bursting events. In the presence of vegetation in the bed, by approaching the sidewall, the distribution of Reynolds shear stresses shows smaller values.

The results of this research reveal that the interaction between vegetation and different sections of a pool (entrance, middle pool and exit sections) has a significant impact on the estimations of Reynolds shear stress, turbulence intensity and shear velocity, as well as the contributions of bursting events.

Application and extension of the results of the present study may help to better understand the key parameters in fluvial hydraulics, including the resistance to flow and sediment transport parameters. This may help river engineers to decrease the cost of projects. A variety of studies on pool bed forms must be carried out in order to improve our knowledge about the interaction of flow in pools under different conditions. For example, the effects of a moving bedform of a pool on the flow structure can be investigated by considering different sand granularities in a pool, or the impact of different types and densities of vegetation on the flow structure in a pool can also be studied. Numerical modeling of the flow over a pool bed in the presence and absence of vegetation is also suggested.

Author Contributions: P.P., experimental works, methodology, software; writing—original draft; H.A., supervision, writing—review, methodology, validation and editing. J.S., writing—review, methodology, validation and editing. H.R.R., methodology and validation. A.R.E., methodology and validation. All authors have read and agreed to the published version of the manuscript.

Funding: This research received no external funding.

Institutional Review Board Statement: Not applicable.

Informed Consent Statement: Not applicable.

Data Availability Statement: Data are contained within the article.

Conflicts of Interest: The authors declare no conflict of interest.

References

1. Tabarestani, E.S.; Afzalimehr, H.; Pham, Q.B. Flow structure investigation over a pool-rife sequence in a variable width river. *Acta Geophys.* **2022**, *70*. [CrossRef]
2. Keller, E. Areal sorting of bed-load material: The hypothesis of velocity reversal: Reply. *Geol. Soc. Am. Bull.* **1972**, *83*, 915–918. [CrossRef]
3. Clifford, N.; Richards, K. The reversal hypothesis and teh maintenance of riffle-pool sequences: A review and field appraisal. In *Lowland Floodplain Rivers: Geomorphological Perspectives*; Wiley: Hoboken, NJ, USA, 1992; pp. 43–70.

4. Clifford, N. Morphology and stage-dependent flow structure in a gravel-bed river. In *Coherent Flow Struct. Open Channels*; Wiley: Chichester, UK, 1996; pp. 545–566.
5. Thompson, D.M. The role of vortex shedding in the scour of pools. *Adv. Water Resour.* **2006**, *29*, 121–129. [CrossRef]
6. Booker, D.; Sear, D.; Payne, A. Modelling three-dimensional flow structures and patterns of boundary shear stress in a natural pool–riffle sequence. *Earth Surf. Process. Landf. J. Br. Geomorphol. Res. Group* **2001**, *26*, 553–576. [CrossRef]
7. Wilkinson, S.N.; Keller, R.J.; Rutherfurd, I.D. Phase-shifts in shear stress as an explanation for the maintenance of pool–riffle sequences. *Earth Surf. Process. Landf.* **2004**, *29*, 737–753. [CrossRef]
8. MacWilliams, M.L., Jr.; Wheaton, J.M.; Pasternack, G.B.; Street, R.L.; Kitanidis, P.K. Flow convergence routing hypothesis for pool-riffle maintenance in alluvial rivers. *Water Resour. Res.* **2006**, *42*. [CrossRef]
9. Sawyer, A.M.; Pasternack, G.B.; Moir, H.J.; Fulton, A.A. Riffle-pool maintenance and flow convergence routing observed on a large gravel-bed river. *Geomorphology* **2010**, *114*, 143–160. [CrossRef]
10. Chao, L.; Shan, Y.-Q.; Yang, K.-J.; Liu, X.-N. The characteristics of secondary flows in compound channels with vegetated floodplains. *J. Hydrodyn. Ser. B* **2013**, *25*, 422–429.
11. MacVicar, B.J.; Rennie, C.D. Flow and turbulence redistribution in a straight artificial pool. *Water Resour. Res.* **2012**, *48*. [CrossRef]
12. Fazel Najafabadi, E.; Afzalimehr, H.; Sui, J. A comparison of two-dimensional and three-dimensional flow structures over artificial pool-riffle sequences. *Can. J. Civ. Eng.* **2017**, *44*, 1084–1098. [CrossRef]
13. Song, T.; Chiew, Y. Turbulence measurement in nonuniform open-channel flow using acoustic Doppler velocimeter (ADV). *J. Eng. Mech.* **2001**, *127*, 219–232. [CrossRef]
14. Okamoto, T.-A.; Nezu, I. Spatial evolution of coherent motions in finite-length vegetation patch flow. *Environ. Fluid Mech.* **2013**, *13*, 417–434. [CrossRef]
15. Huai, W.-X.; Zhang, J.; Katul, G.G.; Cheng, Y.-G.; Tang, X.; Wang, W.-J. The structure of turbulent flow through submerged flexible vegetation. *J. Hydrodyn.* **2019**, *31*, 274–292. [CrossRef]
16. Huai, W.; Zeng, Y.; Xu, Z.; Yang, Z. Three-layer model for vertical velocity distribution in open channel flow with submerged rigid vegetation. *Adv. Water Resour.* **2009**, *32*, 487–492. [CrossRef]
17. Jahadi, M.; Afzalimehr, H.; Ashrafizaadeh, M.; Kumar, B. A numerical study on hydraulic resistance in flow with vegetation patch. *ISH J. Hydraul. Eng.* **2022**, *28*, 243–250. [CrossRef]
18. Nosrati, K.; Afzalimehr, H.; Sui, J. Interaction of Irregular Distribution of Submerged Rigid Vegetation and Flow within a Straight Pool. *Water* **2022**, *14*, 2036. [CrossRef]
19. Wang, J.; He, G.; Dey, S.; Fang, H. Influence of submerged flexible vegetation on turbulence in an open-channel flow. *J. Fluid Mech.* **2022**, *947*, A31. [CrossRef]
20. Shahmohammadi, R.; Afzalimehr, H.; Sui, J. Impacts of Turbulent Flow over a Channel Bed with a Vegetation Patch on the Incipient Motion of Sediment. *Can. J. Civ. Eng.* **2018**, *45*, 803–816. [CrossRef]
21. D'Ippolito, A.; Calomino, F.; Penna, N.; Dey, S.; Gaudio, R. Simulation of accelerated subcritical flow profiles in an open channel with emergent rigid vegetation. *Appl. Sci.* **2022**, *12*, 6960. [CrossRef]
22. Afzalimehr, H.; Riazi, P.; Jahadi, M.; Singh, V.P. Effect of vegetation patches on flow structures and the estimation of friction factor. *ISH J. Hydraul. Eng.* **2021**, *27*, 390–400. [CrossRef]
23. Stephan, U.; Gutknecht, D. Hydraulic resistance of submerged flexible vegetation. *J. Hydrol.* **2002**, *269*, 27–43. [CrossRef]
24. Nepf, H.; Ghisalberti, M. Flow and transport in channels with submerged vegetation. *Acta Geophys.* **2008**, *56*, 753–777. [CrossRef]
25. Luhar, M.; Rominger, J.; Nepf, H. Interaction between flow, transport and vegetation spatial structure. *Environ. Fluid Mech.* **2008**, *8*, 423–439. [CrossRef]
26. Nepf, H.M.; Vivoni, E. Flow structure in depth-limited, vegetated flow. *J. Geophys. Res. Ocean.* **2000**, *105*, 28547–28557. [CrossRef]
27. Afzalimehr, H.; Subhasish, D. Influence of bank vegetation and gravel bed on velocity and Reynolds stress distributions. *Int. J. Sediment Res.* **2009**, *24*, 236–246. [CrossRef]
28. Fazlollahi, A.; Afzalimehr, H.; Sui, J. Effect of slope angle of an artificial pool on distributions of turbulence. *Int. J. Sediment Res.* **2015**, *30*, 93–99. [CrossRef]
29. MacVicar, B.; Obach, L. Shear stress and hydrodynamic recovery over bedforms of different lengths in a straight channel. *J. Hydraul. Eng.* **2015**, *141*, 04015025. [CrossRef]
30. Nezu, I.; Rodi, W. Open-channel flow measurements with a laser Doppler anemometer. *J. Hydraul. Eng.* **1986**, *112*, 335–355. [CrossRef]
31. Afzalimehr, H.; Rennie, C.D. Determination of bed shear stress in gravel-bed rivers using boundary-layer parameters. *Hydrol. Sci. J.* **2009**, *54*, 147–159. [CrossRef]
32. Afzalimehr, H.; Moghbel, R.; Gallichand, J.; Sui, J. Investigation of turbulence characteristics in channel with dense vegetation. *Int. J. Sediment Res.* **2011**, *26*, 269–282. [CrossRef]
33. MacVicar, B.; Roy, A. Sediment mobility in a forced riffle-pool. *Geomorphology* **2011**, *125*, 445–456. [CrossRef]
34. Carling, P.A.; Orr, H.G. Morphology of riffle–pool sequences in the River Severn, England. *Earth Surf. Process. Landf. J. Br. Geomorphol. Res. Group* **2000**, *25*, 369–384. [CrossRef]
35. Przyborowski, Ł.; Łoboda, A.M.; Bialik, R.J. Effect of two distinct patches of Myriophyllum species on downstream turbulence in a natural river. *Acta Geophys.* **2019**, *67*, 987–997. [CrossRef]

36. Mayaud, J.R.; Wiggs, G.F.; Bailey, R.M. Dynamics of skimming flow in the wake of a vegetation patch. *Aeolian Res.* **2016**, *22*, 141–151. [CrossRef]
37. Robinson, S.K. Coherent motions in the turbulent boundary layer. *Annu. Rev. Fluid Mech.* **1991**, *23*, 601–639. [CrossRef]
38. Poggi, D.; Katul, G.; Albertson, J. Momentum transfer and turbulent kinetic energy budgets within a dense model canopy. *Bound.-Layer Meteorol.* **2004**, *111*, 589–614. [CrossRef]
39. Dey, S.; Nath, T.K. Turbulence characteristics in flows subjected to boundary injection and suction. *J. Eng. Mech.* **2010**, *136*, 877–888. [CrossRef]
40. Parvizi, P.; Afzalimehr, H.; Singh, V.P. Impact of pool and vegetated bottom on turbulant flow structure. *Int. J. Hydraul. Eng.* **2021**, *10*, 8–18.
41. Afzalimehr, H.; Anctil, F. Accelerating shear velocity in gravel-bed channels. *Hydrol. Sci. J.* **2000**, *45*, 113–124. [CrossRef]
42. Kironoto, B.; Graf, W.H.; Reynolds. Turbulence characteristics in rough non-uniform open-channel flow. *Proc. Inst. Civ. Eng.-Water Marit. Energy* **1995**, *112*, 336–348. [CrossRef]
43. Afzalimehr, H. Effect of non-uniformity of flow on velocity and turbulence intensities over a cobble-bed. *Hydrol. Process. Int. J.* **2010**, *24*, 331–341. [CrossRef]
44. Najafabadi, E.F.; Afzalimehr, H.; Rowiński, P.M. Flow structure through a fluvial pool-riffle sequence–Case study. *J. Hydro-Environ. Res.* **2018**, *19*, 1–15. [CrossRef]
45. Coles, D. The law of the wake in the turbulent boundary layer. *J. Fluid Mech.* **1956**, *1*, 191–226. [CrossRef]
46. Thornton, C.I.; Abt, S.R.; Morris, C.E.; Fischenich, J.C. Calculating shear stress at channel-overbank interfaces in straight channels with vegetated floodplains. *J. Hydraul. Eng.* **2000**, *126*, 929–936. [CrossRef]
47. MacVicar, B.J.; Rennie, C.D. Lateral distribution of turbulence and secondary currents in non-uniform open channel flow. In Proceedings of the 33rd IAHR Congress: Water Engineering for a Sustainable Environment Hydraulics, Vancouver, BC, Canada, 9–14 August 2009; pp. 1908–1915.

water

MDPI

Article

Effects of Submerged Vegetation Arrangement Patterns and Density on Flow Structure

Mahboubeh Barahimi and Jueyi Sui *

School of Engineering, University of Northern British Columbia, Prince George, BC V2N 4Z9, Canada
* Correspondence: jueyi.sui@unbc.ca; Tel.: +1-250-960-6399

Abstract: Aquatic vegetation appears very often in rivers and floodplains, which significantly affects the flow structure. In this study, experiments have been conducted to investigate the effects of submerged vegetation arrangement patterns and density on flow structure. Deflected and non-bending vegetation is arranged in square and staggered configurations in the channel bed of a large-scale flume. Results showed that the staggered configuration leads to intensified streamwise velocity, turbulence kinetic energy (TKE), and Reynolds shear stress (RSS) compared to the square configuration. When vegetation density is low ($\lambda = 0.04$ and $\lambda = 0.07$), the produced wake in the rear of the vegetation is more expansive than that with high vegetation density ($\lambda = 0.09$ and $\lambda = 0.17$) because the velocity in the center of four vegetation elements is lower than that in the middle of two vegetation elements with low vegetation density. Results of TKE in the wake zone of the deflected vegetation indicate that the maximum root-mean-square velocity fluctuations of flow occur at the sheath section ($z/H = 0.1$) and the top of the vegetation ($z/H = 0.4$). In the wake zone behind the vegetation elements, the maximum value of the RSS occurred slightly above the interface between deflected vegetation and the non-vegetation layer, showing the Kelvin–Helmholtz instability that is associated with inflectional points of the longitudinal velocity. Within the range of vegetation density in this study ($0.04 < \lambda \approx< 0.23$), as the vegetation density increases, the negative and positive values of RSS throughout the flow depth increase.

Keywords: deflected vegetation; non-bending vegetation; vegetation density; Reynolds shear stress; turbulence kinetic energy; flow structure

Citation: Barahimi, M.; Sui, J. Effects of Submerged Vegetation Arrangement Patterns and Density on Flow Structure. *Water* **2023**, *15*, 176. https://doi.org/10.3390/w15010176

Academic Editor: Giuseppe Oliveto

Received: 2 December 2022
Revised: 25 December 2022
Accepted: 29 December 2022
Published: 1 January 2023

1. Introduction

In natural rivers and streams, various arrangement patterns of vegetation can be seen very often in the channel bed, along river banks or on flood plains. Different characteristics of vegetation elements such as vegetation density, shape, and flexibility affect the bending degree of flexible vegetation and have different impacts on flow structure [1,2].

Vegetation creates the ecological habitat and plays an active role in maintaining and protecting biological diversity [3] by providing food and shelter for fish and many other aquatic creatures [4,5]. There is an interaction between vegetation and bed deformation. On the one hand, vegetation influences flow structure, sediment erosion, and deposition [6–8]. On the other hand, as a result of sediment erosion and deposition, organic materials attached to sediment particles spread throughout the river bed and affect vegetation growth and spread [9]. Due to a decrease in flow velocity caused by vegetation in the channel bed, erosion rates decrease [10]. Chen et al. (2011) showed that both the length and depth of scour holes decrease with the increase in vegetation density [11]. Net deposition increased with the distance from the leading edge of vegetation, associated with a decrease in vertical velocity and TKE [12].

Vegetation on riverbanks is a crucial factor in reducing flood damage and coastal erosion by increasing bank stability and damping waves [13–17]. Yue et al. (2020) reported that vegetation roots and sand–root composites provide effective reinforcement to unconsolidated banks, controlling bank erosion and thus reinforcing the stability of banks [18].

In the past decades, to better understand the hydrodynamics in the presence of vegetation in rivers, many research works have been conducted. One of the main concerns is the high turbidity of flowing water in rivers. Water turbidity can negatively affect aquatic creatures. Vegetation is considered as a great measure to reduce resuspension and damping of waves and induce deposition, since additional drag resulted from vegetation reduces the mean flow velocity and bed shear stress within vegetated regions compared to that of the bare channels [19–21]. Ros et al. (2014) found that resuspended sediment concentrations decreased as the flexible canopy density increased [22]. On the other hand, Serra et al. (2018) [23] and Zhang et al. (2018) [24] pointed out that vegetation can promote near-bed turbulence, which will cause enhanced resuspension. Tinoco and Coco (2018) demonstrated a positive correlation between turbulent kinetic energy levels and the vegetation array density since the turbulence kinetic energy (TKE) is the primary driver of resuspension [25]. Although it is commonly thought that the presence of vegetation in rivers can lead to the decrease in resuspension and increase in deposition of sediment, the scour zones around vegetation elements indicate that the presence of vegetation can cause the erosion of the bed [26,27]. These various phenomena are because of different vegetation arrangement and densities. As claimed by Nepf (2012), with a low vegetation arrangement density, the near-bed turbulence can be higher than that over the neighbor bare bed [28].

The vegetation distribution can be classified as either dense or sparse by using $C_D ah$, in which C_D is the vegetation drag coefficient, a is vegetation density m^{-1}, and h is the vegetation bending height. A vegetation patch can be considered as dense if $C_D ah > 0.1$ [29]. In dense vegetation, the turbulence near the channel bed reduces, contributing to sediment retention. For sparse vegetation, $C_D ah < 0.1$, and the turbulence near the bed with the presence of sparse vegetation will increase as stem density increases.

Because of vegetation's positive impacts on water quality, habitat, and channel stability, researchers now advocate replanting and restoring projects in rivers, especially in agricultural waterways, floodways, and emergency spillways. The expansion of vegetation in fluvial systems may worsen the flood impact since highly dense vegetation reduces the channel's capacity and width. Therefore, an accurate and critical assessment of the vegetation density and distribution pattern through reduction of bulk velocity is crucial in sustainable restoration projects. Results of this study will provide vital information for river management, channel restoration, and rehabilitation of fluvial environments through understanding the effect of various vegetation densities, arrangement patterns, and morphology.

2. Materials and Methods

2.1. Flume, ADV, and SonTek IQ Used for This Study

Experiments have been carried out in a large-scale outdoor flume. This flume is 38 m long, 2.0 m wide, and 1.3 m deep, as shown in Figure 1. The longitudinal slope of the flume bed was 0.2%. Two water depths of 20 cm and 30 cm have been used for this experimental study by adjusting the tailgates at the end of the flume. These water depths were chosen based on a real situation in nature since the submerged vegetation typically grows in shallow regions of rivers. Flowing water was supplied by a pump and three valves that feed the upstream holding tank. Water in the holding tank upstream of the main channel was maintained at a constant water level. The desired constant flow rate, which is 100 cm^3/s in this study, was obtained by adjusting these three valves. The holding tank has a volume of 90 m^3 to keep a constant water level during each experimental run.

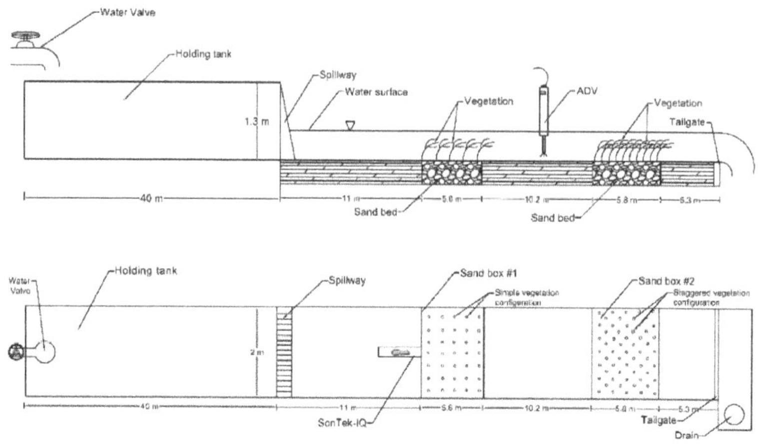

Figure 1. The layout of the experimental flume (vertical and plan views).

The aspect ratio W/H is defined as the ratio of the flume width to water depth. For both water depths of 20 cm and 30 cm used in this experimental study, the flume is classified as a wide flume since the aspect ratio is greater than 5 to 10. This means that in this flume, the effects of the side walls of the channel and the secondary currents can be ignored in the center zone of the flume [30].

There were two sandboxes which are spaced 10.2 m from each other. These sandboxes were 2 m wide and 0.3 m deep. The upstream sandbox was 5.6 m long and the downstream one was 5.8 m long.

In this experimental study, a down-looking Acoustic Doppler Velocimeter (ADV) 10-MHz, developed by Nortek, was used to measure the instantaneous three-dimensional velocity components with a sampling rate of 25 Hz and a sampling volume of 0.25 cc (Figure 2a). The duration of each measurement was 2 min, acquiring 3000 instantaneous velocity data at each measurement point. The vertical intervals between two consecutive points for each velocity profile were 10 mm. The signal-to-noise ratio (SNR) was recorded in the ADV file and used for assessing the strength of the received acoustic signal against the ambient electronic noise level of the ADV [31]. To obtain high-quality data from the ADV, SNR values should be greater than 5 dB for measurements of the mean flow velocity and greater than 15 dB for the instantaneous velocity or turbulence quantities. The filtering method of Goring and Nikora (2002) [32] and Wahl (2002) [33] was selected in this study. The WinADV software was used for data filtering. One of the real-time outputs provided by the ADV is a statistic correlation to assess the quality of the velocity measurements. If the average correlation was less than or equal to 70%, the measured velocity data were filtered out.

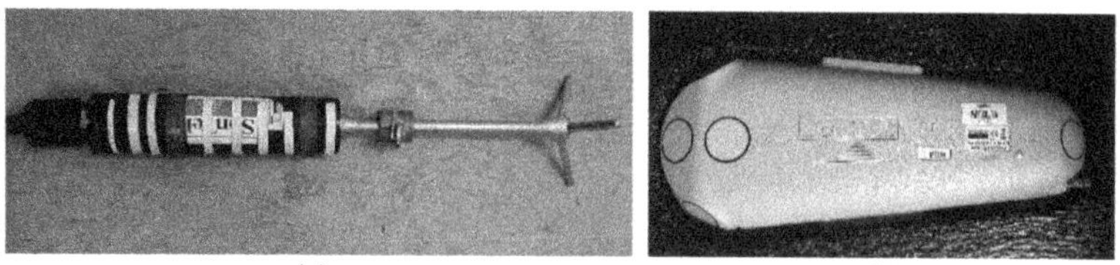

(a) (b)

Figure 2. (**a**) ADV used in this study, (**b**) Sontek IQ.

Measurements at any depth less than 10 cm from the water surface could not be used due to limitations of the ADV since the distance of the sampling volume in this ADV is located 10 cm below the transmitter. It means that the collected data are from the near bed up to 10 cm below the water surface where the ADV was operational. The fluctuating turbulence near the channel bed is more important than that near water surface. Since most sediment transport caused by drag force resulted from vegetation elements and bed shear stress occurs closer to bed [34,35], one can conclude that the limitation of the ADV for measuring the fluctuations from water surface to 10 cm below it does not affect the results much.

After removing spikes using the WinADV software, velocity fluctuation in the longitudinal (x), lateral (y), and vertical (z) directions were calculated as follows:

$$\begin{aligned} u' &= u_i - u \\ v' &= v_i - v \\ w' &= w_i - w \end{aligned} \tag{1}$$

where u, v, and w are the time-averaged velocities that correspond to the directions x, y, and z, respectively; u_i, v_i, and w_i are the instantaneous velocities that correspond to the directions x, y, and z, respectively. The x axis is aligned with the direction of the mean flow. The y axis is the spanwise direction, and the z axis is vertical, with $z = 0$ at the channel bed, and positive upward.

The equilibrium state of the scour process in vegetated channels will achieve after 48 h [36]. To make sure that the exact flow rate has been obtained over the duration of 48 h, a SonTek-IQ Plus was used (Figure 2b). This precise and robust apparatus was also used to measure the average velocity and water depth with advanced post-processing functions [37]. The SonTek IQ is a semi-rectangular shape designed to mount on the channel bottom. Because of the sleek silhouette of the SonTek IQ, its impact on flow is minimal. The surface slope $\partial H/\partial x$ was measured using a staff gauge installed in the middle of the sandbox to verify the water depth manually.

2.2. Sediment Used in Experiments

The sandbox is filled with non-uniform sediment with a median particle size (D_{50}) of 0.50 mm. The standard deviation ($\sigma = \sqrt{D_{84}/D_{16}}$) was used to analyze the uniformity of the distributions where D_{84} and D_{16} are 84% and 16% finer particle diameters, respectively. The smaller the value of σ, the more well-sorted the sediment is [38]. The standard deviation for the sand with median grain size of 0.50 mm in this study is 1.97. Based on that, the sand size used in this experiment is non-uniform. The grain size distribution was obtained using a mechanical shaker and seven different-sized sieves. Figure 3 shows the grain size distributions of the bed material.

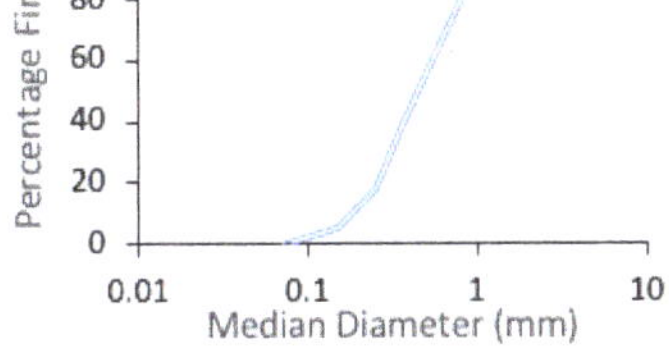

Figure 3. Grain size distribution of bed materials.

The following equation proposed by Hager (1999) is used for determining the roughness coefficient of the channel bed [39]:

$$n_b = 0.039D_{50}^{(1/6)} \tag{2}$$

where n_b is the roughness coefficient of the channel bed and D_{50} is median grain size of sediment particle. Therefore, the roughness coefficient of sand bed, n_b, is estimated as 0.011 for D_{50} = 0.0005 m.

2.3. Vegetation Settings

The model flexible vegetation elements used in this study are made of plastic material. The selected artificial vegetation flexibility is commensurate with the geometry and flexural rigidity of typical aquatic vegetation growing in natural rivers. Each vegetation element consisted of five blades attached to it. Every vegetation element was attached to a grid mesh panel with the spacing distances respectively of 15 cm and 25 cm in a square configuration, and 10.61 cm and 17.68 cm in a staggered configuration. Then, the grid mesh panel with vegetation elements was placed and buried 10 cm below the sand bed surface. Afterward, the surface of the sandbox with vegetation was carefully leveled. By doing so, the vegetation elements were fully stabilized in the sand, representing a natural situation with roots in channel bed.

Figure 4 shows the positions for measurement using an ADV (ADV positions) in channel bed with two different vegetation arrangement patterns, namely, the square and staggered configurations. Measurements at 24 ADV positions around vegetation elements provide robust information for detecting flow structure and turbulence. To determine the wake structure behind each vegetation element, the velocity profile was taken at three points in the wakes of some vegetation elements, as shown in Figure 4.

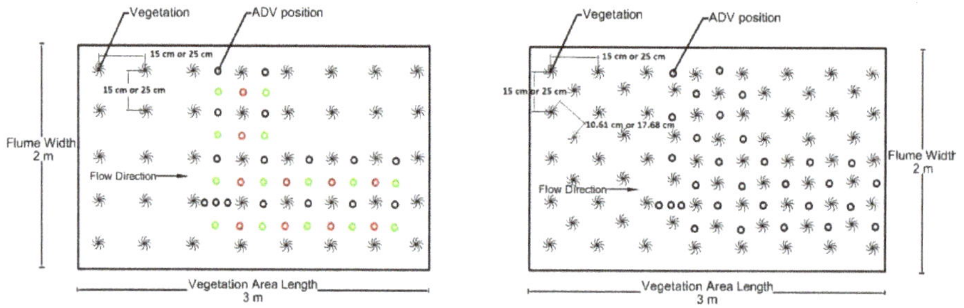

Figure 4. Schematic top view of a vegetative zone in the flume. Black pronged shape shows the positions of individual vegetation elements. Black circles are ADV measurement locations in the wake of each vegetation element. Red circles show the ADV location between two vegetation elements, and green circles show the ADV location in the center of four vegetation elements which have been placed in a square pattern. The spacing distance between adjacent vegetation elements is shown in the figure.

The flowing water from upstream passes through the submerged vegetation patch. After a certain distance from the upstream edge of the submerged vegetation patch, the flow will be fully developed. Upstream of submerged vegetation, flow follows boundary layer conditions. Once the flow approaches submerged vegetation, this condition turns into a mixing layer. In the mixing layer flow, the shear layer known as Kelvin–Helmholtz vortices will be developed and reaches an equilibrium in size depending on the vegetation density and submergence ratio [40]. The flow within the submerged vegetation is fully developed when equilibrium is achieved. In this study, all velocity profiles are collected using the ADV in the fully developed flow inside of the vegetation patch, which is different from the fully developed flow in channels without the presence of vegetation.

In the presence of a finite vegetation patch, channel resistance and conveyance are modified, at least locally, resulting in a deviation from uniform flow conditions [41]. In addition, the channel has a longitudinal slope of 0.2% and the bed material is non-uniform sand, which leads to non-uniform flow in the experiment.

The canopy zone can be divided into two sub-zones: the longitudinal exchange zone ($z < h_p$) and vertical exchange zone ($z > h_p$) (where z is vertical distance from the bed and h_p is penetration depth). The penetration depth ($z = h_p$) is defined as the point into the canopy where shear stress decays to 10% of its maximum value. As canopy density increases ($ahC_D > 0.2$), the penetration depth decreases [42]. Note that C_D is the drag coefficient of vegetation, a is vegetation density a (m^{-1}), and h is the bending height of vegetation. The canopy becomes less ventilated since eddies no longer enter the bed [43]. The designated momentum in the longitudinal exchange zone is a balance of pressure gradient or bed slope and vegetative drag. Turbulence in this region is generated at stem wakes and represents the stem morphology. However, in the vertical exchange zone, flow is affected by momentum balance, contributes to scalar exchange, and turbulence is generated by the KH instabilities [44]. In other words, the difference in drag magnitude between the non-vegetated and vegetated zones leads to the Kelvin–Helmholtz (KH) vortices occurring at the interface between vegetated zones and water. The KH vortices can promote the mass and momentum transport within and over canopies [44,45]. For instance, turbulent mass exchange across the canopy–water interface can regulate the nutrients and contaminants.

To investigate the influence of vegetation density on flow structure, artificial vegetation with densities of 16, 32, 36, and 72 stems/m^2 were used in this study. It may be sufficient to parameterize vegetation based on stem diameter, density, and the number of plants per area for flow without leaf and non-bending vegetation [46,47].

The vegetation density a (m^{-1}) was determined by dividing the projected vegetation area by the vegetation volume (Equation (3)) as [48]:

$$a = \frac{A}{V} = \frac{n\overline{A}_i}{WhL} \tag{3}$$

where n is the amount of vegetation in the area of (W * L), W is the channel width, $\overline{A}_i$ is the mean frontal vegetal area, h is the vegetation bending height, and L is the length of channel in which n was counted. Considering five blades per vegetation element, the vegetation density a varies from 0.256 m^{-1} to 1.2 m^{-1}. Finally, a non-dimensional measure of the canopy density $\lambda = ah$, known as the roughness density, was calculated. The frontal area of vegetation was determined using an image analysis software. This software was designed to distinguish between black and white zones to calculate the silhouette of the vegetation. According to Belcher et al. (2003), there is a scale to distinguish sparse and dense vegetation [29]. In a sparse regime ($\lambda = ah < 0.1$), the vegetation drag is small compared to the bed roughness. Therefore, flow velocity acts following the boundary layer profile. In this regime, the turbulence near the bed will increase as the stem density increases. On the other hand, in dense vegetation regime ($\lambda = ah > 0.1$), the vegetation drag is clearly high compared to the bed stress. An increase in the vegetation density will lead to a decreased near-bed turbulence and increased sedimentation. The vegetation density in this study is summarized in Table 1. One can see from Table 1 that the range of the canopy density $\lambda = ah$ is $0.04 < \lambda \approx< 0.23$. Some researchers found that, for $0.1 < \lambda \approx< 0.2$, the eddies in the mixing layer penetrate toward the bed. In this study, the vortices (eddies) in the middle layer of the flow, named the mixing layer, penetrate toward the bed and are responsible for turbulence patterns across the vegetation, benefiting the resuspension of sediment [49,50]. As a result, no penetration depth needs to be calculated in the present study because the eddies reach the bed.

Table 1. Vegetation density parameters in this study.

Configuration of Vegetation Elements	Vegetation Density $a(m^{-1})$	Canopy Density ($\lambda = ah$)
Square non-bending Square deflected	0.624	0.12 0.09
Staggered non-bending Staggered deflected	1.2	0.23 0.17
Square non-bending Square deflected	0.256	0.05 0.04
Staggered non-bending Staggered deflected	0.506	0.1 0.07

Some researchers used the inflexible and idealized cylinders to represent vegetation to investigate the complicated flow structure caused by vegetation; it is unable to fully predict the behavior of natural vegetation due to its differences in roughness, flexibility, and drag coefficient. Besides, flow structure around a single cylinder or vegetation element cannot be generalized for a vegetation patch since a group of cylinders or vegetation elements interact with each other and on flow structure through sheltering effect, blocking effects, and flow separation. Compared to a single cylinder, turbulent fluctuations in the wakes of upstream elements introduce additional kinetic energy to the boundary layer of each element, delaying separation and reducing drag coefficient [47,51]. Therefore, properly characterizing morphological properties of vegetation is essential for studying the hydrodynamics of vegetated streams. To investigate the effect of the flexibility and height of vegetation on flow, both deflected and non-bending morphology of vegetation were used in this study (Figure 5). In one of these vegetation settings, all vegetation elements were placed in a fully non-bending setting to represent the stiff and rigid vegetation patch (Figure 5a). In this way, the vegetation exhibits no deformation during the experiments, representing the real situation of reeds and sedges in natural rivers. In another setting, all vegetation elements were deflected, representing the flexible vegetation in streams (Figure 5b). The height of non-bending and deflected vegetation elements in 20 cm water depth was 12.6 cm and 8 cm, respectively, and 19 cm and 12 cm in the 30 cm water depth. As you saw in Figure 1, the length of each sandbox is around 5.5 m. The 3 m long vegetation region was located in the middle of the sandbox. The vegetation has a certain degree of flexibility and can swing under flow, but it does not deform. The morphological properties of the vegetation such as the vegetation deflected height are related to some other hydraulic properties of flow such as flow depth and velocity.

(a)

(b)

Figure 5. ADV position between two vegetation elements in (**a**) non-bending vegetation setting; (**b**) deflected vegetation setting.

The ratio of the flow depth to the vegetation bending height is defined as the degree of submergence ($Sr = H/h$) [52]. Thus, the degree of submerged vegetation in this study were selected as 1.58 for non-bending and 2.5 for deflected vegetation arrangement, respectively.

Each vegetation element has the width of 10 mm at the bottom and 22 mm at the top, respectively. Thus, average width of each element was considered as 16 mm. In the present study, for the case of the densest vegetation configuration, the ratio of the total vegetation thickness to the channel width, D/W, was smaller than 0.5; therefore, the effect of channel blockage on the wake structure can be negligible.

The vegetation Reynolds number associated is defined as:

$$Re_d = \frac{Ud}{v} \tag{4}$$

where U is the mean flow velocity, d is the stem diameter, and v is the kinematic viscosity of water.

$$F_r = \frac{U}{\sqrt{gH}} \tag{5}$$

where g is the gravitational acceleration and H is the water depth. Both calculated Reynolds numbers and Froude numbers indicated that the flow was fully turbulent and subcritical for all cases; therefore, no dependence on Fr number was expected.

To start each experimental run, one valve with the low discharge (5 *L/s*) was gradually opened while the tailgates downstream were closed to avoid sediment being washed away. From the holding tank, water was gently discharged through the spillway into the flume. To maintain the desired flow rate, all three valves were fully opened once the desired water depth was reached. In this study, 32 experimental runs have been conducted including two different submergence ratios, four different vegetation densities with two different layouts including square and staggered configurations for two flow depths of 20 cm and 30 cm. Some of the measured hydraulic data for the flow depth of 20 cm are presented in Table 2.

Table 2. Some data for the flow depth of 20 cm.

Configuration	Density (Stem/m^2)	Flexibility	U (cm/s)	Re_d	λ	u_* (cm/s)	h/H
square	16	deflected	15.15	2423.52	0.0487	3.08	2
		non-bending	17.31	2770.35	0.0359	2.94	1.3
	36	deflected	14.87	2378.56	0.0874	3.42	2
		non-bending	17.58	2813.30	0.1186	3.49	1.3
staggered	32	deflected	14.73	2356.67	0.0961	2.56	2
		non-bending	18.11	2897.52	0.0709	3.16	1.3
	72	deflected	12.46	1993.90	0.168	2.68	2
		non-bending	16.93	2708.75	0.228	3.49	1.3

3. Results and Discussions

3.1. Velocity

In this study, the values of shear velocity (u_*) were obtained using the boundary layer method [53]:

$$u_* = \frac{(\delta_* - \theta)\, u_{max}}{C\delta_*} \tag{6}$$

where $u_{\max}$ is maximum streamlined velocity, and C is an empirical constant that was found to be equal to 4.4 in laboratory experiments [53]. The parameter δ_* is the boundary layer displacement thickness and indicates the distance by which the external streamlines are shifted owing to the formation of the boundary layer. It is impossible to present a boundary layer thickness in an unambiguous way because the effect of viscosity in the boundary layer decreases asymptotically outwards [54]. In order to avoid utilization of an arbitrary boundary layer thickness, it is necessary to consider the boundary layer displacement thickness, δ_* [55]:

$$\delta_* = \int_0^H \left(1 - \frac{u}{u_{max}}\right) dz \tag{7}$$

where u is the mean point velocity at a distance z measured from the reference level. Furthermore, the momentum thickness (θ) in Equation (6) indicates the loss of momentum in the boundary layer as compared with potential flow and is defined as:

$$\theta = \int_0^H \frac{u}{u_{max}} \left(1 - \frac{u}{u_{max}}\right) dz \tag{8}$$

The exact values of δ_* and θ depend upon the distribution of downstream velocity at the cross-section normal to the flow.

In a channel with the presence of vegetation, there is an inner layer called the emergent zone that is controlled by stem scale turbulence. Above that layer, there is a layer with Kelvin–Helmholtz (KH) vorticities that dominate mass and momentum exchange. The logarithmic layer refers to the upper layer of turbulent flow where the velocity profile follows a log shape [56]. Kazem et al., 2021, reported these three layers were present in all cases of their experiments [57]. The logarithmic profile may be described by the Karman–Prandtl equation:

$$u = u_* \left(Ln \frac{z - \delta_*}{z_0} \right) / \kappa \tag{9}$$

where u_* is the shear velocity, z_0 is the roughness height, and κ is the von Karman constant, which is 0.41. The value of δ_* was determined to be 0.02 m through Equation (7) for non-bending vegetation and 0.03 for deflected vegetation. In this study, z_0 is assumed to be equal to D_{50}.

Results of the velocity showed there are significant differences by changing the density, morphology, and layout of vegetation. In addition, velocity is affected by both the measurement position and water depth using an ADV.

Figure 6 shows velocity profiles in the presence of a deflected and non-bended vegetation patch in the channel bed. One can see from this figure that these velocity profiles deviate from the logarithmic law distribution (the Karman–Prandtl Equation) and are confined to the upper part of the flow in the presence of the vegetation patch. In addition, there exists a very good correlation-ship between u/u_* and $(z - \delta_*)/z_0$. Note: a logarithmic profile in the flow's inner layer cannot be developed due to the presence of vegetation [58].

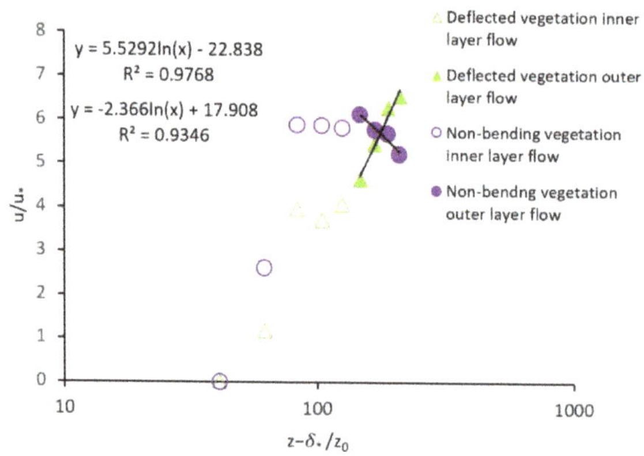

Figure 6. One sample of fitting Logarithmic Law distribution on the upper layer of velocity profile in a deflected and non-bending vegetation (staggered arrangement, flow depth: 30 cm).

3.1.1. Effects of Vegetation Density on Streamwise Velocity

The velocity profiles for the different vegetation densities in the fully developed region are compared, as shown in Figure 7. One can observe from Figure 7 that with the increase in the vegetation density (a = 0.624 m^{-1}), the flow velocity within the canopy decreases, and correspondingly the flow velocity above the canopy increases.

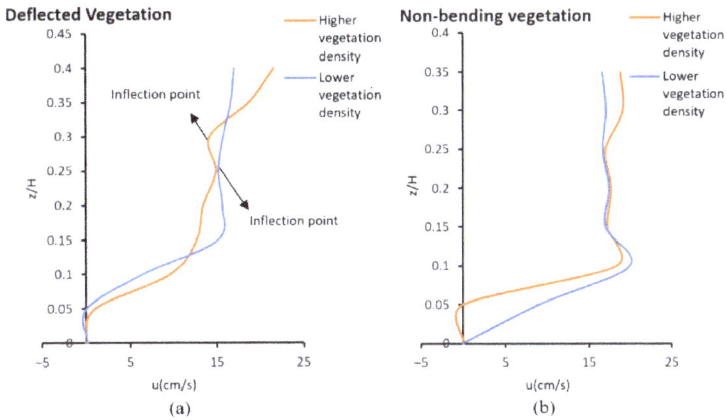

Figure 7. Velocity profiles with high vegetation density (a = 0.624 m^{-1}) and lower vegetation density (a = 0.256 m^{-1}): (**a**) deflected vegetation; (**b**) non-bending vegetation (water depth: 20 cm and square arrangement of vegetation in the wake behind vegetation (black circles in Figure 4).

In the rear of the vegetation wake zone, vegetation creates resistance to flow, and causes flow separation and a decrease in flow velocity near the bed. This phenomenon is the main reason for sediment retention especially behind the deflected vegetation patch with high density. Although this statement is generally true, scour holes around vegetation stems have been observed in many studies and are dependent on vegetation density. Vegetation deflection can be viewed as a passive "drag-reduction" strategy exhibited by vegetation. As shown in Figure 7, the inflection points at the top of the deflected vegetation in the wake zone behind the vegetation are sensible (see arrows in Figure 7a). This finding is in good agreement with that of other researchers that velocity profiles in flows with submerged vegetation contain an inflection point near the top of the vegetation [42,59]. There is an increase in velocity on top of the canopy of the deflected vegetation at z/H = 0.4 compared to the inner layer of vegetation. The difference between the drag magnitude in the non-vegetated zone and that in the vegetated zone causes the Kelvin–Helmholtz (KH) vortices at the interface between vegetation and non-vegetation layer. The KH vortices promote mass and momentum transport both within and over canopies [44,45,47]. The KH instabilities significantly affect the large-scale turbulence structures and the momentum transfer between the non-vegetated and vegetated regimes. The effects of KH instabilities show their effect as an inflection point in velocity profiles [42,59]. Decreasing deflected vegetation spacing (i.e., increasing canopy density) largely retards streamwise velocity at (z/H ≅ 0.3), slightly below the inflection point (see Figure 7a). The inflectional region tends to disappear when the canopy becomes sparser resulting in an increase of the shear length scale associated with the velocity field. Non-bending vegetation lacks this.

On the other hand, in the presence of the non-bending vegetation in the channel bed, for the sparsest vegetation (a = 0.256 m^{-1}), the highest velocity occurs near the bed and lowest velocity near water surface. The peak velocity occurs at the depth of z/H = 0.1, that is, the sheath section where the frontal width is minimal. Because the sheath section is more porous than the middle vegetated layer, it can handle larger flows [11]. A high negative velocity gradient happened from z/H = 0.1 to z/H = 0.15. Then, a decreasing trend of velocity to the water surface is noticeable.

According to Figure 7, it is suggested that the dense deflected vegetation ($\lambda \geq 0.1$) results in the decrease in sediment transport in streams by reducing the velocity near the bed more than non-bending vegetation and sparse densities. Therefore, it is suggested that dense vegetation provides better protection for beds subject to erosion and scour.

3.1.2. Effects of Water Depth of Streamwise Velocity

In the presence of non-bending vegetation in the channel bed with the flow depth of 20 cm, a high velocity gradient has been observed from the channel bed z/H = 0 to the depth of z/H = 0.1 where a peak velocity was reached. After this depth, a decreasing trend of velocity toward the water surface (when 0.1 < z/H < 1.0) is noticeable (Figure 8). However, for the deeper flow of 30 cm, the peak velocity occurs at a larger distance from the channel bed but closer to the water surface. It has also been noticed that, from the depth of z/H = 0.1 to the water surface, the value of flow velocity does not change. The dip phenomenon (the position at which the maximum velocity appears below the water surface) was observed for the case of the non-bending vegetation.

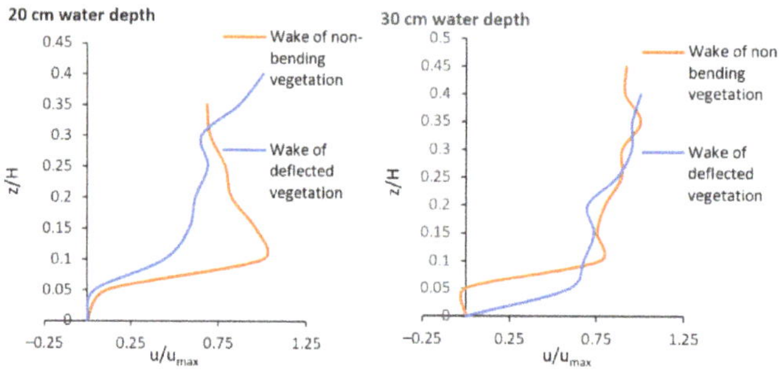

(a) (b)

Figure 8. Comparison of velocity profiles in the wake zone behind vegetation elements arranged in a square configuration, (**a**) water depth = 20 cm; (**b**) water depth = 30 cm.

In the wake zone of deflected vegetation, the velocity in the inner layer of 20 cm-depth flow is less than that in the 30 cm-depth flow, and the velocity in the upper layer in the 20 cm-depth flow is more than that in the 30 cm-depth flow. It can be concluded that, by keeping the same submergence ratio, the shallower the flow depth, the lower the flow velocity in the inner layer in the wake zone of deflected vegetation, and the higher the flow velocity in the outer layer.

Thus, both the flow depth and morphology of vegetation have a substantial effect on the velocity profile. For the non-bending vegetation case, a peak velocity value appears near the bed in shallow water, implying that the non-bending vegetation acts like an emergent vegetation. Therefore, fine sediment particles cannot be easily retained behind the vegetation patch due to high flow velocities. In contrast, in the presence of deflected vegetation in a channel bed, the velocity near bed is smaller than that of non-bending vegetation mainly due to the sheltering effect behind vegetation (see Section 3.2. Turbulence Kinetic Energy for more information). In a deeper flow (30 cm), the peak velocity is shifted toward the water surface compared to that in a shallower flow (20 cm) with high velocity near the channel bed. As a consequence, sediment deposition is more probable in a deeper flow comparing to that in a shallower flow.

3.1.3. Effects of Square Arrangement on Velocity Profile

Figure 9 shows the dimensionless time-averaged streamwise velocity distribution between two vegetation elements (red circles in Figure 4) and in the center of four vegetation elements (green circles in Figure 4) with a density of a = 0.624 m^{-1} in the flow with the depth of 20 cm compared to that in the flow with the depth of 30 cm.

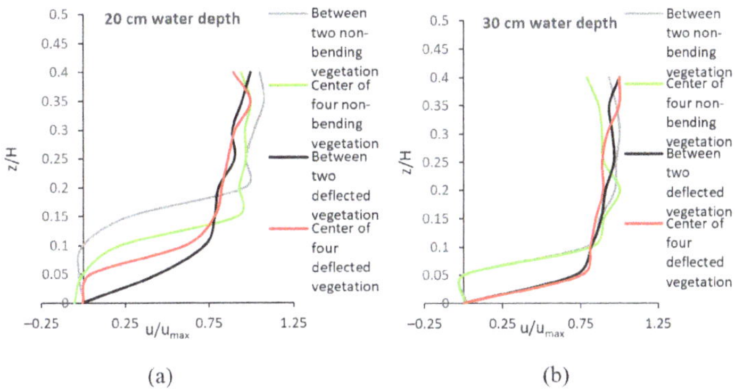

Figure 9. Dimensionless time-averaged streamwise velocity distribution in the presence of square vegetation configuration with a vegetation density of a = 0.624 m^{-1}. ADV locations: between two vegetation elements (red circles in Figure 4) and in the center of four vegetation elements (green circles in Figure 4): (**a**) water depth = 20 cm; (**b**) water depth = 30 cm.

In the presence of dense non-bending vegetation with a square arrangement pattern in the channel bed, for the flow with a depth of 20 cm, the difference between mean streamwise velocities in the middle of two vegetation elements and those in the wake of vegetation elements is 27.39%; and the difference between mean streamwise velocities in the center of four vegetation elements and those in the wake of vegetation elements is 30.64%. In the presence of dense deflected vegetation in the channel bed, these values are 26.79% and 28.08%, respectively. For the flow with a depth of 30 cm, the same trend was calculated. According to these results, in the presence of dense vegetation of $\lambda \geq 0.1$, the velocity in the middle of two vegetation elements is lower than that in the center of four vegetation elements. Accordingly, in vegetation with a high density, the width of the wake behind the vegetation is narrow, leading to the increase in the flow velocity in the center of the vegetation elements. In other words, the vegetation with a high density reduces the flow cross-sectional area locally, and thus results in a narrow wake behind the vegetation which diminishes faster within a shorter distance compared to that for the case of sparse vegetation. In contrast, in the presence of sparse non-bending vegetation with low density (i.e., $\lambda < 0.1$) in the flow with a depth of 20 cm, the difference between the mean streamwise velocities in the middle of two vegetation elements and those in the wake of vegetation elements is 20.56%; and the difference between the mean streamwise velocities in the center of four vegetation elements and those in the wake of vegetation elements is 12.05%. For the case of deflected vegetation, these differences are 34.48% and 33.75%, respectively. The same trend was observed in the flow with a depth of 30 cm. This finding shows that, in the presence of sparse vegetation in the bed, the mean streamwise velocity in the middle of two vegetation elements is higher than that in the center of four vegetation elements. This effect indicates the presence of a wide wake behind each vegetation element that attenuates the velocity at the center of four vegetation elements.

3.1.4. Effects of Staggered Arrangement on Velocity Profile

Figure 10 shows the velocity profiles in the presence of both deflected and non-bending vegetation arranged in a staggered layout in flows with different depths of 20 cm and 30 cm, respectively. The characteristics of velocity profiles of flow in the presence of vegetation with a staggered arrangement follow the same trend as those of velocity profiles in the wake of vegetation arranged in a square configuration, as discussed above. However, velocity profiles are more inflectional in vegetation with a staggered arrangement compared to square arrangement. As one can see from Figure 10, the velocity profile for the case of the staggered vegetation arrangement in the wake of deflected vegetation has an "S" shape. In the middle part close to the leaf zone of vegetation, the flow velocity reaches its maximum. In this zone, the drag force causes an appreciable retardation of the mean velocity; in the flow with a depth of 30 cm, this zone is located at $0.1 < z/H < 0.15$, and in the flow with a depth of 20 cm, this zone is located at $0.15 < z/H < 0.25$. As the velocity increases with the distance above the vegetation zone, the drag resulted from vegetation gradually diminishes, and the shear stresses decline until zero at the free water surface.

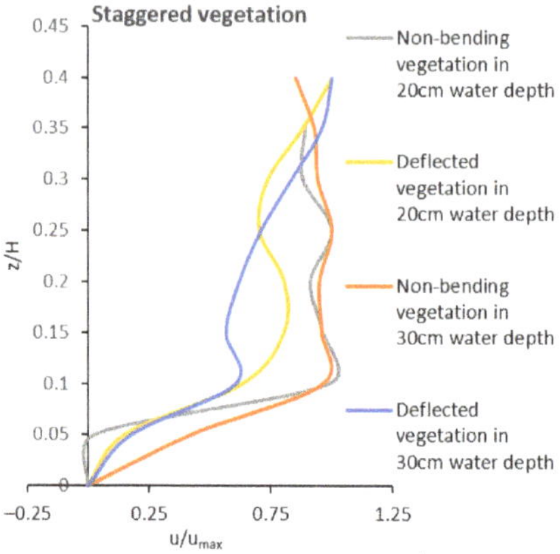

Figure 10. Velocity profiles in the wake zone behind non-bending vegetation elements and deflected vegetation elements, arranged in the staggered configuration in flow with the depths of 20 cm and 30 cm.

3.1.5. Velocity Contours in Square Arrangement

In Figure 11, two-dimensional and three-dimensional velocity contours are presented near the bed (at a depth of $z/H = 0.1$) and at the top of deflected vegetation (at a depth of $z/H = 0.4$). The depression spot in the contours in Figure 11a illustrates the wake zone of vegetation in the separation flow zone where sediment deposition was observed. Near the bed, velocity is higher between rows of vegetation elements. These graphs demonstrate that velocity in the center of the square formed by four vegetation elements is lower than that in the middle two vegetation elements. This figure confirms the results of Section 3.1.3. Some spikes in Figure 11b indicate high velocities at the interface between the deflected vegetation and non-vegetation layer as inflection points in velocity profiles compared to its neighboring zone.

3.2. Turbulence Kinetic Energy (TKE)

The generation of vortices in the stem wake zone behind vegetation drains energy from the mean flow and feeds it into turbulent kinetic energy (TKE). In this case, turbulent energy is produced at the same rate as the work done by the flow against vegetation drag [60]. Most sediment transport models are based on bed shear stress in a bare channel since turbulence is related to bed stress. However, in vegetated channels, the turbulence level is related to the vegetative drag and has little or no link to the bed shear stress [47].

The local turbulent kinetic energy was defined as

$$\mathrm{TKE} = 1/2\left(\overline{u'^2} + \overline{v'^2} + \overline{w'^2}\right) \tag{10}$$

where $\overline{u\prime}$, $\overline{v\prime}$, and $\overline{w\prime}$ represent the root-mean-square velocity fluctuations (RMS), indicating the mean energy per unit mass related to turbulent eddies in streamwise, lateral, and vertical directions, respectively.

$$\begin{cases} \mathrm{RMS}(u') = \overline{u'} = \sqrt{\overline{u'^2}} \\ \mathrm{RMS}(v') = \overline{v\prime} = \sqrt{\overline{v\prime^2}} \\ \mathrm{RMS}(w') = \overline{w\prime} = \sqrt{\overline{w\prime^2}} \end{cases} \quad (11)$$

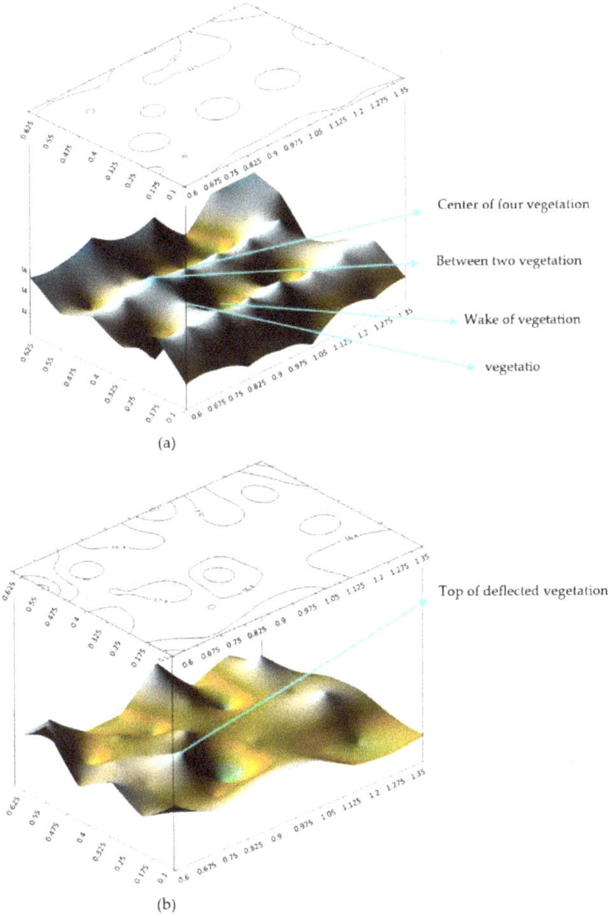

Figure 11. Contours and 3D velocity graphs in vegetation with density of λ = 0.09 at (**a**) near the bed at the depth of z/H = 0.1; (**b**) top of deflected vegetation at the depth of z/H = 0.4.

Compared to that TKE measured between two vegetation elements and those in the center of squares formed by four vegetation elements (vegetation-free streamline), as shown in Figure 12, the TKE was significantly enhanced by vegetation. In this study, this trend will be discussed in greater detail.

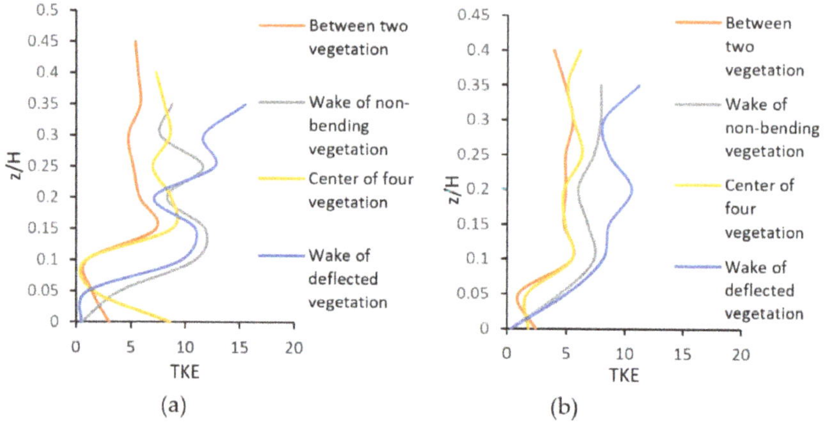

Figure 12. Comparison of Turbulence Kinetic Energy (TKE) between deflected vegetation morphology and non-bending vegetation morphology at different positions: (**a**) higher vegetation density (a = 0.624 m^{-1}); (**b**) lower vegetation density (a = 0.256 m^{-1}).

The TKE in Figure 12 in the wake of the vegetation depicts that the maximum $RMS(u')$, $RMS(v')$, and $RMS(w')$ occur either at the sheath section of the vegetation (at the depth of $z/H = 0.1$) or above the top of the vegetation (at the depth of $z/H \geq 0.4$).

In the sheath section, the frontal projected area is small, and flow can mostly pass through the sheath section. Furthermore, stem scale turbulence was boosted at the sheath section. Turbulence generated by individual stem wakes was attributed to the enhancement of TKE behind vegetation at the depth of about $z/H = 0.1$. The near-bed turbulence kinetic energy is enhanced by both bed-generated and vegetation-generated turbulence in the presence of submerged vegetation. In the region slightly far from the channel bed to the vegetation top, the presence of the von Karman vortex street results in the enhancement of TKE comparing to that in unvegetated channels. Due to the increased turbulent kinetic energy (TKE) associated with the von Karman vortex street, which inhibits sediment deposition, the location of the vortex formation indicates an important transition between zones where net deposition increases and zones where net deposition diminishes compared to those without the presence of vegetation in the bed [61].

On the other hand, there are no obstructions to the flow at the top of the vegetation. Thus, the TKE terms reach their maximum values. This result confirms that of Afzalimehr et al. (2011) [62]. During the transition from the vegetated zone to the unvegetated zone, the kinetic energy has the opportunity to enhance itself.

While in the profiles at the points either between two vegetation elements or in the center of the square formed by four vegetation elements, the TKE near the bed was greater than the TKE behind vegetation elements. As shown in Figure 12, TKE near the bed ($z/H = 0$) in the wake of vegetation elements starts at zero; however, a TKE between two vegetation elements and in the center of square formed by four vegetation elements (vegetation-free streamlines) does not equal zero (TKE > 0). This is a well-known effect of the submerged vegetation on turbulence, and it is called the sheltering or dampening effect. A sheltering effect occurs when two bodies are positioned so that one is located in the wake of the upstream body [63]. The downstream body experiences a lower approaching velocity than that for the upstream body, resulting in a lower drag force. Depending on how vegetation elements are arranged, the sheltering effect can be relevant to vegetation-covered flows [51].

As shown in Figure 12, the TKE profiles for vegetation densities of a = 0.624 m^{-1} and a = 0.256 m^{-1} indicate a greater effect of this phenomenon in denser vegetation due to a shorter distance between vegetation elements (or smaller spacing distance between vegetation elements). Therefore, the sheltering effect is more evident in denser vegetation comparing to that in sparse vegetation. This effect can enhance sediment deposition and protect the bed from erosion [64,65]. Nosrati et al., 2022 [66], found that the bending deformation of vegetation results in a significant reduction in the spacing distance between vegetation elements, causing an intensified sheltering effect and a lower form drag force. The TKE decreases significantly near the bed ($0 < z/H < 0.05$) in the wake zone of denser vegetation. In addition, with the decrease in the flow depth from 30 cm to 20 cm, the TKE near the bed decreases. Shahmohammadi et al. (2018) [8] found that the TKE in bare channels is greater compared to the upstream-approaching flow from a vegetated channel. Results of the present study are in good agreement with that of Shahmohammadi et al. (2018) [8]: namely, as the vegetation density decreases, the TKE near the bed ($z/H = 0$) increases.

One can see from Figure 13, in the wake zone behind each vegetation element, the TKE value at the top of the deflected vegetation has a high value, confirming the presence of KH vortices at that flow depth. This result is in agreement with that of Kazem et al. (2021) [67]. In addition, near the channel bed, the high value of TKE took place in the pathway between vegetation elements. This high value of TKE should be the main cause of sediment transport in vegetated channels. It has been noticed from Figure 13b that the TKE has a low value in front of each vegetation element near the bed that was discussed above as a sheltering effect.

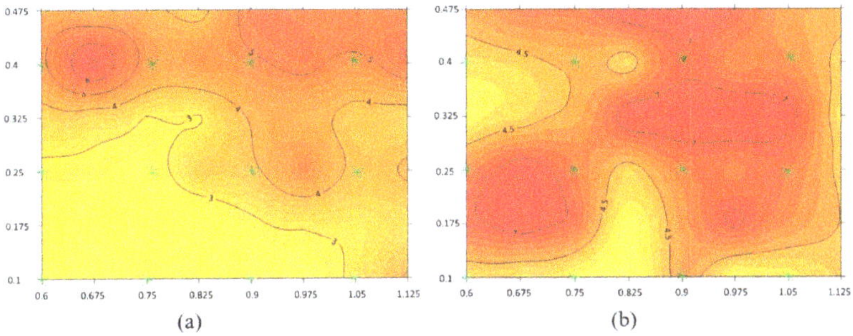

Figure 13. Contour of Turbulence Kinetic Energy (TKE) for deflected vegetation case having a square configuration ($\lambda = 0.04$, water depth = 30 cm): (**a**) around the top of vegetation $z/H = 0.41$; (**b**) near the bed $z/H = 0$.

3.3. Shear Stress

In this study, the primary turbulent Reynolds Shear Stress (RSS) τ at a distance from the bed of z was determined from

$$\tau(z) = -\rho < u'w' > \quad (12)$$

where ρ is the mass density of fluid, angle brackets denote the spatial average of flow variables, and u' and w' are instantaneous velocity fluctuations in the longitudinal and vertical directions, respectively. The values of RSS were normalized by the square of shear velocity (u_*^2).

Results show that the presence of vegetation in a channel bed causes a deviation of the RSS distribution from the linear one. It is also noticed that the RSS distribution is influenced by the aspect ratio (W/H) [62].

In Figure 14, in the presence of vegetation elements with a square arrangement ($a = 0.624\ m^{-1}$), the RSS has been displayed at the following locations: in the middle of two vegetation elements (red circles in Figure 4), at the center of the square formed by four vegetation elements (green circles in Figure 4), in the wake zone of vegetation elements (black circles in Figure 4), and in the wake zone of vegetation elements with staggered configuration ($a = 1.2\ m^{-1}$). At the same location as shown in Figure 14, the RSS presents in Figure 15 for vegetation densities of ($a = 0.256\ m^{-1}$) for a square configuration and ($a = 0.506\ m^{-1}$) for a staggered configuration of vegetation elements.

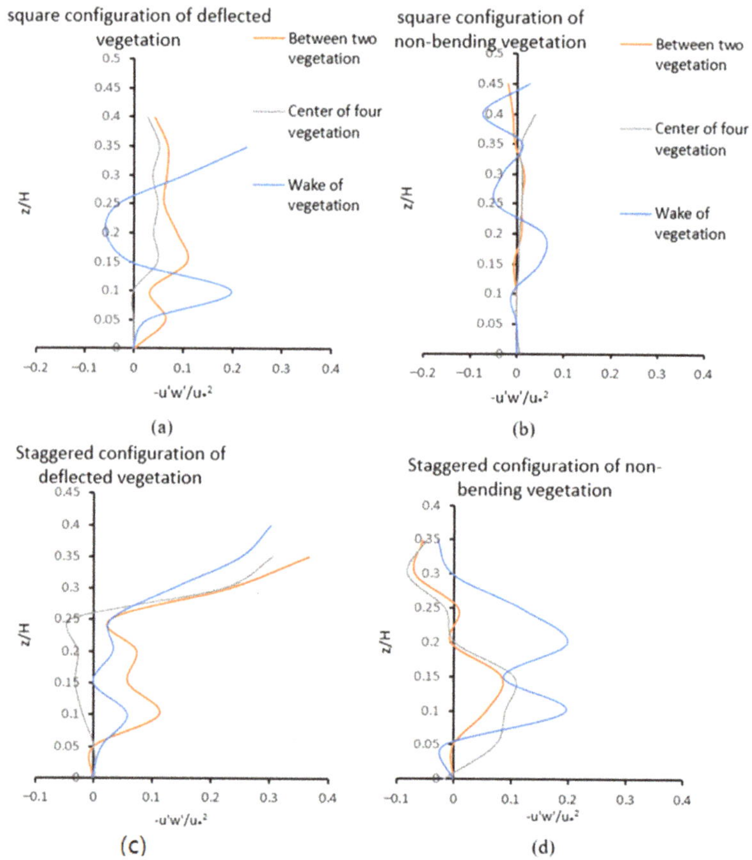

Figure 14. Normalized Reynolds Shear Stress (flow depth = 20 cm): (**a**) Square configuration of deflected vegetation in the wake zone of vegetation, between two vegetation elements and in the center of four vegetation elements with λ = 0.09; (**b**) Square configuration of non-bending vegetation in the wake zone of vegetation, between two vegetation elements and in the center of four vegetation elements with λ = 0.12; (**c**) Staggered configuration of deflected vegetation in wake of three random vegetation elements with λ = 0.17; (**d**) Staggered configuration of non-bending vegetation in wake of three random vegetation elements with λ = 0.23.

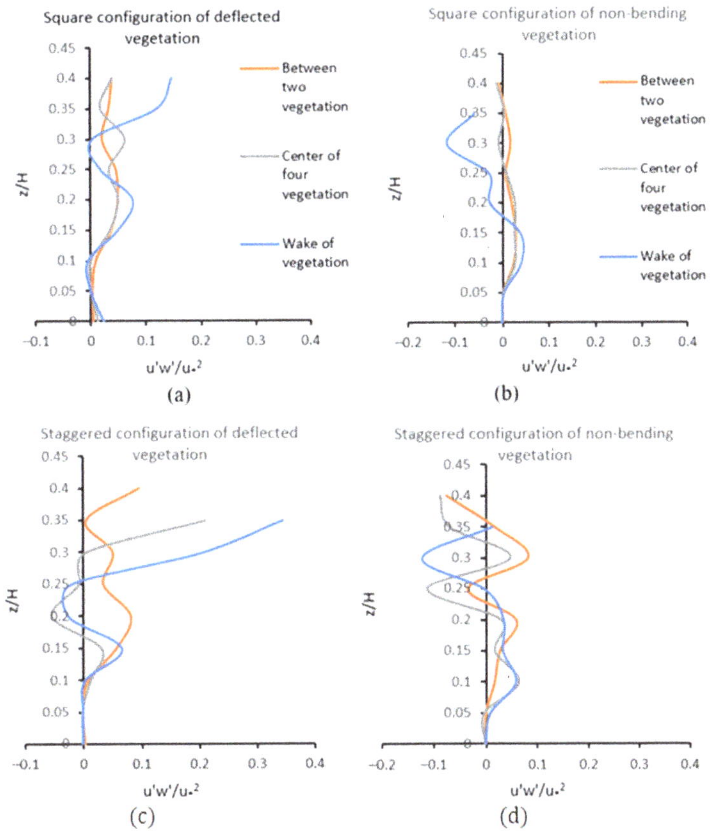

Figure 15. Normalized Reynolds Shear Stress in (water depth = 20 cm): (**a**) Square configuration of deflected vegetation in the wake of vegetation, between two vegetation elements and in the center of four vegetation elements with λ = 0.04; (**b**) Square configuration of non-bending vegetation in the wake of vegetation, between two vegetation elements and in the center of four vegetation elements with λ = 0.05; (**c**) Staggered configuration of deflected vegetation in wake of three random vegetation elements with λ = 0.07; (**d**) Staggered configuration of non-bending vegetation in wake of three random vegetation elements with λ = 0.1.

Results showed that the RSS is highly affected by vegetation density, morphology, and the place for data collection as displayed in Figures 14 and 15. For the case of non-bending vegetation with a square configuration and density of ($a = 0.624\ m^{-1}$), the RSS does not change much throughout the flow depth between two vegetation elements, and at the center of the square formed by four vegetation elements with a slight fluctuation pattern (Figures 14 and 15b). Inside the inner layer of the vegetation, however, more fluctuation in shear stress has been noticed with the negative values of RSS near the water surface. The negative values of RSS near the water surface for the non-bending vegetation case are attributed to turbulent fluxes associated with vegetation morphology and negative streamwise velocity gradients (see Figure 9a).

For the deflected vegetation case (Figure 14a), similar to the non-bending vegetation case (Figure 14b), constant RSS through the water depth was observed between two vegetation elements and at the center of the square formed by four vegetation elements. However, the RSS values for the deflected vegetation case are higher than those for the non-bending vegetation case. For the case of deflected vegetation with a square configuration, at the center of the square formed by four vegetation elements and between two vegetation elements (see Figure 14a), the RSS has a maximum value at a depth of $z/H = 0.15$. However, in the wake zone behind the deflected vegetation, the maximum value of RSS occurred at an elevation slightly higher above the top of the vegetation, indicating the presence of the KH instability at the top of the deflected vegetation and slightly above the top of the vegetation ($z/H \geq 0.4$). As one can see from Figure 14a, above the top of the vegetation, the increasing trend of RSS is continuous. The shift of the maximum RSS above the top of the vegetation is caused by the presence of branches that alter the peak of RSS to a higher location above the top of the vegetation. Near the channel bed, the RSS is close to zero and starts to increase around $z/H = 0.1$, which is associated with stem scale turbulence at this zone. Then, the RSS reaches the peak value at the depth above the vegetation bending height, namely around $z/H = 0.4$. When the vegetation is deflected, the drag discontinuity at the edge of the vegetation produces a shear layer at this interface. The Kelvin–Helmholtz instability forms large coherent vortices within the shear layer, and these structures dominate vertical transport between vegetation and the water column above.

When the vegetation elements are in a staggered layout, the RSS (Figure 14c,d and Figure 15c,d) are similar to those in the wake zone of squared layout elements with intensified RSS values. Thus, the streamwise velocity, TKE, and RSS for the staggered layout vegetation are intensified compared to those for a square configuration.

The canopy morphology and resistance affect the depth to which KH vortices penetrate the canopy. In the present study, the range of λ is $0.04 < ah = \lambda \approx< 0.23$; consequently, the mixing layer eddies penetrate toward the bed and are responsible for turbulence patterns across the vegetation [49]. As a result, no penetration depth needs to be calculated in the present study because the eddies reach the bed.

To assess the effect of the vegetation density on the RSS, comparing Figure 14 to Figure 15, almost the same distribution pattern for the RSS was observed for all four graphs. It is found that the RSS value increases with the decrease in the spacing distance between vegetation elements. In other words, as the density of vegetation increases, both the negative and positive values of RSS throughout the water depth increases. This result is in agreement with that of Barahimi et al., 2018 [68], who concluded there existed greater maximum and smaller minimum values of Reynolds shear stress in dense vegetation comparing to that in sparse vegetation. However, for dense vegetation ($\lambda > 0.1$), as the vegetation density increases, the influence of the bed shear stress decreases. Based on that, the submerged vegetation can be viewed as an extra layer of riverbed, implying that the dense vegetation has shielded riverbed roughness from its effects. As a result, the influence of the bed can be negligible near the bed, and the vegetation density affects the flow structure as a new layer of rigidity [7]. As the trend clearly showed in Figure 14d (the densest vegetation), the RSS near the bed is negative compared to other profiles which have zero or positive RSS near the bed.

4. Conclusions

Based on experiments in a large-scale flume, this study aims to better understand the impact of morphology, density, and arrangement of submerged vegetation on flow velocity, turbulence kinetic energy (TKE), and Reynolds shear stress. Most of the data were collected in the wake zones behind vegetation elements, between two vegetation elements, and at the center of a square formed by four vegetation elements. Results showed that flow depth, density, and morphology of vegetation in the bed had a substantial effect on velocity profiles. The following conclusions were drawn from this study:

In the presence of vegetation in the bed with a high density ($\lambda = 0.09$ and $\lambda = 0.17$), the velocity between two vegetation elements is lower than at the center of square formed by four vegetation elements. In other words, in the presence of vegetation with a high density, the width of the wake zone behind the vegetation element is narrow; therefore, it leads to the increase in velocity at the center of the square formed by four vegetation elements. In other words, the vegetation with a high density reduces the flow cross-sectional area locally, and thus results in a narrow wake behind vegetation which diminishes faster within a shorter distance compared to that for the case of sparse vegetation. On the other hand, for the case of sparse vegetation (λ = 0.04 and λ = 0.07), the mean streamwise velocity between two vegetation elements is higher than that at the center of a square formed by four vegetation elements. This effect indicates the presence of a wide wake behind each vegetation element that attenuates the velocity at the center of four vegetation elements. With the decrease in the spacing distance between the deflected vegetation elements (i.e., increasing canopy density), the streamwise velocity will be largely retarded at the flow depth of $z/H \cong 0.3$ which is slightly below the inflection point. The inflectional region tends to disappear when the vegetation canopy becomes sparser since the shear length scale associated with the velocity field will be increased. This kind of inflection point has not been observed in the non-bending vegetation. Besides, velocity profiles are more inflectional for the case of a staggered arrangement of vegetation elements compared to that for the case of a square arrangement of vegetation elements.

The dense deflected vegetation ($\lambda \geq 0.1$) results in the decrease in sediment transport in streams by reducing the velocity near the bed more than non-bending vegetation and sparse densities. Therefore, it is suggested that dense vegetation provides better protection for beds subject to erosion and scour.

The TKE behind vegetation starts at zero at the bed ($z/H = 0$); however, the TKE between two vegetation elements and at the center of the square formed by four vegetation elements has a value greater than zero (TKE > 0). This is a well-known effect of submerged vegetation on turbulence and is called the sheltering or dampening effect. A greater sheltering effect was observed in denser vegetation due to a shorter distance between vegetation elements. The TKE in the wake of the vegetation depicts that the maximum $\mathrm{RMS}(u')$, $\mathrm{RMS}(v')$, and $\mathrm{RMS}(w')$ occur either at the sheath section of the vegetation (at the depth of $z/H = 0.1$) or above the top of the vegetation (at the depth of $z/H \geq 0.4$). In the sheath section, the frontal projected area is small, and flow can mostly pass through the sheath section. Furthermore, stem scale turbulence was boosted at the sheath section. In the region slightly far away from the channel bed to the vegetation top, the presence of the von Karman vortex street results in the enhancement of TKE comparing to that in unvegetated channels.

In the wake zone behind the deflected vegetation, the maximum value of RSS occurred at an elevation slightly higher above the top of the vegetation, indicating the presence of the KH instability at the top of the deflected vegetation and slightly above the top of the vegetation ($z/H \geq 0.4$). Above the top of the vegetation, the increasing trend of RSS is continuous. The shift of the maximum RSS above the top of the vegetation is caused by the presence of branches that alter the peak of RSS to a higher location above the top of the vegetation. Within the range of vegetation density in this study ($0.04 < \lambda \approx< 0.23$), as the vegetation density increases, the negative and positive values of RSS throughout the flow depth increase. However, for dense vegetation ($\lambda > 0.1$), as the vegetation density increases, the influence of the bed shear stress decreases. Based on that, the submerged vegetation can be viewed as an extra layer of riverbed, implying that the dense vegetation has shielded riverbed roughness from its effects. When the vegetation elements are in a staggered layout, the RSS are similar to those in the wake zone of squared layout elements with intensified RSS values.

In the presence of non-bending vegetation in the channel bed with the flow depth of 20 cm, a high velocity gradient appears from the depth of $z/H = 0$ to $z/H = 0.1$, reaching a peak velocity at the depth of $z/H = 0.1$, and a decreasing trend of velocity toward the water surface is noticeable. However, for the deeper flow of 30 cm, the peak velocity occurs at a higher location close to the water surface. The constant value of velocity from the depth of $z/H = 0.1$ to the water surface has been observed. The dip phenomenon was observed in the non-bending vegetation case.

Author Contributions: Conceptualization, M.B. and J.S.; methodology, M.B. and J.S.; software, M.B.; validation, M.B. and J.S.; formal analysis, M.B. and J.S.; investigation, M.B. and J.S.; resources, M.B. and J.S.; data curation, M.B.; writing—original draft preparation, M.B.; writing—review and editing, M.B. and J.S.; supervision, J.S.; project administration, M.B. and J.S.; funding acquisition, J.S. All authors have read and agreed to the published version of the manuscript.

Funding: The Natural Sciences and Engineering Research Council of Canada (NSERC) under the Discovery Grant Program RGPIN-2019-04278.

Institutional Review Board Statement: Not applicable.

Informed Consent Statement: Not applicable.

Data Availability Statement: The data presented in this study are available on request.

Conflicts of Interest: The authors declare no conflict of interest.

References

1. Fathi-Moghadam, M.; Drikvandi, K.H.; Lashkarara, B.; Hammadi, K. Determination of Friction Factor for Rivers with Nonsubmerged Vegetation in Banks and Floodplains. *Sci. Res. Essays* **2011**, *6*, 4714–4719.
2. Kouwen, N. Modern Approach to Design of Grassed Channels. *J. Irrig. Drain. Eng.* **1992**, *118*, 733–743. [CrossRef]
3. Nedjimi, B.; Beladel, B.; Guit, B. Biodiversity of Halophytic Vegetation in Chott *Zehrez* Lake of Djelfa (Algeria). *Am. J. Plant Sci.* **2012**, *3*, 1527–1534. [CrossRef]
4. Waycott, M.; Longstaff, B.; Mellors, J. Seagrass Population Dynamics and Water Quality in the Great Barrier Reef Region: A Review and Future Research Directions. *Mar. Pollut. Bull.* **2005**, *51*, 343–350. [CrossRef] [PubMed]
5. Short, F.T.; Short, C.A.; Novak, A.B. Seagrasses. In *The Wetland Book*; Finlayson, C.M., Milton, G.R., Prentice, R.C., Davidson, N.C., Eds.; Springer: Dordrecht, The Netherlands, 2016; pp. 1–19. ISBN 978-94-007-6173-5.
6. López, F.; García, M. *Open-Channel Flow Through Simulated Vegetation: Turbulence Modeling and Sediment Transport*; US Army Engineer Waterways Experiment Station: Vicksburg, MS, USA, 1997.
7. Neumeier, U.; Ciavola, P. Flow Resistance and Associated Sedimentary Processes in a Spartina Maritima Salt-Marsh. *J. Coast. Res.* **2004**, *202*, 435–447.
8. Shahmohammadi, R.; Afzalimehr, H.; Sui, J. Impacts of Turbulent Flow over a Channel Bed with a Vegetation Patch on the Incipient Motion of Sediment. *Can. J. Civ. Eng.* **2018**, *45*, 803–816. [CrossRef]
9. Vandenbruwaene, W.; Temmerman, S.; Bouma, T.J.; Klaassen, P.C.; de Vries, M.B.; Callaghan, D.P.; van Steeg, P.; Dekker, F.; van Duren, L.A.; Martini, E.; et al. Flow Interaction with Dynamic Vegetation Patches: Implications for Biogeomorphic Evolution of a Tidal Landscape: FLOW INTERACTION WITH DYNAMIC PATCHES. *J. Geophys. Res. Earth Surf.* **2011**, *116*, F01008. [CrossRef]
10. Trimble, S.W. Stream Channel Erosion and Change Resulting from Riparian Forests. *Geology* **1997**, *25*, 467–469. [CrossRef]
11. Chen, S.C.; Kuo, Y.M.; Li, Y.H. Flow Characteristics within Different Configurations of Submerged Flexible Vegetation. *J. Hydrol.* **2011**, *398*, 124–134. [CrossRef]
12. Zhang, J.; Lei, J.; Huai, W.; Nepf, H. Turbulence and Particle Deposition Under Steady Flow Along a Submerged Seagrass Meadow. *J. Geophys. Res. Oceans* **2020**, *125*, e2019JC015985. [CrossRef]
13. Kabiri, F.; Afzalimehr, H.; Sui, J. Flow structure over a wavy bed with vegetation cover. *Int. J. Sediment Res.* **2017**, *32*, 186–194. [CrossRef]
14. Pollen, N.; Simon, A. Estimating the Mechanical Effects of Riparian Vegetation on Stream Bank Stability Using a Fiber Bundle Model: Modeling root reinforcement of stream banks. *Water Resour. Res.* **2005**, *41*, W07025. [CrossRef]
15. Afzalimehr, H.; Dey, S. Influence of Bank Vegetation and Gravel Bed on Velocity and Reynolds Stress Distributions. *Int. J. Sediment Res.* **2009**, *24*, 236–246. [CrossRef]
16. Barbier, E.B.; Hacker, S.D.; Kennedy, C.; Koch, E.W.; Stier, A.C.; Silliman, B.R. The Value of Estuarine and Coastal Ecosystem Services. *Ecol. Monogr.* **2011**, *81*, 169–193. [CrossRef]
17. Arkema, K.K.; Griffin, R.; Maldonado, S.; Silver, J.; Suckale, J.; Guerry, A.D. Linking Social, Ecological, and Physical Science to Advance Natural and Nature-Based Protection for Coastal Communities: Advancing Protection for Coastal Communities. *Ann. N. Y. Acad. Sci.* **2017**, *1399*, 5–26. [CrossRef]
18. Yu, G.A.; Li, Z.; Yang, H.; Lu, J.; Huang, H.Q.; Yi, Y. Effects of Riparian Plant Roots on the Unconsolidated Bank Stability of Meandering Channels in the Tarim River, China. *Geomorphology* **2020**, *351*, 106958. [CrossRef]
19. Sanchez-Gonzalez, J.; Sanchez-Rojas, V.; Memos, C. Wave Attenuation Due to Posidonia Oceanica Meadows. *J. Hydraul. Res.* **2011**, *49*, 503–514. [CrossRef]
20. Hansen, J.; Reidenbach, M. Wave and Tidally Driven Flows in Eelgrass Beds and Their Effect on Sediment Suspension. *Mar. Ecol. Prog. Ser.* **2012**, *448*, 271–287. [CrossRef]
21. Schleiss, A.J.; Franca, M.J.; Juez, C.; De Cesare, G. Reservoir Sedimentation. *J. Hydraul. Res.* **2016**, *54*, 595–614. [CrossRef]
22. Ros, À.; Colomer, J.; Serra, T.; Pujol, D.; Soler, M.; Casamitjana, X. Experimental Observations on Sediment Resuspension within Submerged Model Canopies under Oscillatory Flow. *Cont. Shelf Res.* **2014**, *91*, 220–231. [CrossRef]

23. Serra, T.; Oldham, C.; Colomer, J. Local Hydrodynamics at Edges of Marine Canopies under Oscillatory Flows. *PLoS ONE* **2018**, *13*, e0201737. [CrossRef] [PubMed]
24. Zhang, Y.; Tang, C.; Nepf, H. Turbulent Kinetic Energy in Submerged Model Canopies Under Oscillatory Flow. *Water Resour. Res.* **2018**, *54*, 1734–1750. [CrossRef]
25. Tinoco, R.O.; Coco, G. Turbulence as the Main Driver of Resuspension in Oscillatory Flow Through Vegetation. *J. Geophys. Res. Earth Surf.* **2018**, *123*, 891–904. [CrossRef]
26. Bouma, T.J.; van Duren, L.A.; Temmerman, S.; Claverie, T.; Blanco-Garcia, A.; Ysebaert, T.; Herman, P.M.J. Spatial Flow and Sedimentation Patterns within Patches of Epibenthic Structures: Combining Field, Flume and Modelling Experiments. *Cont. Shelf Res.* **2007**, *27*, 1020–1045. [CrossRef]
27. Temmerman, S.; Bouma, T.J.; Van de Koppel, J.; Van der Wal, D.; De Vries, M.B.; Herman, P.M.J. Vegetation Causes Channel Erosion in a Tidal Landscape. *Geology* **2007**, *35*, 631. [CrossRef]
28. Nepf, H.M. Hydrodynamics of Vegetated Channels. *J. Hydraul. Res.* **2012**, *50*, 262–279. [CrossRef]
29. Belcher, S.E.; Jerram, N.; Hunt, J.C.R. Adjustment of a Turbulent Boundary Layer to a Canopy of Roughness Elements. *J. Fluid Mech.* **2003**, *488*, 369–398. [CrossRef]
30. Chow, V.T. *Open-Channel Hydraulics*; McGraw-Hill Civil Engineering Series; Blackburn Press: Blackburn, UK, 2009; ISBN 1-932846-18-2.
31. Sontek, A.D.V. *Operation Manual, Firmware Version 4.0*; Sontek: San Diego, CA, USA, 1997.
32. Goring, D.G.; Nikora, V.I. Despiking Acoustic Doppler Velocimeter Data. *J. Hydraul. Eng.* **2002**, *128*, 117–126. [CrossRef]
33. Wahl, T.L. Discussion of "Despiking Acoustic Doppler Velocimeter Data" by Derek G. Goring and Vladimir I. Nikora. *J. Hydraul. Eng.* **2003**, *129*, 484–487. [CrossRef]
34. Lowe, R.J.; Koseff, J.R.; Monismith, S.G. Oscillatory Flow through Submerged Canopies: 1. Velocity Structure. *J. Geophys. Res.* **2005**, *110*, C10016. [CrossRef]
35. Abdolahpour, M.; Ghisalberti, M.; Lavery, P.; McMahon, K. Vertical Mixing in Coastal Canopies: Vertical Mixing in Coastal Canopies. *Limnol. Oceanogr.* **2017**, *62*, 26–42. [CrossRef]
36. Nabaei, S.F.; Afzalimehr, H.; Sui, J.; Kumar, B.; Nabaei, S.H. Investigation of the Effect of Vegetation on Flow Structures and Turbulence Anisotropy around Semi-Elliptical Abutment. *Water* **2021**, *13*, 3108. [CrossRef]
37. *SonTek-IQ Series Intelligent Flow Featuring Smart Puls HD User's Manual*; Xylem Inc: San Diego, CA, USA, 2017; p. 135.
38. Blott, S.J.; Pye, K. GRADISTAT: A Grain Size Distribution and Statistics Package for the Analysis of Unconsolidated Sediments. *Earth Surf. Process. Landf.* **2001**, *26*, 1237–1248. [CrossRef]
39. Hager, W.H. *Wastewater Hydraulics: Theory and Practice*, 2nd ed.; Springer: Berlin, Germany; London, UK, 2010; ISBN 978-3-642-11383-3.
40. Huai, W.; Zhang, J.; Katul, G.G.; Cheng, Y.; Tang, X.; Wang, W. The Structure of Turbulent Flow through Submerged Flexible Vegetation. *J. Hydrodyn.* **2019**, *31*, 274–292. [CrossRef]
41. Siniscalchi, F.; Nikora, V.I.; Aberle, J. Plant Patch Hydrodynamics in Streams: Mean Flow, Turbulence, and Drag Forces: Plant patch hydrodynamics. *Water Resour. Res.* **2012**, *48*, W01513. [CrossRef]
42. Aberle, J.; Järvelä, J. Hydrodynamics of Vegetated Channels. In *Rivers–Physical, Fluvial and Environmental Processes*; Rowiński, P., Radecki-Pawlik, A., Eds.; Springer International Publishing: Cham, Switzerland, 2015; pp. 519–541. ISBN 978-3-319-17718-2.
43. Nikora, V. Hydrodynamics of Aquatic Ecosystems: An Interface between Ecology, Biomechanics and Environmental Fluid Mechanics. *River Res. Appl.* **2010**, *26*, 367–384. [CrossRef]
44. Nepf, H.M.; Vivoni, E.R. Flow Structure in Depth-Limited, Vegetated Flow. *J. Geophys. Res. Oceans* **2000**, *105*, 28547–28557. [CrossRef]
45. Raupach, M.R.; Finnigan, J.J.; Brunei, Y. Coherent Eddies and Turbulence in Vegetation Canopies: The Mixing-Layer Analogy. *Bound.-Layer Meteorol.* **1996**, *78*, 351–382. [CrossRef]
46. Petryk, S.; Bosmajian, G. Analysis of Flow through Vegetation. *J. Hydraul. Div.* **1975**, *101*, 871–884. [CrossRef]
47. Nepf, H.M. Drag, Turbulence, and Diffusion in Flow through Emergent Vegetation. *Water Resour. Res.* **1999**, *35*, 479–489. [CrossRef]
48. Nezu, I.; Sanjou, M. Turburence Structure and Coherent Motion in Vegetated Canopy Open-Channel Flows. *J. Hydro-Environ. Res.* **2008**, *2*, 62–90. [CrossRef]
49. Nepf, H.; Ghisalberti, M. Flow and Transport in Channels with Submerged Vegetation. *Acta Geophys.* **2008**, *56*, 753–777. [CrossRef]
50. Huai, W.; Li, S.; Katul, G.G.; Liu, M.; Yang, Z. Flow Dynamics and Sediment Transport in Vegetated Rivers: A Review. *J. Hydrodyn.* **2021**, *33*, 400–420. [CrossRef]
51. Etminan, V.; Lowe, R.J.; Ghisalberti, M. A New Model for Predicting the Drag Exerted by Vegetation Canopies: New model for vegetation drag. *Water Resour. Res.* **2017**, *53*, 3179–3196. [CrossRef]
52. Nepf, H.M. Flow and Transport in Regions with Aquatic Vegetation. *Annu. Rev. Fluid Mech.* **2012**, *44*, 123–142. [CrossRef]
53. Afzalimehr, H.; Rennie, C.D. Determination of Bed Shear Stress in Gravel-Bed Rivers Using Boundary-Layer Parameters. *Hydrol. Sci. J.* **2009**, *54*, 147–159. [CrossRef]
54. Schlichting, H.; Gersten, K. *Boundary-Layer Theory*; Springer: Berlin/Heidelberg, Germany, 2000; ISBN 978-3-642-85831-4.
55. Clauser, F.H. The Turbulent Boundary Layer. In *Advances in Applied Mechanics*; Elsevier: Amsterdam, The Netherlands, 1956; Volume 4, pp. 1–51. ISBN 978-0-12-002004-1.

56. Dyer, K.R. Sediment Processes in Estuaries: Future Research Requirements. *J. Geophys. Res.* **1989**, *94*, 14327–14339. [CrossRef]
57. Kazem, M.; Afzalimehr, H.; Sui, J. Characteristics of Turbulence in the Downstream Region of a Vegetation Patch. *Water* **2021**, *13*, 3468. [CrossRef]
58. Shi, Z.; Pethick, J.S.; Burd, F.; Murphy, B. Velocity Profiles in a Salt Marsh Canopy. *Geo-Mar. Lett.* **1996**, *16*, 319–323. [CrossRef]
59. Afzalimehr, H.; Barahimi, M.; Sui, J. Non-Uniform Flow over Cobble Bed with Submerged Vegetation Strip. *Proc. Inst. Civ. Eng.-Water Manag.* **2017**, *172*, 86–101. [CrossRef]
60. Raupach, M.; Shaw, R. Averaging Procedures for Flow within Vegetation Canopies. *Bound Lay Met* **1982**, *22*, 79–90. [CrossRef]
61. Chen, Z.; Ortiz, A.; Zong, L.; Nepf, H. The Wake Structure behind a Porous Obstruction and Its Implications for Deposition near a Finite Patch of Emergent Vegetation. *Water Resour. Res.* **2012**, *48*, W09517. [CrossRef]
62. Afzalimehr, H.; Moghbel, R.; Gallichand, J.; Sui, J. Investigation of Turbulence Characteristics in Channel with Dense Vegetation. *Int. J. Sediment Res.* **2011**, *26*, 269–282. [CrossRef]
63. Raupach, M.R. Drag and Drag Partition on Rough Surfaces. *Bound.-Layer Meteorol.* **1992**, *60*, 375–395. [CrossRef]
64. Leonard, L.A.; Croft, A.L. The Effect of Standing Biomass on Flow Velocity and Turbulence in Spartina Alterniflora Canopies, Estuarine. *Coast. Shelf Sci.* **2006**, *69*, 325–336. [CrossRef]
65. Neumeier, U.; Amos, C.L. The Influence of Vegetation on Turbulence and Flow Velocities in European Salt-Marshes. *Sedimentology* **2006**, *53*, 259–277. [CrossRef]
66. Nosrati, K.; Afzalimehr, H.; Sui, J. Drag Coefficient of Submerged Flexible Vegetation Patches in Gravel Bed Rivers. *Water* **2022**, *14*, 743. [CrossRef]
67. Kazem, M.; Afzalimehr, H.; Sui, J. Formation of Coherent Flow Structures beyond Vegetation Patches in Channel. *Water* **2021**, *13*, 2812. [CrossRef]
68. Brahimi, M.; Afzalimehr, H. Effect of Submerged Vegetation Density on Flow under Favorable Pressure Gradient. *SN Appl. Sci.* **2019**, *1*, 57. [CrossRef]

water

Article

Numerical Investigation of Hydrodynamics in a U-Shaped Open Channel Confluence Flow with Partially Emergent Rigid Vegetation

Zhengrui Shi * and Sheng Jin

School of Civil and Hydraulic Engineering, Dalian University of Technology, 2 Ling Gong Road, Dalian 116024, China
* Correspondence: thomasczr0614@mail.dlut.edu.cn

Abstract: The effects of partially emergent rigid vegetation on the hydrodynamics of a curved open-channel confluence flow were simulated using OpenFOAM. The numerical model using the Volume of Fluid method and the RNG k-ε turbulence model in the Reynolds-averaged Navier–Stokes equations was first validated by existing experimental data with good agreement. Then the characteristics of hydrodynamics were analyzed in aspects of separation zone, water level, streamwise velocities, secondary flows, bed shear stress and flow resistance. Some main conclusions can be drawn from the results. Compared to the non-vegetated cases, the separation zones in vegetated cases are smaller in both length and width. With higher vegetation Solid Volume Fraction (SVF), the separation zone is divided into two parts, a smaller one right after the confluence point and a larger one on the second half of the curved reach after the confluence. The main circulation cell shrinks and the circulation near the concave bank moves towards the channel midline. The differences in velocities and bed shear stress between the convex and concave banks become larger with a higher SVF. Under the same SVF, a larger vegetation density has more disturbance on the tributary than a larger stem diameter.

Keywords: U-shape open channel; confluence flows; emergent rigid vegetation; separation zone; bed shear stress; OpenFOAM

Citation: Shi, Z.; Jin, S. Numerical Investigation of Hydrodynamics in a U-Shaped Open Channel Confluence Flow with Partially Emergent Rigid Vegetation. *Water* **2022**, *14*, 4027. https://doi.org/10.3390/w14244027

Academic Editors: Jueyi Sui and Hossein Afzalimehr

Received: 11 November 2022
Accepted: 6 December 2022
Published: 9 December 2022

1. Introduction

Meandering open-channel confluence flows are ubiquitous in natural and artificial rivers. Confluences play a critical role in river networks and irrigation systems. Interlaced channels form intricate flow structures in confluence areas, where strong physical gradients develop. The complex hydrodynamics further affect the local transport of pollutants and sediment, as well as aquatic vegetation seeds. According to the investigation carried out by Roberts [1], flow structures in confluence flows can be divided into six zones, which are described as stagnation zone, deflection zone, separation zone, acceleration zone, recovery zone and shear layer (shown in Figure 1). Flow velocities are much smaller in the separation zone so, there is a higher possibility that aquatic vegetation grows in separation zones. Moreover, the vegetation's obstruction has an influence on flow structures in return resulting in a more complex hydro-environment. Due to the above, it is quite favorable for understanding flow behaviors and sedimentation issues at river confluences to study curved open-channel confluence flows with partial vegetation.

Former investigations of curved open-channel confluence flows and rigid vegetated open-channel flows are reviewed in these two paragraphs, respectively. Amounts of previous studies in confluence flows were carried out in the past two decades. Some of the studies focused on the flow behaviors and bed morphology of the natural river confluences. These studies were carried out by using different kinds of numerical models or analytical methods based on the tested data of rivers to investigate the effects between the confluence flows and bed morphology formation [2–7]. The studies took the bed discordance into consideration and found that the discordance of the bed brought the deflection to the confluence flow and changed the lateral pressure gradients resulting in the changes of secondary flow strength in the local area of confluences. Some investigations paid attention to the effects of confluence angle, flow discharge ratio and the confluence location on the flow patterns in experimental curved confluence flows [8–13]. These works gained some findings that with the increase in tributary flow discharge ratio and confluence angle the main flows were deflected more obviously towards the convex bank and the separation zone became larger in both length and width. A few researchers studied the effects of obstructions in curved confluence flows [14]. The studies mentioned above investigated the influence on open-channel confluence flows caused by bed morphology, confluence angle, and discharge ratio. Nevertheless, few of these studies took the effects of vegetation obstructions into consideration, which is a common occurrence in both natural rivers and artificial channels.

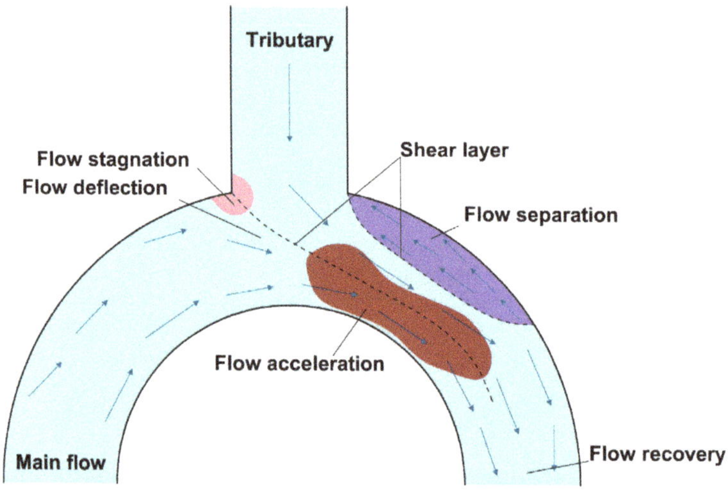

Figure 1. Various zones at the confluence of curved open-channel with a tributary.

The studies of rigid vegetated flows focused on straight or curved open channels in the past twenty years. Nepf et al. [15–20] studied rigid vegetation in flow in a laboratory straight open channel to investigate the generation of shear layer caused by vegetation blockages and the turbulence in vegetated flows. As for the numerical studies, a large number of simulations were carried out in two-dimensional models by adopting shallow water equations [21–25]. In most 2D simulations, the effects of vegetation usually were defined as bed friction or roughness. Even though these 2D models showed the capabilities in modeling the velocity distributions of the vegetated flow, the neglect of three-dimensional effects led to the deficiency in turbulent flow and secondary flow structures. Moreover, the generalized vegetation roughness on the channel bed also resulted in an over-prediction of bed shear stress. Some investigations paid attention to the flow of vegetation in curved open channel flows and straight confluence flows [26–30]. These studies used the experimental method or 3-D numerical models to investigate the velocity distributions, secondary flow structures and bed shear stress of the vegetated flows. The studies demonstrated that the vegetation stems in the channel obviously compressed the flow area and pushed the secondary flow cells away from the vegetation array. In some situations, with a large density of vegetation covering the channel, the secondary flow cells were broken into small cells or even disappeared. These studies focused on meandering open channel flows with vegetation and provided analyses of the secondary flow structures. However, the flow structures in the curved open channel confluence flows are more complex than that of curved channels.

The former studies of vegetated flows focused on relatively simple flow fields with straight or curved channels. Few researchers have paid attention to the effects of vegetation on curved open channel confluence flows. Moreover, with the low velocities in the separation zone along the concave bank after the confluence, the sedimentation of silts and seeds is highly possible to occur in this area. Thus, an investigation of the effects of vegetation on separation zones, secondary flows and bed shear stress can help to understand the complex hydrodynamics in the meandering confluences and guide interests on these issues.

In this study, a numerical simulation using the VOF method and the RNG k-ε turbulence model is applied on OpenFOAM (Open Field Operation and Manipulation) to simulate the flow in a U-shaped open-channel confluence with partially emergent rigid vegetation regions near the concave bank after the confluence point. Then the hydrodynamic characteristics are analyzed in the aspect of the separation zone, water level, streamwise velocities, secondary flows and bed shear stress, to obtain the effects of rigid emergent vegetation on the flow regime of curved open-channel confluence flows.

2. Materials and Methods

In this paper, the simulation is carried out by using the incompressible multiphase flow solver interFoam [31] with the RNG k-ε turbulence model [32]. The Navier–Stokes equations in Cartesian coordinates are defined as follows:

Continuity equation

$$\frac{\partial u_i}{\partial x_i} = 0 \tag{1}$$

Momentum equation

$$\frac{\partial \rho u_i}{\partial t} + \frac{\partial}{\partial x_j}(\rho u_j u_i) = -\frac{\partial p}{\partial x_i} + \frac{\partial}{\partial x_j}\left(\tau_{ij} + \tau_{t_{ij}}\right) + \rho g_i + f_{\sigma i} + f_i \tag{2}$$

where ρ represents the fluid density, u represents the velocity, p represents the pressure, τ_{ij} and $\tau_{t_{ij}}$ are the viscous and turbulent stresses, g_i is the component of gravitational acceleration, $f_{\sigma i}$ is the surface tension, f_i represents the component of external forces such as the vegetation drag forces.

InterFoam solver uses the VOF method to simulate the water free-surface. In the VOF method, a volume fraction α is defined to distinguish the two fluid phases in the computational domain. With the volume fraction α the fluid density ρ is defined as follows:

$$\rho = \alpha \rho_1 + (1 - \alpha)\rho_2 \tag{3}$$

In this equation, α is 1 when inside the fluid 1 with the density ρ_1 and α is 0 when inside the fluid 2 with the density ρ_2. The value of α varies between 0 and 1 at the water free-surface. To capture the free-surface, an equation regarded as the conservation of mixture components along the path of a fluid parcel is defined for α:

$$\frac{\partial \alpha}{\partial t} + \frac{\partial (\alpha u_j)}{\partial x_j} = 0 \tag{4}$$

The surface tension $f_{\sigma i}$ in the momentum equation is also related to α [33]:

$$f_{\sigma i} = \sigma \kappa \frac{\partial \alpha}{\partial x_i} \tag{5}$$

where σ is the surface tension constant and κ is the curvature. The curvature κ is approximated as follows [34]:

$$\kappa = -\frac{\partial n_i}{\partial x_i} = -\frac{\partial}{\partial x_i}\left(\frac{\partial \alpha / \partial x_i}{|\partial \alpha / \partial x_i|}\right) \tag{6}$$

The turbulence model RNG k-ε equations are shown below:

$$\frac{\partial \rho k}{\partial t} + \frac{\partial}{\partial x_j}(\rho k u_j) = \frac{\partial}{\partial x_j}\left[\left(\mu + \frac{\mu_t}{\sigma_k}\right)\frac{\partial k}{\partial x_j}\right] + P_k - \rho\varepsilon \tag{7}$$

$$\frac{\partial \rho \varepsilon}{\partial t} + \frac{\partial}{\partial x_j}(\rho \varepsilon u_j) = \frac{\partial}{\partial x_j}\left[\left(\mu + \frac{\mu_t}{\sigma_\varepsilon}\right)\frac{\partial \varepsilon}{\partial x_j}\right] + C_{1\varepsilon}\frac{\varepsilon}{k}P_k - C_{2\varepsilon}^{*}\rho\frac{\varepsilon^2}{k} \tag{8}$$

$$\mu_t = \rho C_\mu \frac{k^2}{\varepsilon} \tag{9}$$

$$P_k = \mu_t S^2 \tag{10}$$

$$S = \sqrt{2S_{ij}S_{ij}} \tag{11}$$

$$S_{ij} = \frac{1}{2}\left(\frac{\partial u_i}{\partial x_j} + \frac{\partial u_j}{\partial x_i}\right) \tag{12}$$

$$C_{2\varepsilon}^{*} = C_{2\varepsilon} + \frac{C_\mu \eta^3}{1+\beta\eta^3}\left(1 - \frac{\eta}{\eta_0}\right) \tag{13}$$

$$\eta = S\frac{k}{\varepsilon} \tag{14}$$

In the equations, k is the turbulent kinetic energy, ε is the turbulence dissipation rate, μ is the kinematic viscosity, μ_t is the turbulence viscosity, σ_k and σ_ε are the turbulent Prandtl numbers, S represents the modulus of the mean rate of the strain tensor, S_{ij} is the strain-rate tensor, P_k is the production of turbulence kinetic energy. The value of constant coefficients C_μ, $C_{1\varepsilon}$, $C_{2\varepsilon}$, σ_k, σ_ε, η_0, β is shown in Table 1.

Table 1. Value of constant coefficients in the RNG k-ε turbulence model.

Constant Coefficient	Value
C_μ	0.0845
$C_{1\varepsilon}$	1.42
$C_{2\varepsilon}$	1.68
σ_k	0.71942
σ_ε	0.71942
η_0	4.38
β	0.012

The liner central differencing scheme is used for the interpolation. The implicit Euler scheme was applied to discretize the temporal terms. The gradient terms were obtained by second order Gaussian integration. The convection terms were discretized by the van-leer scheme, a second order Total Variation Diminishing (TVD) limited scheme. The upwind scheme was employed for kinetic energy and turbulent dissipation. The PIMPLE algorithm, which is a combination of the PISO algorithm and SIMPLE algorithm, was applied for the pressure-velocity coupling at each time step iteration. The under-relaxation method was applied to reduce solution oscillations and keep the computation stable. The relaxation factor for velocities was 0.7 while it was 0.3 for the pressure, turbulence kinetic energy and turbulence dissipation rate.

As for the boundary conditions, the velocities were set based on the water flow discharges for the inflow boundary. The outflow boundary was set as a zero-gradient boundary condition. The free-surface boundary was defined as an atmospheric boundary condition. The standard wall functions were applied for the wall boundaries on the channel bed, side banks and the surfaces of the vegetation stems.

3. Model Validations and Simulation Cases Set Up

There are few available experimental investigations on curved open channel confluence flow with rigid vegetation, in this chapter, an experiment of a U-shape open channel flow through partially rigid vegetation and another hydraulic experiment in the same U-shape open channel with a tributary were chosen to be simulated to complete the validation.

3.1. Curved Open Channel Flow with Partially Rigid Vegetation

The experiment of a U-shaped open-channel flow through partially rigid vegetation was carried out by Huai et al. [27] at the State Key Laboratory of Water Resources and Hydropower Engineering Science, Wuhan University, Wuhan, China. The experimental flume with a 1 m wide and 0.25 m high rectangular cross-section consisted of a 4 m straight inflow reach, a 180° curved reach, and a 4 m straight outflow reach. The radius R of the centerline in the curved reach was 2 m. A 0.25 m wide band was placed along the convex bank from 2 m after the inlet to 2 m before the outlet. To simulate the rigid vegetation, reinforcing steel bars with 0.006 m diameter were perpendicularly planted on the band with a 0.05 m interval space. The height of the steel bars was 0.15 m. The water permeability of stems was neglected. The inflow discharge was 0.03 m^3/s and the water depth measured at the outlet was 0.148 m. The sketch of this experiment is shown in Figure 2a. The mean streamwise velocity profiles of test lines (r=1.65 m, 2.00 m, and 2.30 m, respectively) on the three chosen test cross sections (the entrance, middle and exit of the curved reach) were sampled to compare with the experimental data.

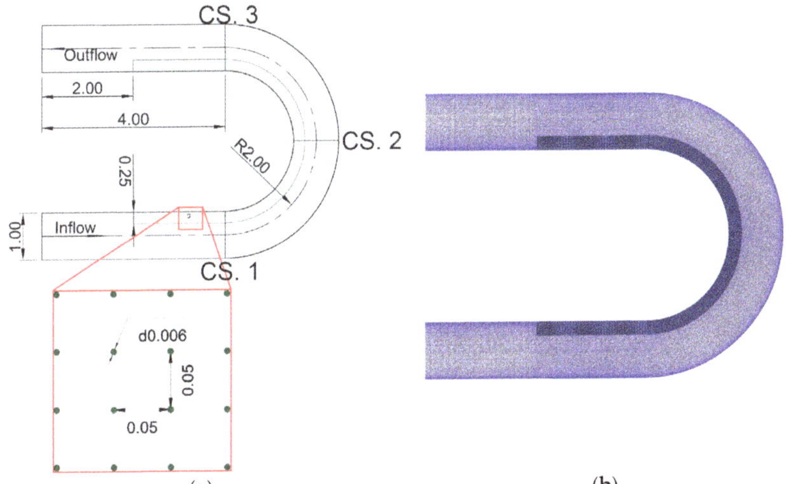

Figure 2. (**a**) The schematic of a U-shape open-channel with an emergent rigid vegetation region on the convex bank and the detailed view of the vegetation region (units: m); (**b**) The local refined calculating mesh.

The calculating mesh was created by the commercial software ICEM-CFD. According to a mesh in-dependency analysis, a structured 6.94-million-cell mesh was determined with local refinement in the vegetation region on the convex bank (shown in Figure 2b). The minimum mesh size was 0.0015 m for the first layer of mesh at the wall boundaries of the surface on vegetation stems. This was to keep the y-plus value within an appropriate range ($30 \le y+ \le 300$) to ensure the effectiveness of the standard wall function method. The boundary conditions were the same as those described in Section 2. The volume fraction α of the part below the water level in the computational domain was set to 1 for the initialization of the calculation. The time step was controlled at 0.00175 s so that the Courant number was kept below 1.

The comparison of the mean streamwise velocity profiles on the test lines of the three cross-sections between the experimental data and simulation results is shown in Figure 3. The simulation results were basically consistent with the experimental data. Figure 4 shows the comparison of secondary flow structures on the test cross-sections between experimental data and simulation results. The secondary flow structures of the simulation results were similar to those of the experimental data.

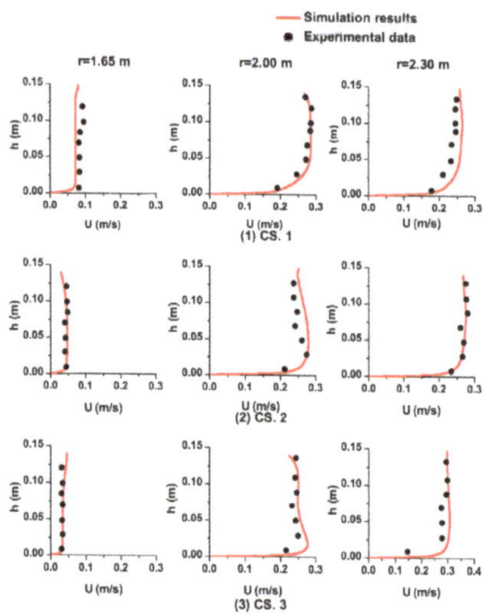

Figure 3. Comparisons of the mean streamwise velocity profiles of the test lines between the simulation results and experimental data.

To evaluate the agreement between the simulation results and the experiment data, the indexes root-mean-squared error (RMSE) and coefficient of determination (R^2) were calculated on the three tested cross-sections. The formulas of RMSE and R^2 are described as below:

$$RMSE = \sqrt{\frac{\sum_{i=1}^{n}(\hat{y}_i - y_i)^2}{n}} \quad (15)$$

$$\overline{y} = \frac{1}{n}\sum_{i=1}^{n} y_i \quad (16)$$

$$R^2 = 1 - \frac{\sum_{i=1}^{n}(y_i - \hat{y}_i)^2}{\sum_{i=1}^{n}(y_i - \overline{y})^2} \quad (17)$$

where n represents the number of values in the experiment data, y_i is the measured value, $\hat{y}_i$ is the calculated value, and $\overline{y}$ represents the average value of y_i. The RMSE value can show the dispersion degree of the data. A smaller RMSE value indicates a better prediction of the model. Moreover, the R^2 value is a quantized form for the goodness of fit. The value of R^2 is usually between 0 and 1, a better fit is gained when the value is closer to 1. Table 2 shows the sectional RMSE and R^2 of mean streamwise velocities.

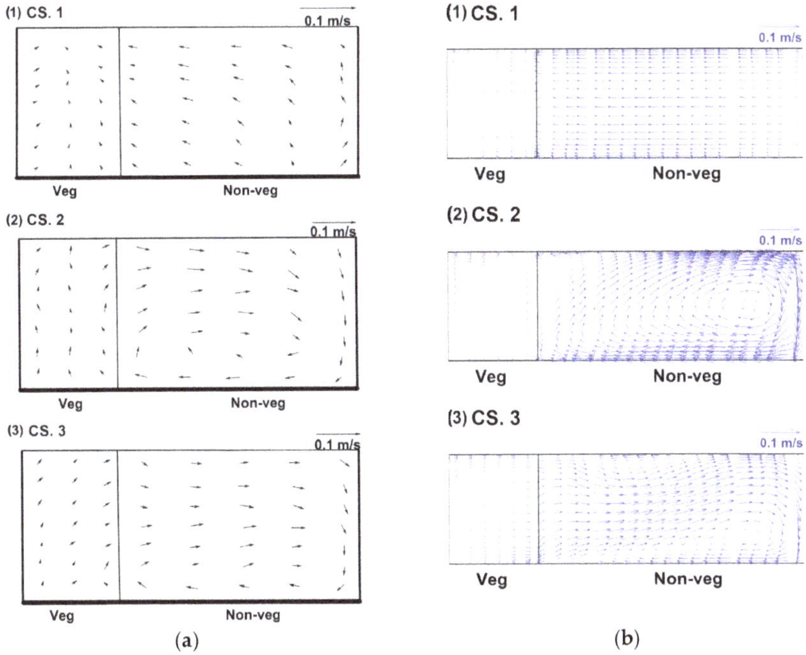

Figure 4. Comparison of secondary flow structures between the experimental data and simulation results on the test cross-sections: (**a**) Secondary flow vectors on the three test cross-sections of experimental data; (**b**) Secondary flow vectors on the three test cross-sections of simulation results.

Table 2. RMSE and R^2 values of mean streamwise velocities on the three cross-sections.

Section	RMSE	R^2
1	0.01828	0.9458
2	0.01518	0.977
3	0.01791	0.9077

The RMSE values on the three tested cross-sections were quite small and the values of R^2 were close to 1, which indicated the reliability of the model in predicting the mean streamwise velocities. Especially, the RMSE value of CS. 2 was smaller and the R^2 value was larger, which means that the model was better in the curved reach than at the entrance and exit of the curved reach. This was similar to the results gained by Shaheed et al. in modeling a curved open channel flow by using k-ε models. Basically, the validation demonstrated the capability of the numerical model in simulating flows in curved open channels with partially rigid vegetation.

3.2. Curved Open Channel Confluence Flow

The physical modeling investigation of a curved open channel confluence flow was performed by Sui et al. [12]. The main channel was the same as the U-shaped channel described above in Section 3.1. The tributary channel was a 3.5 m-long straight flume with a 0.3 m wide and 0.25 m high rectangular cross-section. The confluence was located at a 90° cross-section of the curved reach and the confluence angle between the tributary and the main channel was also 90°. As for the hydraulic conditions, the main channel inflow discharge Q_M was 0.03 m^3/s and the tributary discharge Q_T was 0.018 m^3/s. The water level measured at the entrance of the curved reach was 0.182 m. Test lines (r = 1.71 m, 2.05 m, and

2.34 m, respectively) on the two cross sections (CS. 1 $\varphi = 60°$ and CS. 2 $\varphi = 120°$) were chosen to validate the experimental data. The schematic of the experiment is shown in Figure 5a.

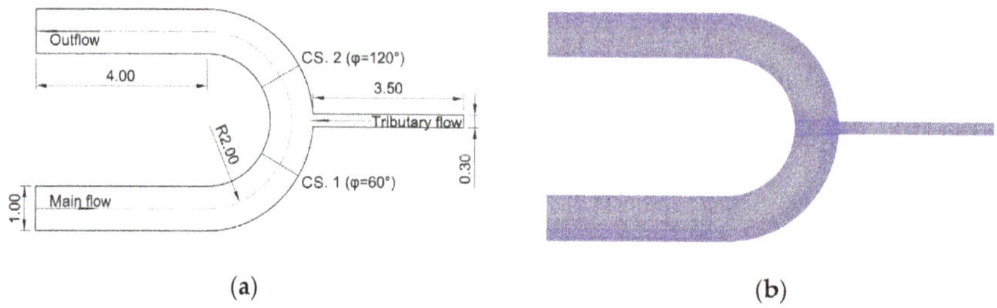

Figure 5. (**a**) The schematic of a U-shape open channel with a tributary (units: m); (**b**) The local refined calculating mesh.

The calculating mesh owned 1.15 million structured cells with local refinement at the confluence point (shown in Figure 5b). The minimum mesh size was 0.003 m for the first layer of mesh at the wall boundaries. The volume fraction α of the part below the water level in the computational domain was set to 1 for the initialization of the calculation. The time step was controlled at 0.0123 s to keep the Courant number no more than 1.

The comparison of the mean streamwise velocity profiles on the test lines of the two cross-sections between the experimental data and simulation results is shown in Figure 6. Good agreements were obtained. Table 3 shows the RMSE and R^2 values on the two tested cross-sections. The reliability of the model was proved. Figure 7 displays the comparison of free-surface velocity vectors between experimental data and simulation results. The simulation results showed the same trends for the velocity vectors as the experimental data. The comparison of results confirmed the ability of the numerical model to simulate confluence flows in curved open channels with a separation zone.

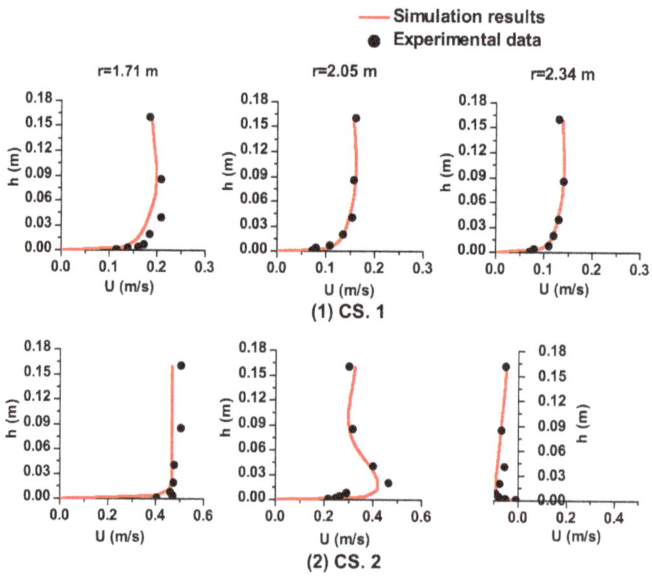

Figure 6. Comparisons of the mean streamwise velocity profiles of the test lines between the simulation results and experimental data.

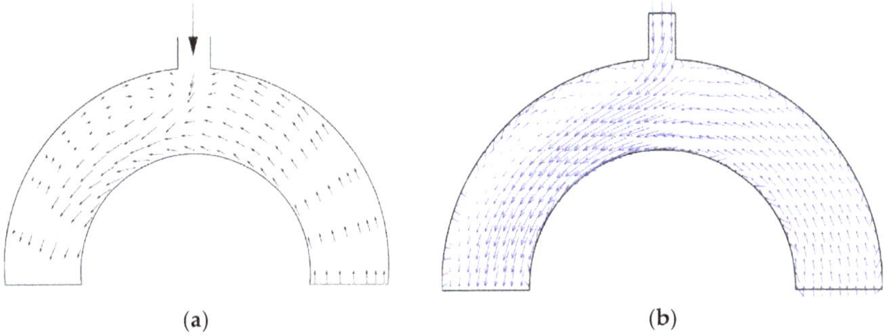

Figure 7. Comparison of free-surface velocity vectors between the experimental data and simulation results: (**a**) Free-surface velocity vectors of experimental data; (**b**) Free-surface velocity vectors of simulation results.

Table 3. RMSE and R^2 values of mean streamwise velocities on the two cross-sections.

Section	RMSE	R^2
1	0.01574	0.9034
2	0.03356	0.9798

3.3. Numerical Simulation Cases Set Up

Based on the numerical model validations above, the capability of the model to simulate the curved open channel confluence flows through rigid vegetation was verified. Then the model was adopted to study the effects of partially emergent rigid vegetation on a U-shaped open-channel confluence flow.

Based on the validation simulations above, the numerical simulations were carried out in the same U-shaped channel as in the experiment described in Section 3.2. Likewise, the main channel inflow discharge Q_M and water level h_0 were also the same as the second validation case. This was for controlling variables so as to compare the results with different tributary flow discharge ratios. Then two different discharge ratios λ ($\lambda = Q_T/Q_M$, Q_T represents the tributary flow discharge) were chosen in the simulations. As for the vegetated cases, considering the higher possibility of vegetation growing in the low velocities zone along the concave bank, a 0.25 m-wide vegetation region was set along the concave bank after the confluence point. The vegetation stems with no water permeability considered were all 0.25 m height to keep the stems sticking out of the water free-surface. The stems were arranged in parallel. Two different stem diameters d and interval spaces Δx were set in different cases. Three cross-sections (CS. 1: the cross-section right before the confluence point; CS. 2: the 135° cross-section of the curved reach; CS. 3: the exitance section of the curved reach) in the main channel were monitored. The mesh used in simulations had about 3.68 million structured grids with local refinement along the concave bank after the confluence point. The minimum size of the grid near the wall boundary was also 0.0015 m. The computing time step was set to be about 0.00175 s. The details of all the cases are shown in Table 4. The simulation schematic and the calculating mesh are shown in Figure 8.

Table 4. Detailed information list of simulation cases.

Cases	Q_M (m^3/s)	Q_T (m^3/s)	λ	h_0 (m)	d (m)	Δx (m)	Solid Volume Fraction
1	0.03	0.009	0.3	0.182	-	-	-
2	0.03	0.018	0.6	0.182	-	-	-
3	0.03	0.009	0.3	0.182	0.006	0.05	1.14
4	0.03	0.018	0.6	0.182	0.006	0.05	1.14
5	0.03	0.009	0.3	0.182	0.006	0.025	4.55
6	0.03	0.018	0.6	0.182	0.006	0.025	4.55
7	0.03	0.009	0.3	0.182	0.012	0.05	4.55
8	0.03	0.018	0.6	0.182	0.012	0.05	4.55

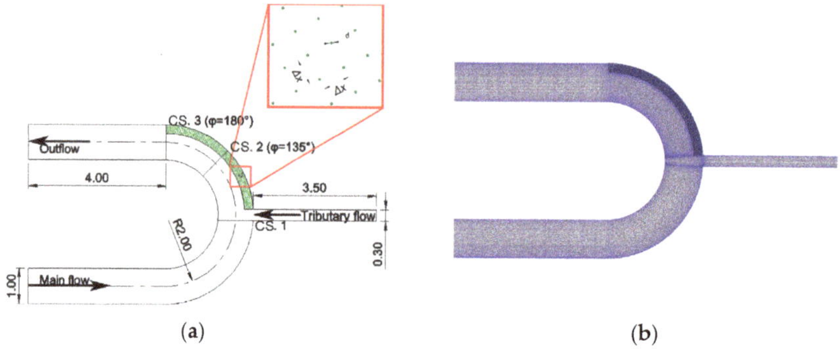

Figure 8. (**a**) The schematic of a U-shape open-channel with a tributary and an emergent rigid vegetation region on the concave bank and the detailed view of the vegetation region (units: m); (**b**) The local refined calculating mesh.

4. Results and Discussion

In this section, the flow analysis was performed from the following aspects of streamwise velocities, separation zone, water level, cross-section secondary flows, bed shear stress and flow resistance.

4.1. Steam-Wise Velocities and Separation Zone

Figure 9 below displays the contours and 2-D streamlines of dimensionless streamwise velocities ($U^* = U/U_{Average}$, $U_{Average}$ represents the averaged velocity on the cross-section along the curved channel) on the vertical section near the water surface in the curved reach after the confluence point of all the cases. In the non-vegetated cases, the velocities on the convex bank were significantly greater than those on the concave bank, and the velocity difference between the convex bank and concave bank increased with the increase in tributary discharge. Because the tributary flow on the concave bank brought a lateral impact to the main flow and deflected the main flow to the convex bank after the confluence point. A strong shear layer with a large velocity gradient and momentum exchange was generated. So, a larger tributary discharge caused a much stronger flow deflection and resulted in a larger velocity difference between the convex bank and the concave bank. Moreover, the strong shear layer divided the main flow into two parts, a separation zone with smaller reversed velocities near the concave bank and a flow acceleration zone with the largest velocities near the convex bank, respectively. Thus, the border of the separation zone could be obtained by sampling the points with zero streamwise velocities. In Figure 9, the black solid lines represent the sampled separation borders and the dashed lines represent the edge of the vegetation region in vegetated cases. From the streamlines in Case 1 and Case 2, a circulation was formed beside the concave bank after the confluence point. The circulation extended in both length and width as the tributary discharge increased. Moreover, the border of the separation zone just passed through the center of circulation.

The size of the separation zone also increased in both length and width with the increase in tributary flow discharge. As for the vegetated cases, the flow velocities decreased obviously in the vegetation region due to the vegetation blockage. With higher Solid Volume Fraction (SVF), the velocities in the vegetation region decreased more obviously. The circulations were also much smaller than those in the non-vegetated cases and the separation zones became smaller in length and width as well. Especially, in the cases with higher SVF, the position of the separation zone was different from the cases with low SVF. In Cases 1–4, there was only one large separation zone located right after the confluence point. However, in Cases 5–8, a relatively smaller separation zone was located at the channel confluence while another large separation zone was captured near the downstream of the curved channel. According to the 2-D streamlines, the streamlines near the concave bank became disordered as the vegetation resistance increased. In cases 1–4, with smaller resistance, the tributary flow rushed into the main channel and formed a circulation after the confluence point. While in cases 5–8, the tributary flow was disturbed by the large resistance and divided into two parts. A little part whirled at the corner right after the confluence point forming the little separation zone and the other part flowed through the stem space and slowed down gradually downstream which formed the large separation. Moreover, it could be observed that in the non-vegetated cases the main flow streamlines deflected in a similar direction to the separation zone extensions. However, in the vegetated cases the main flow streamlines deflected along the edge of the vegetation region.

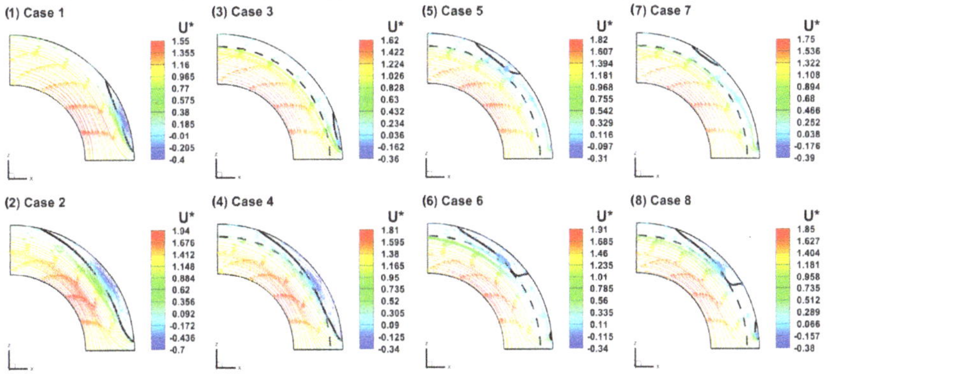

Figure 9. Contours and 2-D streamlines of dimensionless streamwise velocities on the vertical section near the water surface in the curved reach after the confluence point. (The black dashed line and black solid line in the figure represent the edge of vegetation region and the border of separation zone, respectively).

Figure 10 shows the contours of the radial gradient of streamwise velocities on the vertical section near the water surface in the curved reach after the confluence point. It is obvious that a shear layer with a large negative velocity gradient was formed between the midline and concave bank. In the shear layer, the velocities changed fast and the momentum exchange was violent. In the non-vegetated cases, the shear layer was located right beside the border of the separation zone. With the tributary discharge increased, the value of the velocity gradient increased and the range of the shear layer became larger. In the vegetated cases, due to the blockage of stems, the velocity gradient in the vegetation region was relatively small. The shear layer is located beside the edge of the vegetation region and deflected slightly towards the midline. This phenomenon is consistent with the deflection of streamlines in Figure 9. What is more, the velocity gradient at the beginning of the vegetation region was quite large. This is because the tributary flow with large velocities knocked into the vegetation stems generating a sharp decline in velocities.

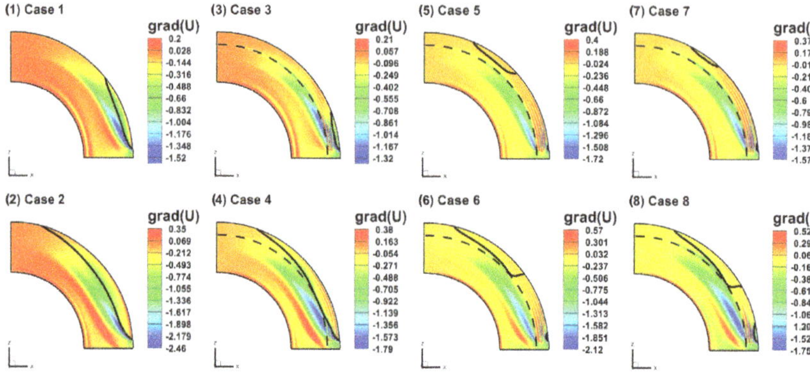

Figure 10. Contours of radial gradient of streamwise velocities on the vertical section near the water surface in the curved reach after the confluence point. (The black dashed line and black solid line in the figure represent the edge of vegetation region and the border of separation zone, respectively).

4.2. Water Level

Figure 11 below shows the dimensionless water level ($h^* = h/h_0$) contours in the main channel of all the cases. According to the water level contours, for all the cases, the lowest water level is located near the convex bank in the curved reach after the confluence point due to the centrifugal force. This character is basically consistent with that in curved open channel flow. In the non-vegetated cases, the water level near the concave bank in the separation zone fell visibly as the tributary discharge ratio increased. That is because the tributary flow diverted water from the concave bank into the central channel. As the tributary discharge increased, backwater occurred upstream of the confluence, which caused the water level to increase sharply; the waterhead before and after the confluence was much larger. The lowest water level points also moved downstream as the tributary discharge increased. For the vegetated cases, the water level at the confluence point rose due to the blockage of the vegetation. The downstream water level varied obviously along the channel width near the convex bank while the water level of the concave bank became relatively flat. The low water level area near the convex bank shrunk compared to that in the non-vegetated case because the vegetation region blocked the flows and narrowed the main flow area. With the increase in SVF, the upstream water level became higher and the waterhead between upstream and downstream became larger. Especially with the same SVF ($SVF = 4.55\%$), the upstream water level in Case 5 was a little bit higher than that in Case 7, and a similar result was also found between Case 6 and Case 8. This phenomenon demonstrates that under the same solid volume fraction and water barrier area a larger vegetation density brings more resistance to the flow than a larger stem diameter.

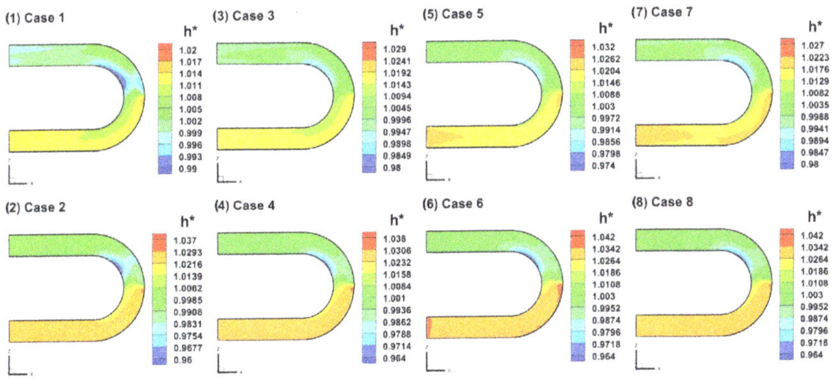

Figure 11. Contours of dimensionless water level in the main channel of all the cases.

4.3. Cross-Section Secondary Flow

To totally study the effects of discharge ratios and vegetation on secondary flow structures, three typical cross-sections were chosen. The three cross-sections were the upstream cross-section right before the confluence point, the 135° intersecting surface in the curved reach and the exit section of the channel bend, respectively (shown in Figure 8a).

Figures 12–14 below exhibit the contours of dimensionless streamwise velocity ($U^* = U/U_{Average}$, $U_{Average}$ represents the averaged velocity in the cross-section) distribution and 2-D velocity vectors on the three cross-sections of all the cases. According to the velocity contours, the maximum streamwise velocity occurred near the convex bank, which was consistent with Figure 9. On the upstream cross-section of the confluence point, with the tributary discharge ratio increased, the maximum velocity regions moved close to the convex bank. The dimensionless velocity near the convex bank increased while on the concave bank the dimensionless velocity decreased. There were only minor changes in velocity distribution between vegetated cases and non-vegetated cases on CS. 1, which demonstrated that the vegetation region after the confluence point had less influence on the upstream cross-section than the tributary flows. In CS. 2 and CS. 3, the dimensionless velocity on the concave bank became obviously much smaller in the vegetated cases, the minimum velocity region basically overlapped with the vegetation region. From the velocity vectors on the cross-sections, due to the centrifugal force, a clockwise main circulation cell was located near the channel bed of the convex bank, which was in tune with the curved open channel flow. As the tributary discharge and vegetation SVF increased, the scope of main circulation cells apparently shrunk. On the upstream cross-section right before the confluence point, due to the tributary flow, there was no circulation cell on the concave bank. In CS. 2, besides the main circulation cell, there was another reversed smaller circulation on the concave bank. The concave bank circulation was formed by the effects of turbulence flow and the driving forces brought by the main circulation cell. In the vegetated cases, the main circulation cell near the convex bank shrunk and the concave-bank circulation moved towards the channel midline. This phenomenon was much more apparent in the cases with higher SVF. Moreover, in the vegetation region, due to the obstruction of stems, the anisotropy of turbulence flow was enhanced so that the vectors became smaller and more chaotic. Moreover, in CS. 3, the main circulation cells near the bed of the convex bank were much smaller than those in CS. 1 and CS. 2. The circulations near the concave bank also shrunk significantly, which in the vegetated cases were hardly observed.

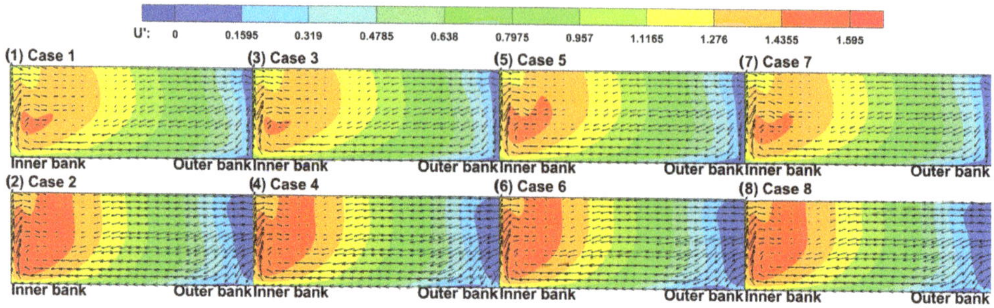

Figure 12. The contours of dimensionless streamwise velocity distribution and 2-D vectors on CS. 1 of all the cases.

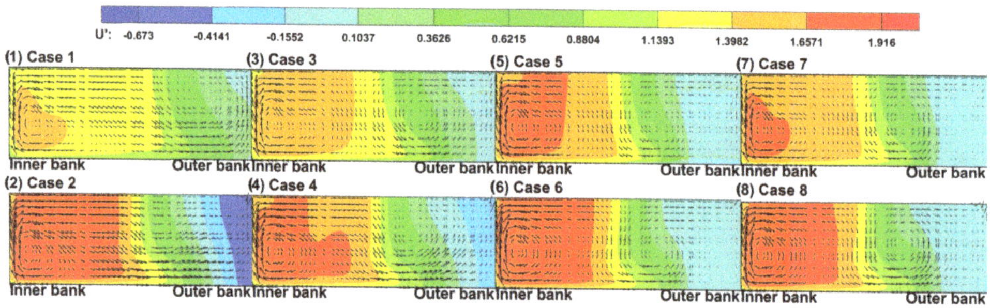

Figure 13. The contours of dimensionless streamwise velocity distribution and 2-D vectors on CS. 2 of all the cases.

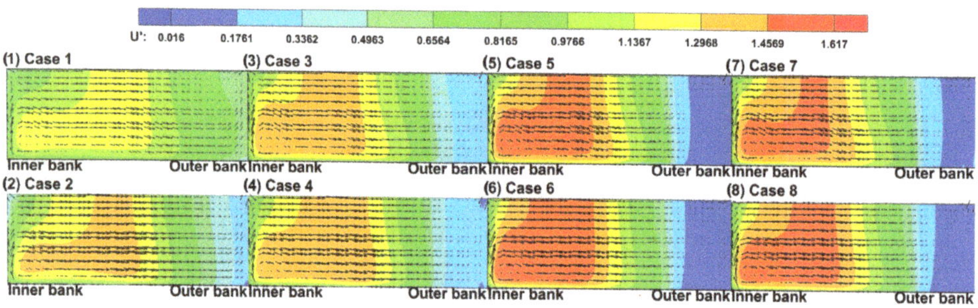

Figure 14. The contours of dimensionless streamwise velocity distribution and 2-D vectors on CS. 3 of all the cases.

The secondary flow strength is described by a dimensionless coefficient S_{xy}, which was introduced by Shukry [35] in his studies on the streamflow at the riverbank. The coefficient S_{xy} represents the kinetic energy ratio of the lateral current and mainstream on a given cross-section, and the formula is defined as follows:

$$S_{xy} = \frac{\int ({U_r}^2 + {U_v}^2) dA}{\int ({U_s}^2 + {U_r}^2 + {U_v}^2) dA} \tag{18}$$

where U_s, U_r and U_v represents the streamwise velocity, radial velocity and vertical velocity on the cross-section, respectively.

Figure 15 shows the secondary flow strength on the cross-sections along the curved reach of all the cases. According to the S_{xy} lines, the secondary flow strength before the confluence point increased gradually along the channel and was almost the same in all the cases with different flow discharge ratios and different vegetation densities. This phenomenon demonstrates the tributary and vegetation had a tiny effect on the upstream secondary flows which increased due to the centrifugal effects brought by the curvature. The largest S_{xy} point is located at the 90° cross-section with a sharp increase around the confluence point. This is because the tributary flow brought a strong lateral current to the main flow and the turbulence characteristics were strengthened in the confluence area. Moreover, in the cases with a larger tributary discharge ratio ($\lambda = 0.6$) the maximum secondary flow strength was much larger than those in the cases with a smaller discharge ratio ($\lambda = 0.3$). After the confluence point, the secondary flow strength declined dramatically and reached the minimum value at the 120° cross-section. That is because the flow acceleration zone with the narrowest main flow width was located in this area and the streamwise velocities reached the maximum value. The kinetic energy of the mainstream was much larger than the that of lateral current. After the 120° cross-section, in the non-vegetated cases, the secondary flow strength increased gradually along the channel. While in the vegetated cases, due to the blockage brought by the vegetation, the width of the main flow was narrowed along the channel and the secondary flow strength hardly changed until the 180° cross-section. What is more, as the vegetation resistance blocked the flow on the concave bank, the secondary strength in the vegetated cases was larger at the 90° cross-section and smaller in the downstream curved reach than that in the non-vegetated cases.

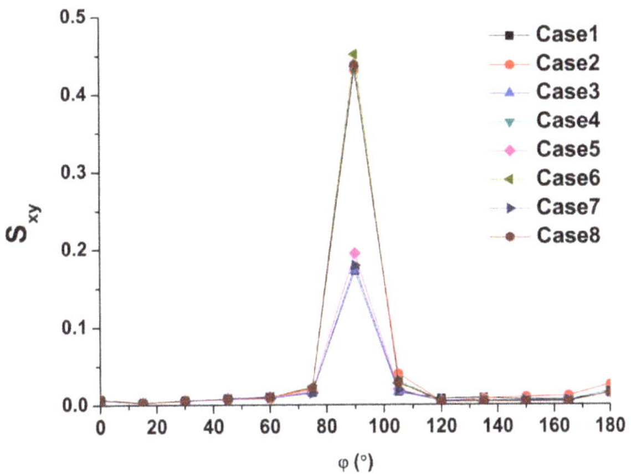

Figure 15. Secondary flow strength (S_{xy}) along the curved reach in the main channel.

4.4. Bed Shear Stress

The dimensionless bed shear stress, i.e., friction coefficient C_f can be defined as the following expression:

$$C_f = \frac{\tau}{\rho U_\infty{}^2} \tag{19}$$

where τ represents the bed shear stress, ρ represents the water density, and U_∞ represents the averaged streamwise velocity at the channel exit.

Figure 16 displays the contours of bed shear stress friction coefficient C_f of the main channel in all the cases. According to the investigation carried out by Kashyap et al. [36], for the high curvature flatbed channels ($R/B \leq 3$), the region of maximum bed shear stress is located near the convex bank upstream after the curved reach entrance, and gradually moves towards the concave bank downstream. While in this paper, the results were quite different from those in curved open channels. The region of maximum bed shear stress was always located near the convex bank along the whole curved reach. The tributary flow deflected the main flow from the concave bank to the convex bank and induced a large velocity difference between the inner side and outer side of the curved channel. The near-bed vertical velocity gradient was much larger on the convex bank than that on the concave bank. As the same as the streamwise velocities, with the increase in tributary discharge, the difference in bed shear stress between the convex and concave banks was obviously larger. For vegetated cases, due to the resistance brought by stems, the velocities near the convex bank became much larger, and so was the near-bed velocity gradient. Then the bed shear stress in the non-vegetated region was also much larger than in the vegetation region. With the increase in SVF, the maximum bed shear stress became larger. Especially, with the same SVF ($SVF = 4.55\%$), the maximum bed shear stress in Case 5 and Case 6 was a little bit bigger than that in Case 7 and Case 8. The reason leading to these results is similar to the streamwise velocity gradient. The bed shear stress is in proportion to the near-bed vertical velocity gradient [37]. Moreover, in the non-vegetated cases, the maximum bed shear stress near the convex bank basically occurred in the flow acceleration zone which was on the opposite of the separation zone. While in the vegetated cases, the maximum bed shear stress region almost covered the convex bank after the confluence point. Furthermore, near the concave bank, there was a dramatic decline in the friction coefficient. Moreover, in the non-vegetated cases, the region of decline was associated with the tributary discharge, with the increase in tributary discharge, the decline region moved gradually from the concave bank to the midline. While in the vegetated cases, the dramatic decline of bed shear stress basically occurred around the edge of the vegetation region.

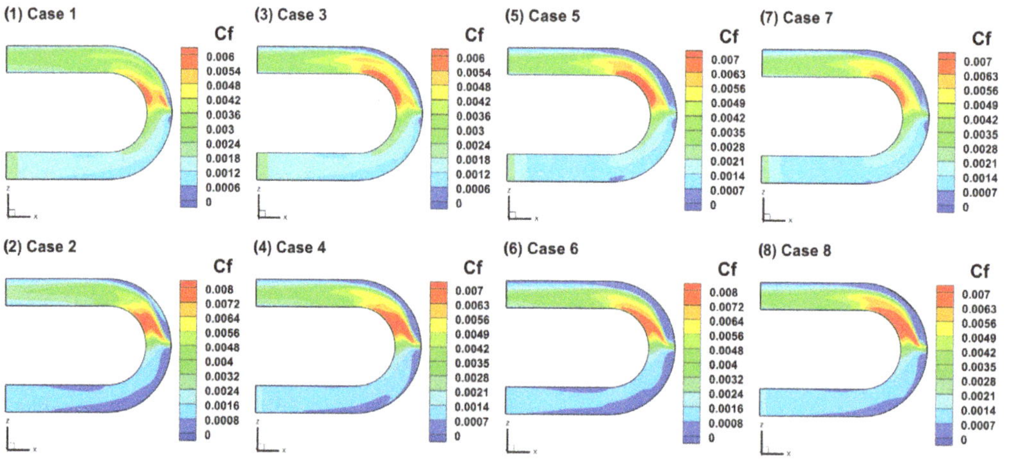

Figure 16. Contours of dimensionless bed shear stress friction coefficient C_f in the main channel of all the cases.

4.5. Flow Resistance

The flow resistance brought by rigid vegetation is often quantified by the vegetation drag coefficient, which is a normalized form of drag force. The drag coefficient of a single vegetation stem is defined below:

$$C_{di} = \frac{F_{di}}{0.5A\rho {U_{ref}}^2} \tag{20}$$

where F_{di} represents the drag force of the stem, A is the projection area of the stem, ρ is the fluid density, U_{ref} is the reference velocity approach to the stem. Since the velocity approach to a stem is quite difficult to observe in a vegetation array, the drag coefficient of the stem is also defined as an averaged drag coefficient:

$$C_d = \frac{F_d}{0.5AN\rho U_m{}^2} \tag{21}$$

where F_d represents the total drag force of the vegetation array, A is the projection area of the stem, ρ is the fluid density, N is the number of stems in the array, and U_m is the averaged streamwise velocity on the cross-section right before the vegetation region.

The time-averaged drag coefficients of all the vegetated cases are shown in Figure 17 below. According to Figure 17, the drag coefficients of all the cases were quite small due to the low velocities on the concave bank after the curved open channel confluence. The low velocities in the separation zone brought weak interaction between the flow and stems, and the drag forces in this area were very small. Comparing the drag coefficients in the cases, with the same planting scheme, the drag coefficients in the cases with larger flow discharge were a bit smaller than those in the cases with smaller discharge. Moreover, the drag coefficients decreased obviously with the increase in SVF. Under the same SVF ($SVF = 4.55\%$), the drag coefficients in Case 5 and Case 6 were much smaller than those in Case 7 and Case 8. The results indicate that the vegetation drag coefficient in the curved open channel confluence flow decreases with the increase in stem Reynolds number and vegetation planting density, which is consistent with the findings in Tanino and Nepf [38].

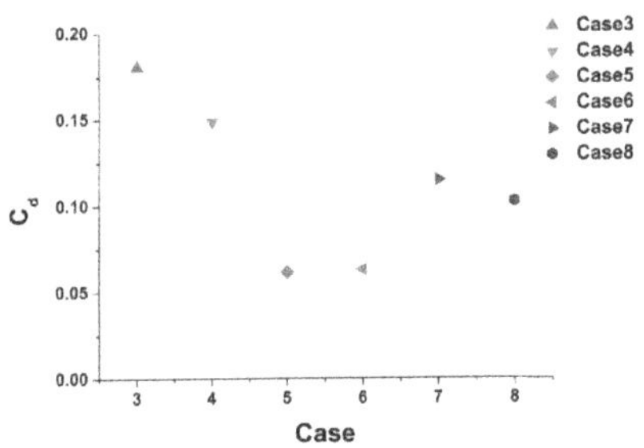

Figure 17. The time-averaged drag coefficients of all the vegetated cases.

Figure 18a,b display the local time-averaged drag coefficients along the curved channel after the confluence. From Figure 18, in each case, the local drag coefficients showed the same variation trend. The largest drag coefficient occurred at the beginning of the vegetation array. The large flow velocities coming from the confluence point knocked into the vegetation with a strong impact. In the middle reach of the vegetation zone, the flow was deflected to the convex bank; the velocities in this area were relatively smaller, thus the smallest drag coefficient located in this area. With the flow recovery before the exit of the curved reach, the drag coefficient became a little bit larger. Comparing the drag coefficients in the cases with the two different discharge ratios, the drag coefficients in the cases with larger tributary discharge were larger at the beginning of the vegetation array and smaller downstream of the confluence. This was because the larger tributary discharge brought a stronger impact to the stems at the beginning of the vegetation region and deflected the main flow downstream more obviously away from the concave bank towards the convex bank.

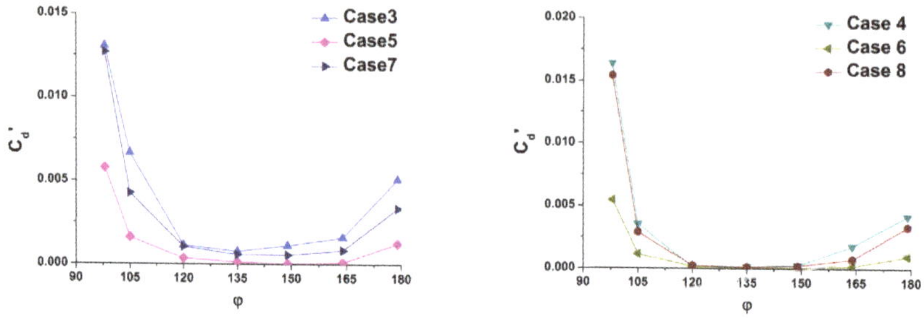

(a) (b)

Figure 18. The local time-averaged drag coefficients along the curved channel after the confluence: (**a**) Cases with tributary discharge ratio $\lambda = 0.3$; (**b**) Cases with tributary discharge ratio $\lambda = 0.6$.

5. Conclusions

The hydrodynamics of a U-shape open channel confluence flow with partially rigid emergent vegetation were simulated by the RNG k-ε numerical model coupled with the VOF method in OpenFOAM. The results of streamwise velocity, separation zone, water level, cross-section secondary flows and bed shear stress were analyzed. Some main findings are summarized as follows:

- The streamwise velocities of the convex bank were significantly greater than those of the concave bank and this velocity difference increased as the tributary discharge increased. The vegetation blocked the tributary flow to the mainstream causing an obvious decrease in velocities in the vegetation region. Moreover, the velocity difference between the convex bank and the concave bank was much larger. This change was more apparent with the higher vegetation solid volume fraction.
- The tributary flow impacted and deflected the main flow producing a separation zone with reversed smaller velocities on the concave bank after the confluence point. With the increase in tributary discharge, the separation zone became larger in both length and width. The vegetation near the concave bank played a role in blockage on the tributary flow and changed the mainstream deflecting direction. Compared to the non-vegetated cases, the separation zone was much smaller in length and width. Especially, with higher vegetation solid volume fraction ($SVF = 4.55\%$), the separation zone was divided into two parts, a smaller one right after the confluence point and a larger one on the second half of the curved reach after the confluence.
- The lowest water level point was located near the convex bank in the curved reach after the confluence point. Moreover, with the tributary discharge increased, the water level before the confluence increased while the water level after the confluence fell visibly. The lowest water level points also moved downstream as the tributary discharge increased. The vegetation brought resistance to the tributary flow causing backwater upstream of the confluence. The water level varied fast near the convex bank and tended to be gentle in the channel midline. With the same SVF ($SVF = 4.55\%$), a larger vegetation density brought more resistance to the tributary flow than a larger vegetation stem diameter.
- On the upstream cross-section right before the confluence point, there was only one main circulation cell located near the channel bed of the convex bank. While on the downstream cross-sections, besides the main circulation cell, there was another reversed smaller circulation on the concave bank. As the tributary discharge and vegetation SVF increased, the main circulation cell apparently shrunk and the concave-bank circulation moved towards the channel midline. Moreover, on the 180° cross-section, the circulation cells were much smaller than those in the curved reach. The maximum secondary flow strength was at the 90° cross-section where the confluence point was located.

- As the same trends of streamwise velocities, the region of maximum bed shear stress was always located near the convex bank along the whole curved reach. Compared to the non-vegetated case, the maximum bed shear stress was much larger in the vegetated case. Moreover, the bed shear stress in the non-vegetated region was also larger than that in the vegetated region. As the vegetation SVF increased, this difference became larger.
- The time-averaged drag coefficients of the vegetation decreased with the increase in stem Reynolds number and vegetation planting density. Along the arc length of the vegetation region, the largest local drag coefficient was located at the beginning of the region and decreased sharply in the middle reach, then it became a little bit larger downstream before the exit of the curved reach.

In this study, a U-shape open channel flow with a tributary on the concave bank was simulated by a numerical model. An emergent rigid vegetation region was set on the concave bank after the confluence point. The effects of tributary discharge ratio and vegetation density on the hydrodynamics were analyzed based on the results. This work provided a reference guide to understanding the dynamics and structures of the weedy interlaced meandering channels. The main implications of the study are to benefit water ecological management and restoration, whether natural rivers or urban artificial canals.

However, there are still lots of problems that are worthy of being investigated in this issue. In this study, the open channel was a 180° curved rectangular-cross-section channel with a tributary located at the 90° point of the curve; the vegetation region was covered with emergent rigid stems. In future studies, the degree of the curve, the shape of the cross-section, the location of the confluence point and the flexibility of the vegetation are to be taken into consideration. Even more, the contaminant transport and sediment deposition in the curved open channel confluence flow through vegetation are also good extensions of this topic.

Author Contributions: Conceptualization, Z.S. and S.J.; methodology, Z.S.; formal analysis, Z.S.; investigation, Z.S.; writing—original draft preparation, Z.S.; writing—review and editing, S.J.; supervision, S.J. All authors have read and agreed to the published version of the manuscript.

Funding: This research was funded by the National Natural Science Foundation of China (Grant No. 51739011), the National Key Research and Development Program of China (Grant No. 2016YFC0402707-03).

Data Availability Statement: Not applicable.

Conflicts of Interest: The authors declare no conflict of interest.

References

1. Roberts, M.V.T. Flow Dynamics at Open Channel Confluent-Meander Bends. Doctoral Thesis, University of Leeds, Leeds, UK, 2004.
2. Bradbrook, K.F.; Lane, S.N.; Richards, K.S. Numerical simulation of three-dimensional time-averaged flow structure at river channel junctions. *Water Resour. Res.* **2000**, *36*, 2731–2746. [CrossRef]
3. Biron, P.M.; Ramamurthy, A.S.; Han, S. Three-dimensional numerical modeling of mixing at river junctions. *J. Hydraul. Eng. -ASCE* **2004**, *130*, 243–253. [CrossRef]
4. Bradbrook, K.F.; Biron, P.M.; Lane, S.N.; Richards, K.S.; Roy, A.G. Investigation of controls on secondary circulation in a simple junction geometry using a three-dimensional numerical model. *Hydrol. Process.* **1998**, *12*, 1371–1396. [CrossRef]
5. Bradbrook, K.F.; Lane, S.N.; Richards, K.S.; Biron, P.M.; Roy, A.G. Role of bed discordance at asymmetrical river junctions. *J. Hydraul. Eng. -ASCE* **2001**, *127*, 351–368. [CrossRef]
6. Miyawaki, S.; Constantinescu, S.G.; Kirkil, G.; Rhoads, B.; Sukhodolov, A. Numerical investigation of three-dimensional flow structure at a river junction. In Proceedings of the 33rd International Association Hydraulic Research Congress, Vancouver, BC, Canada, 9–14 August 2009.
7. Constantinescu, G.; Miyawaki, S.; Rhoads, B.; Sukhodolov, A.; Kirkil, G. Structure of turbulent flow at a river junction with a momentum and velocity ratios close to 1: Insight from an eddy-resolving numerical simulation. *Water Resour. Res.* **2011**, *47*, W05507. [CrossRef]
8. Riley, J.D.; Rhoads, B.L. Flow structure and channel morphology at a natural confluent meander bend. *Geomorphology* **2012**, *163*, 84–98. [CrossRef]

9. Riley, J.D.; Rhoads, B.L.; Parsons, D.R.; Johnson, K.K. Influence of junction angle on three-dimensional flow structure and bed morphology at confluent meander bends during different hydrological conditions. *Earth Surf. Process. Landf.* **2014**, *40*, 252–271. [CrossRef]
10. Shamloo, H.; Pirzadeh, B. Investigation of flow pattern in curved channels with different lateral intake locations. In Proceedings of the 13th Annual & 2nd International Fluid Dynamics Conference, Shiraz University, Shiraz, Iran, 26–28 October 2010.
11. Ghobadian, R.; Tabar, Z.S.; Koochak, P. Numerical study of junction-angle effects on flow pattern in a river junction located in a river bend. *Water SA* **2016**, *42*, 43–51. [CrossRef]
12. Sui, B.; Huang, S.H. Numerical analysis of flow separation zone in a confluent meander bend channel. *J. Hydrodyn.* **2017**, *29*, 716–723. [CrossRef]
13. Montaseri, H.; Asiaei, H.; Baghlani, A. Numerical study of flow pattern around lateral intake in a curved channel. *Int. J. Mod. Phys. C* **2019**, *30*, 1950083. [CrossRef]
14. Mohammadiun, S.; Neyshabouri, S.A.A.S.; Naser, G.; Vahabi, H. Numerical Investigation of Submerged Vane Effects on Flow Pattern in a 90° Junction of Straight and Bend Open Channels. *Iran. J. Sci. Technol. Trans. Civ. Eng.* **2016**, *40*, 349–365. [CrossRef]
15. Nepf, H. Drag, turbulence and diffusivity in flow through emergent vegetation. *Water Resour. Res.* **1999**, *35*, 479–489. [CrossRef]
16. Ghisalberti, M.; Nepf, H. Mixing layers and coherent structures in vegetated aquatic flow. *J. Geophys. Res.* **2002**, *107*, 3-1–3-11. [CrossRef]
17. Ghisalberti, M.; Nepf, H. The limited growth of vegetated shear-layers. *Water Resour. Res.* **2004**, *40*, W07502. [CrossRef]
18. Ghisalberti, M.; Nepf, H. Mass transport in vegetated shear flows. *Environ. Fluid Mech.* **2005**, *5*, 527–551. [CrossRef]
19. Ghisalberti, M.; Nepf, H. The structure of the shear layer over rigid and flexible canopies. *Environ. Fluid Mech.* **2006**, *6*, 277–301. [CrossRef]
20. Nepf, H. Flow and transport in regions with aquatic vegetation. *Annu. Rev. Fluid Mech.* **2012**, *44*, 123–142. [CrossRef]
21. Wu, W.; Shields, F.D.; Bennett, S.J.; Wang, S.S.Y. A depth-averaged, two-dimensional model for flow, sediment transport, and bed topography in curved channels with riparian vegetation. *Water Resour. Res.* **2005**, *41*, W03015. [CrossRef]
22. Zhang, J.; Su, X. Numerical model for flow motion with vegetation. *J. Hydrodyn.* **2008**, *20*, 172–178. [CrossRef]
23. Camporeale, C.; Perucca, E.; Ridolfi, L.; Gurnell, A.M. Modeling the interactions between river morphodynamics and riparian vegetation. *Rev. Geophys.* **2013**, *51*, 379–414. [CrossRef]
24. Le Bouteiller, C.; Venditti, J.G. Sediment transport and shear stress partitioning in a vegetated flow. *Water Resour. Res.* **2015**, *51*, 2901–2922. [CrossRef]
25. Yang, Z.-H.; Bai, F.-P.; Huai, W.-X.; Li, C.-G. Lattice Boltzmann method for simulating flows in open-channel with partial emergent rigid vegetation cover. *J. Hydrodyn.* **2019**, *31*, 717–724. [CrossRef]
26. Li, C.W.; Zeng, C. 3D Numerical modelling of flow divisions at open channel junctions with or without vegetation. *Adv. Water Resour.* **2009**, *32*, 49–60. [CrossRef]
27. Huai, W.-X.; Li, C.-G.; Zeng, Y.-H.; Qian, Z.-D.; Yang, Z.-H. Curved open channel flow on vegetation roughened inner bank. *J. Hydrodyn.* **2012**, *24*, 124–129. [CrossRef]
28. Schnauder, I.; Sukhodolov, A.N. Flow in a tightly curving meander bend: Effects of seasonal changes in aquatic macrophyte cover. *Earth Surf. Process. Landf.* **2012**, *37*, 1142–1157. [CrossRef]
29. Termini, D. Vegetation effects on cross-sectional flow in a large amplitude meandering bend. *J. Hydraul. Res.* **2017**, *55*, 423–429. [CrossRef]
30. Wang, M.; Avital, E.; Korakianitis, T.; Williams, J.; Ai, K. A numerical study on the influence of curvature ratio and vegetation density on a partially vegetated U-bend channel flow. *Adv. Water Resour.* **2020**, *148*, 103843. [CrossRef]
31. InterFoam–OpenFOAMWiki. Available online: https://openfoamwiki.net/index.php/InterFoam (accessed on 10 April 2021).
32. Yakhot, V.; Orszag, S.A. Renormalization group analysis of turbulence. *J. Sci. Comput.* **1986**, *1*, 3–51. [CrossRef]
33. Brackbill, J.; Kothe, D.; Zemach, C. A continuum method for modeling surface tension. *J. Comput. Phys.* **1992**, *100*, 335–354. [CrossRef]
34. Heyns, J.A.; Oxtoby, O.F. Modelling surface tension dominated multiphase flows using the VOF approach. In Proceedings of the 6th European Conference on Computational Fluid Dynamics, Barcelona, Spain, 20–25 July 2014.
35. Shukry, A. Flow around bends in an open flume. *Trans. Am. Soc. Civ. Eng.* **1950**, *115*, 751–779. [CrossRef]
36. Kashyap, S.; Constantinescu, G.; Rennie, C.D.; Post, G.; Townsend, R. Influence of channel aspect ratio and curvature on flow, secondary circulation, and bed shear stress in a rectangular channel bend. *J. Hydraul. Eng.-ASCE* **2012**, *138*, 1045–1059. [CrossRef]
37. Katritsis, D.; Kaiktsis, L.; Chaniotis, A.; Pantos, J.; Efstathopoulos, E.P.; Marmarelis, V. Wall shear stress: Theoretical considerations and methods of measurement. *Prog. Cardiovasc. Dis.* **2007**, *49*, 307–329. [CrossRef] [PubMed]
38. Tanino, Y.; Nepf, H. Laboratory investigation on mean drag in a random array of Rigid, Emergent Cylinders. *J. Hydraul. Eng.* **2008**, *134*, 4–41. [CrossRef]

water

MDPI

Article

Dynamic Roughness Modeling of Seasonal Vegetation Effect: Case Study of the Nanakita River

André Araújo Fortes [1], Masakazu Hashimoto [1,2,*], Keiko Udo [1], Ken Ichikawa [1,3] and Shosuke Sato [1,2]

1 Department of Civil and Environmental Engineering, Tohoku University, Sendai 980-8579, Japan
2 International Research Institute of Disaster Science, Tohoku University, Sendai 980-8572, Japan
3 Fukken Gijyutsu Consultants Co., Ltd., Sendai 980-0012, Japan
* Correspondence: hashimoto@irides.tohoku.ac.jp

Abstract: Hydraulic models of rivers are essential for vulnerability assessment in disaster management. This study simulates the 2019 Typhoon Hagibis at the Nanakita River using a dynamic roughness model. The model estimates the roughness of the river on a pixel level from the relationship between the Manning roughness coefficient and the degree of submergence of vegetation. This degree is defined as the ratio of water depth to plant height. After validating the model, the effect of vegetation on the water level in different seasons from April 2020 to March 2021 was assessed. The vegetation area and height were obtained on a pixel level using unmanned aerial vehicle photogrammetry. The dynamic roughness model showed that the water level profile increased by 7.03% on average. The seasonal effect of vegetation was observed, revealing a strong correlation between variations in the vegetation conditions and water level profile. This approach may help mitigate flood damage by indicating the factors that can increase the risk of flooding.

Keywords: flood vulnerability; hydraulic resistance; Manning roughness coefficient; Typhoon Hagibis; unmanned aerial vehicles; structure from motion; vegetated flow

Citation: Araújo Fortes, A.; Hashimoto, M.; Udo, K.; Ichikawa, K.; Sato, S. Dynamic Roughness Modeling of Seasonal Vegetation Effect: Case Study of the Nanakita River. *Water* **2022**, *14*, 3649. https://doi.org/10.3390/w14223649

Academic Editors: Hossein Afzalimehr and Jueyi Sui

Received: 29 September 2022
Accepted: 9 November 2022
Published: 12 November 2022

1. Introduction

Hydraulic models of rivers are essential for vulnerability assessment in disaster management, especially in areas prone to floods due to cyclones and typhoons. These models are influenced by various parameters, with roughness as the most critical. Roughness has several influencing factors [1], with vegetation as the dominant [2]. Several investigations on the effect of vegetation on flow and roughness have been made [2–12]. Studies have attributed the roughness variation to the degree of submergence of plants [2,5–8] and plant density [2,9–11]. It is important to note that other factors such as the plant morphology and flexibility play an important role in the roughness determination [12]. Physical parameters are the primary determinants of roughness, and the ratio of the average canopy height to the mean flow depth is the most accurate descriptor of roughness [5]; consequently, with depth fluctuations in flood events, the roughness value is a dynamic variable. Hence, constant roughness is not applicable for every type of flow condition [2]. Mohammadi and Kashefipour [13] considered the flow depth and velocity as variables to calculate a dynamic roughness value in the Kārūn River, Iran, without directly considering the vegetation. Yoshida et al. [14] used vegetation information acquired by LiDAR to calculate a distributed roughness based on the vegetative drag force dynamically in a flood model. Variations of vegetation characteristics due to seasonal change alter the flow dynamics of rivers [15], making vegetation control at the riverbank a crucial task.

Vegetation characteristics are often excluded from hydraulic modeling due to high time consumption and financial cost. Unmanned aerial vehicles (UAVs) are useful tools for acquiring the data through river mapping [14,16–19]. Combining high-resolution UAV imagery with artificial neural networks (ANN) is a good strategy to identify vegetation and other features in river environments [17]. In addition, UAV imagery has enabled the creation of three-dimensional (3D) terrain models through structure from the motion (SfM) approach [18], being used to monitor seasonal variations of riparian vegetation height [19]. UAVs with real-time kinematic (RTK) positioning have obviated the need for ground control points to generate accurate digital surface models (DSMs) from 3D point clouds, improving the photogrammetry process. This has further reduced the computational cost and time required to generate 3D point clouds accurate to less than 10 cm [20]. Studies that consider the vegetation-induced roughness in hydraulic modeling from remote-sensed vegetation data, especially height and density, are needed [21]. Furthermore, the pixel-based acquisition of vegetation information through UAV imagery has allowed the application of dynamic vegetative roughness estimation for flood analysis. In addition, affordable UAV imagery and photogrammetry with RTK positioning allow frequent observations and acquisitions of seasonal vegetation information.

With many rivers, Japan is often affected by typhoons and suffers constant river flooding. Yet, the country has not implemented sufficient vegetation management to mitigate typhoon induced inundations. About 90% of Japan's river length is administered by prefectural governments, of which only a few rivers are overseen by management strategies owing to budgetary constraints. On 12 October 2019, Typhoon Hagibis devastated ten prefectures in Japan with record rainfall and wind speeds [22]. According to the Ministry of Land, Infrastructure, Transport and Tourism [23], the event overtopped approximately 130 embankments along several rivers, causing 90 deaths and nine disappearances and damaging approximately 80,000 houses. In Miyagi prefecture, the floods severely damaged the town of Marumori and abutting areas [24]. To ensure the safety of the population, flood vulnerability studies are very important for indicating factors that can increase the flood risk at determined locations [25]. This is mostly important for small- and medium-sized rivers because they tend to flood more severely than large rivers.

The objectives of this study are (1) to estimate the seasonal effect of the riparian vegetation on flow dynamics using a dynamic roughness hydraulic model, (2) to obtain the vegetation conditions throughout the year from 3D point cloud data, (3) to quantify the effect of riparian vegetation on flood vulnerability through a case study of Typhoon Hagibis, and (4) to assess the effect of the seasonal variation of vegetation conditions on the changes in water level using the hydraulic event of Typhoon Hagibis.

2. Study Area

The rivers in Japan are classified into Class A and Class B depending on their dimension as well as their social importance [26]. Class A rivers are managed by the national government, whereas Class B rivers are managed by the prefectural governments. Nanakita River is a Class B river located in Miyagi prefecture; it courses through the cities of Sendai and Tagajo. According to Viet et al. [27], the river has a length of 45 km and a catchment area of 229 km^2, the average annual discharge is 10 m^3/s, and the flood discharge of a 100-year return period is 1650 m^3/s. The bed slope of the Nanakita River is about 0.0016 in the relatively upstream reach and about 0.0003 in the lower reach [28].

The effect of vegetation on the flow dynamics is more visible in Class B rivers owing to their smaller size, and therefore, the Nanakita River is suitable for achieving the research objectives of this study. The Nanakita Dam was constructed in the upstream part of the river basin. According to the Japan Dam Foundation [29], the dam type is a rockfill dam; it has a water surface area of 50 hectares and a capacity of 9,200,000 m^3, with the purpose of flood control, river flow maintenance, and water supply. According to the prefecture's administration, the dam was not assigned the control of the river discharge at the time of Typhoon Hagibis.

The Nanakita River is the only Class B river in the prefecture with a flood forecast system, which is effective in the downstream portion of the river. The upstream region relies on a water level awareness system. The river catchment is a flood-prone area and the river has overflowed several times in the past due to heavy rains. The last flood occurred in 2015 caused by typhoons 17 and 18 [30].

Figure 1a–c show a map of the Nanakita River basin location in Miyagi prefecture, its topography and river system with the location of the dam and five water level gauge stations, and the 2 km stretch of the river chosen for the two-dimensional (2D) hydraulic simulation, respectively. This stretch has a dense population and is frequented by residents, with many houses on the coast. It is densely vegetated, with a predominant population of *Pueraria montana var. Lobata* (Kudzu), *Miscanthus sinensis*, *Phragmites australis*, and *Solidago canadensis*, making it ideal for studying the effects of a typhoon. In addition, although the water level reached a considerable height, no overtopping of the embankments happened during the typhoon event.

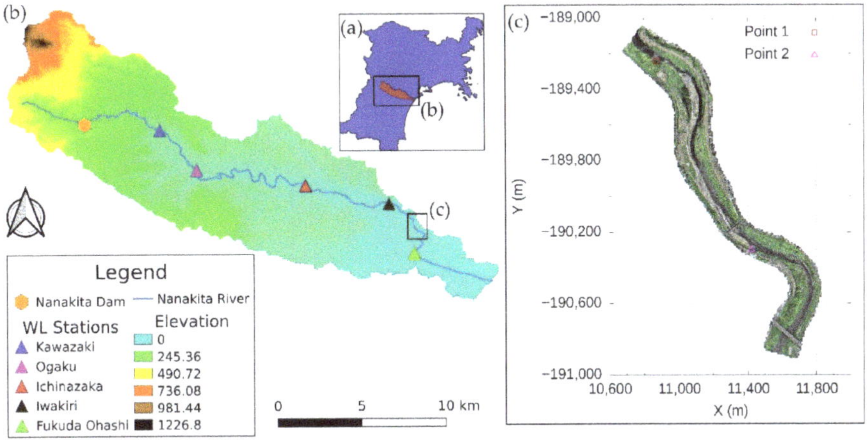

Figure 1. Study area: (**a**) Location of the Nanakita River basin in Miyagi prefecture; (**b**) basin map with the topography used as calculation area for the rainfall-runoff inundation model; (**c**) the 2 km stretch of the Nanakita River obtained from the UAV observation of September 2019 used as calculation area for the 2D hydraulic model.

Thirteen UAV observations took place in the stretch shown in Figure 1c, once in September 2019 and then monthly from April 2020 to March 2021. The prefecture removed the vegetation in October 2019, right after the disaster, as a countermeasure but could not continue the efforts owing to regrowth.

3. Methodology

The 2019 Typhoon Hagibis was simulated with a 2D hydraulic model in the UAV observed area shown in Figure 1c; the model was altered to dynamically calculate the Manning roughness coefficient considering the vegetation height (dynamic roughness model). The discharge of the event in the upstream section of the stretch was first generated by the rainfall–runoff–inundation (RRI) model.

The vegetation conditions in September 2019 and from April 2020 to March 2021 were obtained from the 3D point cloud data files.

The recreation of Typhoon Hagibis using the dynamic roughness model used the vegetation conditions obtained from the September 2019 UAV observation, which contained the pre-event conditions. The results of the simulation were compared with those of a static roughness simulation (2D hydraulic model before the alteration). Both simulations used the same topography and discharge, the only difference being the roughness setting; whereas the dynamic model calculated the Manning roughness coefficient for the entire simulation period, the static model considered it a constant value.

After model validation, the seasonal effect of the vegetation on the water level of the river was assessed. To this end, the same hydrologic event was simulated 12 more times, wherein only the vegetation conditions obtained from the monthly UAV observations from April 2020 to March 2021 were varied.

3.1. UAV Observations

The purpose of UAV photogrammetry was to obtain the vegetation information and determine the seasonal changes of the riparian vegetation on the river floodplains. We captured photographs once in September 2019 and then monthly from April 2020 to March 2021. Thus, we generated 13 orthophotographs and 3D point cloud data files. The UAV (Phantom 4RTK (DJI)) has an RTK module that receives data from satellites and fixed bases and accurately corrects the position of the image in flight in real time. The observations were shot from an altitude of 50 m. The wrap rate was set at 80% overlap and 60% side wrap. To improve the accuracy of the images in the vertical direction, a camera inclined at 70° was mounted for the SfM analysis.

The point cloud had a vertical accuracy of less than 10 cm at the top of the embankment (approximately 8 cm).

Metashape (Agisoft) was used for the SfM analysis, and the point cloud density was 400 points m^{-2}.

3.2. Hydrologic Model

The RRI model was used for the hydrologic simulation; it is a 2D model that can simulate both rainfall–runoff and inundation [31–34]. The model used the following input data: (1) the observed rainfall data collected from radar/rain gauge-analyzed (RA) precipitation [35] with hourly observations spanning 48 h from October 12 to 14, 2019; (2) the MERIT hydro digital elevation model (DEM) [36] with a resolution of 90 m used for the topographic data; and (3) the land-cover map from the MLITT [23] with a resolution of 100 m. In this model, a Manning roughness coefficient value of 0.04 was adopted for the river channel. In the slope area, the values were 0.5, 0.3, and 0.04 for vegetated areas, urban areas, and water bodies, respectively. The soil depth of the vegetated area was assumed to be 1.0 m, whereas that of the other two areas were assumed to be 0.1 m. Simulated and observed water depth in the five water level gauge stations shown in Figure 1b were used to validate the model.

3.3. Two-Dimensional Hydraulic Model

The 2D hydraulic model used for flood inundation was developed by Iwasa and Inoue [37] and Inoue et al. [38], and further used in Hashimoto et al. [39]. The model utilizes the continuity and momentum equations in the x- and y-directions as governing equations, shown in Equations (1)–(3).

$$\frac{\partial h}{\partial t} + \frac{\partial M}{\partial x} + \frac{\partial N}{\partial y} = 0, \tag{1}$$

$$\frac{\partial M}{\partial t} + \frac{\partial uM}{\partial x} + \frac{\partial vM}{\partial y} = -gh\frac{\partial H}{\partial x} - gn^2u\frac{\sqrt{u^2+v^2}}{h^{\frac{1}{3}}}, \tag{2}$$

$$\frac{\partial N}{\partial t} + \frac{\partial uN}{\partial x} + \frac{\partial vN}{\partial y} = -gh\frac{\partial H}{\partial y} - gn^2v\frac{\sqrt{u^2+v^2}}{h^{\frac{1}{3}}}, \tag{3}$$

where h is the depth, x and y are the flux directions, and u and v are the flow velocities in the x and y directions, respectively. $M = u \times h$, $N = v \times h$, n is the roughness factor, and H is the water level.

The topographic data for the model was constructed by interpolating the 21 cross-sections spaced 100 m apart provided by the Miyagi prefecture administration. Thus, a DEM with 23,040 cells with 10 m resolution was generated. The 10 m grid cell was chosen because it accurately represents the vegetation in the UAV-observed area. The hydraulic data used as input was the discharge obtained from RRI model.

3.4. Dynamic Roughness

The hydraulic model was patched with a pixel-based dynamic-roughness calculation routine based on the degree of submergence of vegetation. The concept of the degree of submergence is illustrated in Figure 2, as described in Nikora et al. [5], and is defined by the ratio of the water depth to the vegetation height (Equation (4)).

$$Submergence\ degree = \frac{h_{water}}{h_{vegetation}}, \tag{4}$$

where h_{water} is the water depth of the section, and $h_{vegetation}$ is the vegetation height.

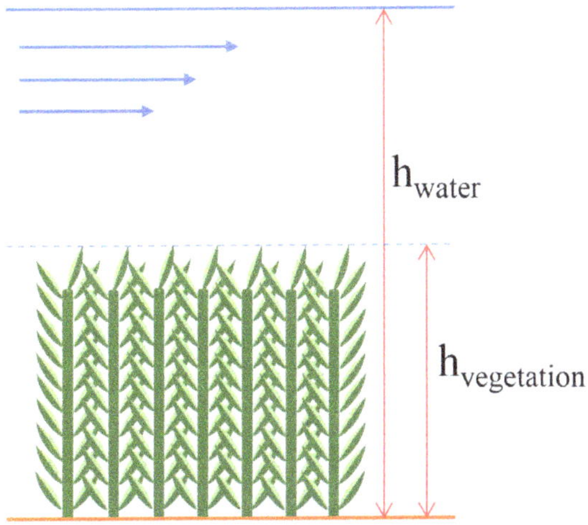

Figure 2. Scheme showing the degree of submergence of the vegetation.

The code adaptation comprehends the creation of a new input file with the location of the grid cells with vegetation and the vegetation height in each cell. The model then calculates the pixel-based roughness at every vegetated grid cell for each loop of the simulation, based on the degree of submergence of vegetation using the formula presented in Equation (5), obtained from the analysis of the relationship between the two parameters by the Japan Institute of Country-ology and Engineering [40]. The data present in [40] was obtained by field experiment, and the equation was obtained from regression analysis using MATLAB. The analysis produced a coefficient of determination (R^2) of 0.88 and a root mean square error (RMSE) of 0.007. Unlike the dynamic roughness calculation method of Mohammadi and Kashefipour [13], which used only water depth and velocity threshold to define roughness, this model directly correlated the hydraulic resistance with the vegetation height.

$$n = 0.084 * \left(\frac{h_{water}}{h_{vegetation}}\right)^{-0.98} + 0.023, \tag{5}$$

Figure 3 shows a representation of the relationship between the Manning roughness coefficient and the degree of submergence of various types of vegetation obtained from the Japan Institute of Country-ology and Engineering [40] and the regression equation. The equation from this relationship was derived from all the points shown in Figure 3, unifying all types of grass vegetation hardness into a single curve.

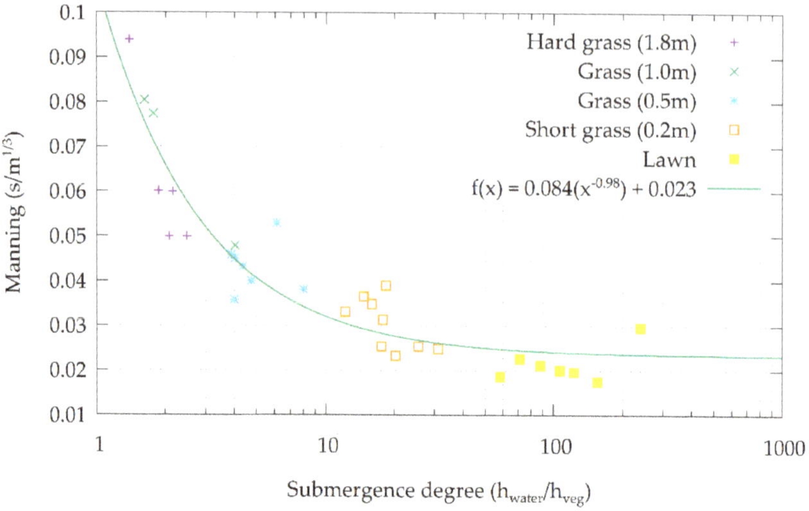

Figure 3. Relationship between the Manning roughness coefficient and the degree of submergence for different types of vegetation hardness and the regression curve.

In the code adaptation, the estimation of the Manning roughness coefficient is only applicable for fully submerged vegetation, when the degree of submergence is higher than 1; therefore, when the vegetation is emerging (degree of submergence less than or equal to 1), the model assumes a no-flow condition, where the water enters the cell and is trapped until it submerges the vegetation. For non-vegetated grid cells, the Manning roughness coefficient was set to a constant value of 0.022, a value recommended by Arcement and Verne [41], considering the surface material to be a smooth, firm soil.

The results from the dynamic roughness model were compared with those from the static model. The latter used constant Manning roughness coefficient values of 0.022 for the channel area and 0.038 for the floodplains; the values were fixed by river management. The value for the channel was set according to the bed material, composed mainly of firm soil and sand, whereas the value for the floodplains was chosen considering the range between 0.025 and 0.050 that represents a large amount of vegetation, according to Arcement and Verne [41]. To identify channel and floodplain, a simulation in the same area was performed with the average yearly discharge of 10 m^3/s. Both simulations were validated by comparing the water level profile in the peak with the floodmark points at specific locations.

3.5. Vegetation Characteristics

A Fortran 90 program was written to obtain digital surface models (DSMs) and RGB orthoimages from each of the UAV 3D point cloud data files. The RGB images were used to obtain the vegetation location with a multi-layer perceptron (MLP) algorithm trained based on the RGB values, and the DSMs were used to obtain the vegetation height for each month.

Because of the change of the vegetation color with the passing of seasons, two MLP models were trained, one for the spring and summer seasons (group 1) and another for the autumn and winter (group 2). Figure 4 depicts a flowchart with the steps for the acquisition of the vegetation location for the spring and summer seasons.

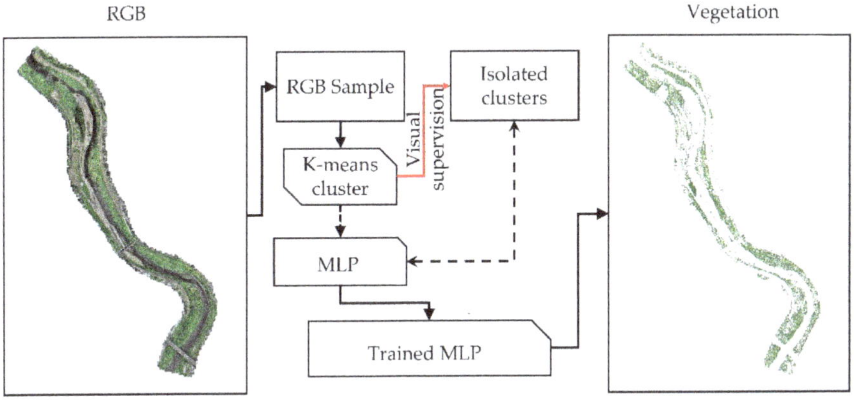

Figure 4. Flowchart of the method for the identification of the vegetated grid cells.

The method of identification of the vegetated pixel was similar to the method applied by Casado et al. [17]: rather than identifying all the features in the river environment, this study focused only in the identification of the vegetation. First, two samples of the RGB images of September 2019 and January 2021 with resolution of 1 m were cropped and submitted to a k-means cluster algorithm. Different from Casado et al. [17], who used the CIELAB color spectrum to cluster the pixels, in this methodology, the RGB color spectrum was used. Each cluster was isolated, and visual supervision was used to obtain the vegetated pixels for the two samples. Once the vegetation was located, the MLP models were trained for each group. Subsequently, the trained models were used to predict the vegetation location in the entire RGB images. The images were re-sampled to a 10 m grid size with ArcMAP 10.5, using the Majority algorithm to match the cell size chosen for the hydraulic simulations. The Majority algorithm determined the new value of the larger cell based on the most popular values between the smaller grid cells in the location. This method produced smoother results than the nearest neighbor method, avoiding the overestimation of the vegetation conditions.

With the vegetated grid cells located, the vegetation height could be calculated. As discussed in Weidner and Förstner [42], the DSM differs from the DEM because it comprehends not only the ground level but the level of the objects above; therefore, by normalizing the DSM with the DEM, the height of the aboveground objects can be determined. Thus, the vegetation height was obtained by this method only at the vegetated grid cells. Differing from the methodology used by van Iersel et al. [19], the normalization was performed with the DEM obtained from the interpolated cross-sections rather than from a digital terrain model (DTM), which is calculated from the DSM by removing the objects above ground level. The accuracy of the DTM depends on the time when the UAV observation is performed, varying with the seasons.

4. Results and Discussion

4.1. Vegetation Conditions

Both MLP models obtained good accuracy and made good predictions of the vegetation location, like the results obtained in [17]. The model accuracy for group 1 was 0.99. Group 2 obtained an accuracy of 0.96, as shown in Table 1, along with other evaluation parameters.

Table 1. Evaluation parameters of the MLP model of both groups.

Parameter	Group 1	Group 2
Accuracy	0.99	0.96
Precision	0.93	0.84
Misclassification	0.01	0.04
Recall	0.98	0.74

The precision and accuracy achieved for group 2 was considerably lower than the accuracy of group 1. The same can be observed for the precision; thus, the MLP trained for group 1 was able to classify the pixels with more correction than group 2. In addition, misclassification of group 2 was higher, meaning that more non-vegetated pixels were wrongfully classified as vegetated pixels in group 2 than in group 1, although, with the misclassification value of 4%, the wrongfully classified pixels were not great in number; therefore, it was considered an acceptable value for this research.

The lower values of accuracy and precision, along with the higher misclassification achieved by the group 2 MLP, occurred because the training sample represented the vegetation in the winter season, when there is a lack of greenness in the plants due to the loss of foliage; this promotes confusion with sand banks and the flood control structures.

The recall shows the percentage of corrected classification of the vegetated pixels. The value of 98% for the MLP trained for group 1 demonstrates that most of the vegetated pixels were correctly identified. For group 2 on the other hand, the recall of 74% shows that a considerable part of the vegetated pixels was not identified by the MLP. These results show that the vegetated area of spring and summer seasons were in closer accordance with the reality, but for autumn and winter seasons, the MLP produced an underestimation of the vegetated area.

Figure 5 illustrates the DSM, DEM, and vegetation height as per the UAV observations from September 2019. From this image, the clear difference between the DSM and the DEM in the vegetated pixels is shown, expressing the vegetation height.

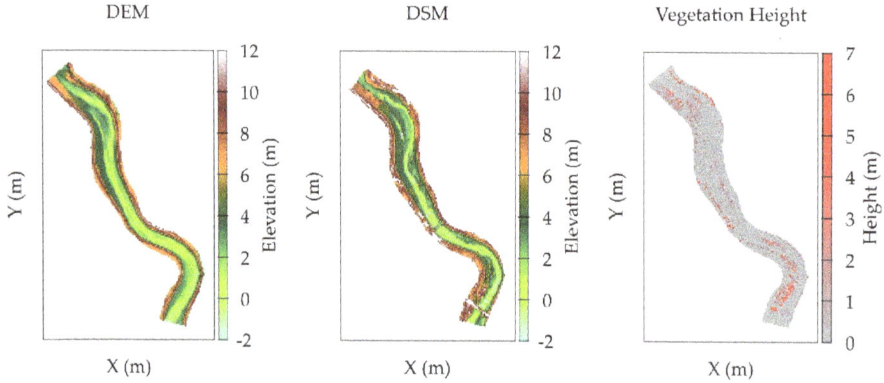

Figure 5. Image of the DEM (**left**), DSM (**center**), and vegetation height (**right**); vegetation height is the result of DSM–DEM for the vegetated grid cells.

The vegetation area and height varied significantly with the passing of seasons. The vegetation area has been shown to follow a clear pattern of variation according to the seasons. The highest area coverage was observed in the summer, when precipitation and temperature are high. In contrast, with low temperature and precipitation, the winter showed the lowest amount of vegetation coverage area. The average height also showed a similar pattern of variation, although less accentuated than the area. Figure 6 shows the values of area and average height for each month from April 2020 to March 2021.

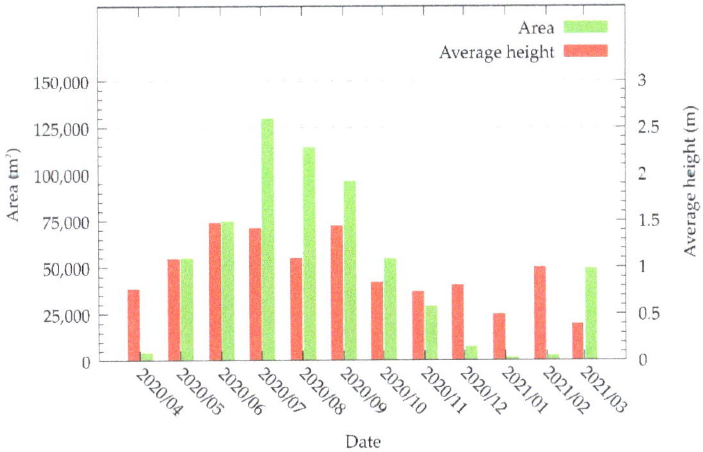

Figure 6. Vegetated area and vegetation average height in the UAV-observed stretch shown in Figure 1c from April 2020 to March 2021.

4.2. Hydrologic Simulation

The results of the RRI model simulation conformed to the observed data. The simulated hydrograph in the outlet revealed a peak discharge of approximately 1250 m^3 s^{-1}, lower than the 100-year return period of 1650 m^3 s^{-1} affirmed by Tanaka et al. [26]. Figure 7 shows the discharge in each station, in the upstream section of the UAV-observed area and the outlet. The comparison between the observed and simulated water level in each station is shown in Figure 8. Nash–Sutcliffe efficiency and RMSE were calculated for each station to validate the model. Except for the Kawazaki station, all stations showed Nash–Sutcliffe values ranging from 0.62 to 0.88. Because the Nash–Sutcliffe efficiency of most of the gauge stations was close to 1, the model could be considered sufficiently accurate.

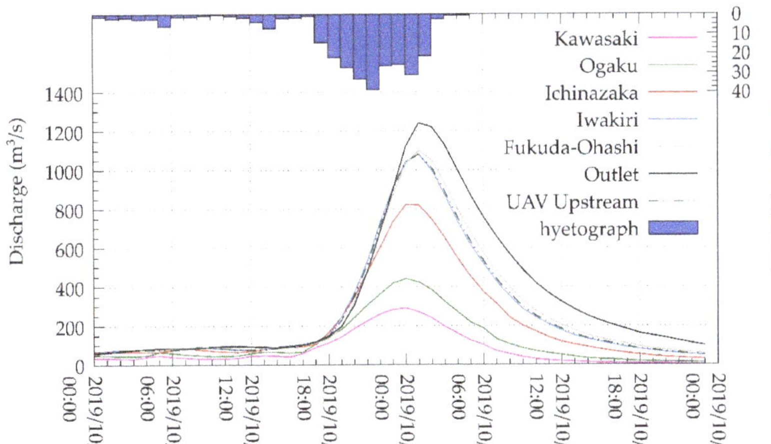

Figure 7. RRI simulated discharges in all the water level gauge stations shown in Figure 1b, in the UAV upstream section, and in the outlet.

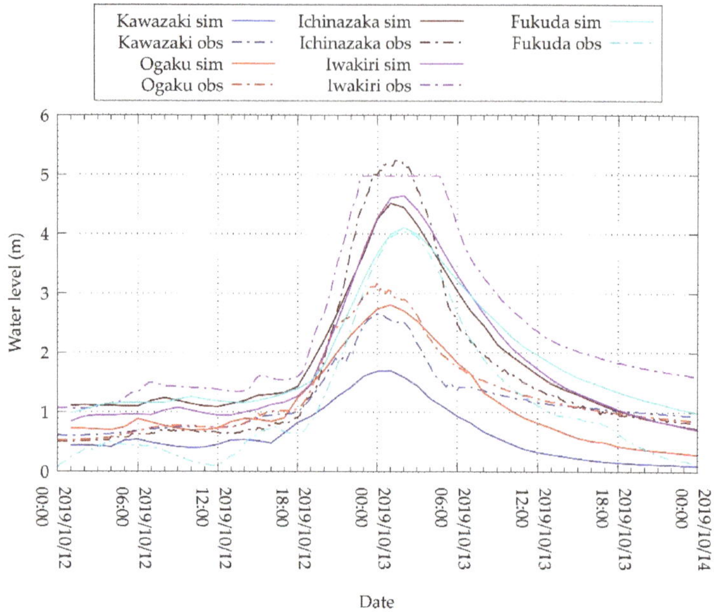

Figure 8. Comparisons between the RRI simulated and observed water levels in the 5 gauge stations.

The results of the simulation were like those of Bhagabati and Kawasaki [32], producing a reasonable discharge while overestimating the flooded area on a flat region of the river basin. The peak inundation (Figure 9) showed a virtual inundation in the residential area west of the Nanakita River between the Iwakiri and Fukuda Ohashi stations. This overestimation occurred because the model did not consider the city's drainage system. In the same stretch, a small flood was also observed because the model did not consider the flood control structures. Thus, a minor reduction in the discharge occurred from the Iwakiri to the following station, Fukuda Ohashi. This result contradicts those of Nastiti et al. [31], which produced a realistic inundated area but unrealistic discharge.

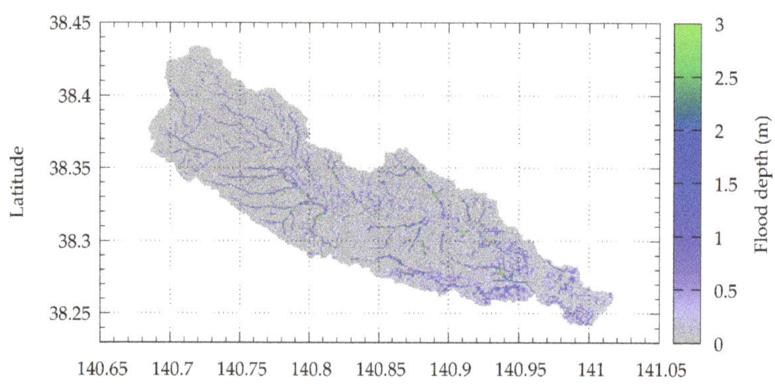

Figure 9. RRI-simulated peak inundation caused by the typhoon event in the river basin shown in Figure 1b.

4.3. Hydraulic Simulation

Considering the vegetation conditions of September 2019, the dynamic roughness model simulated a higher water level profile than that of the static roughness model. Both results were compared with the observed peak water level at specific locations. The dynamic roughness model obtained a more precise result, obtaining an RMSE of 0.17 m compared with that of 0.67 m produced by the static roughness simulations. As proposed by Ebrahimi et al. [2], the adoption of a non-constant Manning roughness coefficient routine improved the quality of the results, which are also in agreement with the results obtained by Mohammadi and Kashefipour [13]. The increased higher accuracy obtained from the dynamic roughness model demonstrates that, as in Yoshida et al. [14], the distributed vegetation in the reach can be considered a good parameter to calculate the roughness in the reach scale.

As shown in Figure 10, the water level in the dynamic roughness simulation increased by 7.03% on average compared with that in the static roughness simulation. The absolute water levels increased by 57 cm at 500 m and 39 cm at 2000 m from the downstream end. The water level increased by a maximum height of 61 cm at 700 m from the downstream boundary. These differences in the actual flood levels evidenced using the dynamic roughness calculation routine confirmed the statement of Wu et al. [8] regarding the consequences of using a constant roughness coefficient, demonstrating that the set Manning roughness coefficient value of 0.038 for the floodplain areas in the static roughness simulation was not enough to represent the vegetation effect in this river stretch.

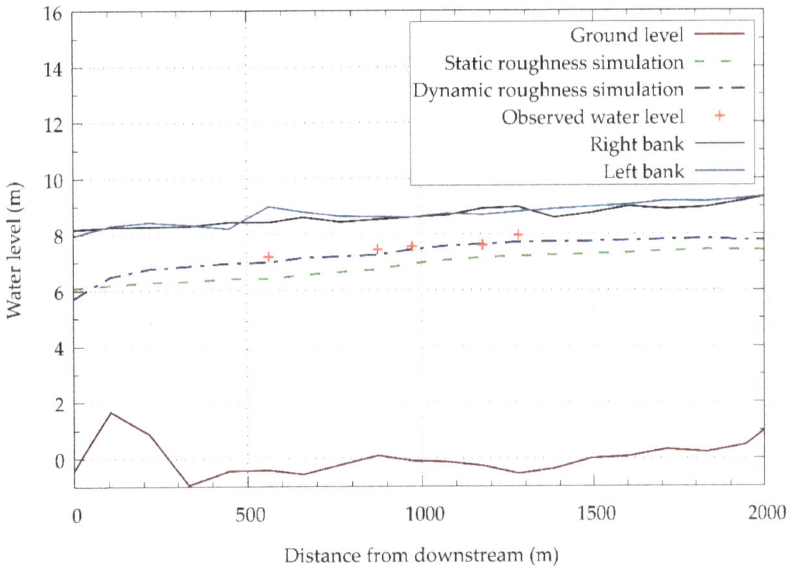

Figure 10. Simulated water level profiles of the simple Manning roughness coefficient setting simulation and the dynamic Manning roughness coefficient calculation simulation.

Measuring the increase in the water level is important for the prediction of areas with greater likelihood of overtopping, aiding in flood vulnerability assessment. The results of the dynamic roughness adaptation in the 2D model prove its usefulness in urban flood control and inundation prediction. The proposed model is better suited for preparedness once it has predicted a worsened and more realistic scenario.

Sensitivity analysis of the Manning value in the floodplain areas was performed in the static roughness model aiming to achieve a roughness value that better represents the roughness caused by the vegetation and other surfaces. The comparison between observed

The dynamic and static roughness models obtained RMSEs of 0.17 m and 0.67 m, respectively, leading to the conclusion that the dynamic roughness model achieved a more realistic simulation of Typhoon Hagibis than the static roughness model did. This demonstrates that the consideration of a fluctuating Manning roughness coefficient based on the degree of submergence of the plants improved the simulation results, as suggested by previous studies [2,5,8]. As predicted by Wu et al. [8], the use of varied Manning roughness coefficient values provoked considerable discrepancies in the simulated water levels. The largest difference of the simulated water level between the static model and the dynamic model was 61 cm, with the higher water level being predicted by the dynamic roughness model. The dynamic roughness model was shown to improve the safety for flood vulnerability studies in the Nanakita River.

Seasonal variations in vegetation area and height have a clear effect on the flow dynamics of the river. From the results of the simulations, it could be concluded that the water level is proportional to the amount of vegetation in the riparian zones. The highest water level was obtained in the summer, when the vegetation volume is at its peak, and the lowest value during the winter, when there is less vegetation. A low water level was also simulated in the early spring season, in April 2020, owing to the removal of the plants by Miyagi prefecture in October 2019, after the typhoon event.

The vegetation area and average height were demonstrated to have a good correlation with the simulated water levels, with average height achieving the strongest relationship, with an R of 0.91 and R^2 of 0.83. Therefore, considering only the vegetation area to estimate the water level, as does the static roughness model, would be less efficient. Considering the distributed vegetation height together with the vegetation area provides a stronger relationship between the riparian vegetation and the water level.

The method to obtain the vegetation location and its distributed height, with the consideration of the parameters in the dynamic roughness model proposed in this study was proven to be applicable for the purpose of river management in Nanakita River.

In the future, the limitations of the dynamic roughness model will be addressed. The identification of the different types of vegetation in the floodplains will be performed; thus, vegetation flexibility and the flow in emergent conditions will be considered in the model. Therefore, the applicability of the model will be broader, and it can be tested in other locations.

Author Contributions: Conceptualization, A.A.F. and M.H.; methodology, A.A.F., M.H. and K.U.; software, A.A.F. and M.H.; validation, A.A.F., M.H., K.U., K.I. and S.S.; investigation, A.A.F., M.H., K.U. and K.I.; resources, M.H., K.U. and K.I.; writing—original draft preparation, A.A.F.; writing—review and editing, A.A.F., M.H. and K.U.; supervision, M.H. and K.U.; project administration, K.U. and S.S.; funding acquisition, K.U. All authors have read and agreed to the published version of the manuscript.

Funding: This research was supported by a collaboration research fund of International Research Institute of Disaster Science, Tohoku University and Fukken Gijyutsu Consultants Co., Ltd.

Institutional Review Board Statement: Not applicable.

Informed Consent Statement: Not applicable.

Data Availability Statement: Not applicable.

Acknowledgments: The authors acknowledge the support of the Miyagi prefecture for providing the water depth data of the Nanakita River used for the model validation, as well as the cross-sections of the river used to generate the terrain topography.

Conflicts of Interest: The authors declare no conflict of interest.

References

1. Cowan, W.L. Estimating hydraulic roughness coefficients. *Agric. Eng.* **1956**, *37*, 473–475.
2. Ebrahimi, N.G.; Fathi-Moghadam, M.; Kashefipour, S.M.; Saneie, M.; Ebrahimi, K. Effects of flow and vegetation states on river roughness coefficients. *J. Appl. Sci.* **2008**, *8*, 2118–2123. [CrossRef]

3. Darby, S.E. Effect of riparian vegetation on flow resistance and flood potential. *J. Hydraul. Eng.* **1999**, *125*, 443–454. [CrossRef]
4. Sun, X.; Shiono, K.; Rameshwaran, P.; Chandler, J.H. Modelling vegetation effects in irregular meandering river. *J. Hydraul. Res.* **2010**, *48*, 775–783. [CrossRef]
5. Nikora, V.; Larned, S.; Nikora, N.; Debnath, K.; Cooper, G.; Reid, M. Hydraulic resistance due to aquatic vegetation in small streams: Field study. *J. Hydraul. Eng.* **2008**, *134*, 1326–1332. [CrossRef]
6. Wang, P.; Wang, C.; Zhu, D.Z. Hydraulic resistance of submerged vegetation related to effective height. *J. Hydrodynam.* **2010**, *22*, 265–273. [CrossRef]
7. Wilson, C.A.M.E. Flow resistance models for flexible submerged vegetation. *J. Hydrol.* **2007**, *342*, 213–222. [CrossRef]
8. Wu, F.-C.; Shen, H.W.; Chou, Y.-J. Variation of roughness coefficients for unsubmerged and submerged vegetation. *J. Hydraul. Eng.* **1999**, *125*, 934–942. [CrossRef]
9. Devi, T.B.; Kumar, B. Experimentation on submerged flow over flexible vegetation patches with downward seepage. *Ecol. Eng.* **2016**, *91*, 158–168. [CrossRef]
10. Jalonen, J.; Järvelä, J.; Aberle, J. Leaf area index as vegetation density measure for hydraulic analyses. *J. Hydraul. Eng.* **2013**, *139*, 461–469. [CrossRef]
11. Schoneboom, T.; Aberle, J.; Dittrich, A. Hydraulic resistance of vegetated flows: Contribution of bed shear stress and vegetative drag to total hydraulic resistance. In *River Flow*; Dittrich, A., Koll, K., Aberle, J., Geisenhainer, P., Eds.; Bundesanstalt für Wasserbau: Karlsruhe, German, 2010; pp. 269–276. ISBN 978-3-939230-00-7.
12. Caroppi, G.; Järvelä, J. Shear layer over floodplain vegetation with a view on bending and streamlining effects. *Environ. Fluid Mech.* **2022**, *22*, 587–618. [CrossRef]
13. Mohammadi, S.; Kashefipour, S.M. Numerical modeling of flow in riverine basins using an improved dynamic roughness coefficient. *Water Resour.* **2014**, *41*, 412–420. [CrossRef]
14. Yoshida, K.; Maeno, S.; Ogawa, S.; Mano, K.; Nigo, S. Estimation of distributed flow resistance in vegetated rivers using airborne topo-bathymetric LiDAR and its application to risk management tasks for Asahi River flooding. *J. Flood Risk Manag.* **2020**, *13*, e12584. [CrossRef]
15. Song, S.; Schmalz, B.; Xu, Y.P.; Fohrer, N. Seasonality of roughness—The indicator of annual river flow resistance condition in a lowland catchment. *Water Resour. Manag.* **2017**, *31*, 3299–3312. [CrossRef]
16. Carbonell-Rivera, J.P.; Estornell, J.; Ruiz, L.A.; Torralba, J.; Crespo-Peremarch, P. Classification of UAV-based photogrammetric point clouds of riverine species using machine learning algorithms: A case study in the Palancia River, Spain. *Int. Arch. Photogramm. Remote Sens. Spatial Inf. Sci.* **2020**, *XLIII-B2-2020*, 659–666. [CrossRef]
17. Casado, M.R.; Gonzalez, R.B.; Kriechbaumer, T.; Veal, A. Automated identification of river hydromorphological features using UAV high resolution aerial imagery. *Sensors* **2015**, *15*, 27969–27989. [CrossRef]
18. Javernick, L.; Brasington, J.; Caruso, B. Modeling the topography of shallow braided rivers using structure-from-motion photogrammetry. *Geomorphology* **2014**, *213*, 166–182. [CrossRef]
19. van Iersel, W.; Straatsma, M.; Addink, E.; Middelkoop, H. Monitoring height and greenness of non-woody floodplain vegetation with UAV time series. *ISPRS J. Photogramm. Remote Sens.* **2018**, *141*, 112–123. [CrossRef]
20. Forlani, G.; Dall'Asta, E.; Diotri, F.; di Cella, U.M.; Roncella, R.; Santise, M. Quality assessment of DSMs produced from UAV flights georeferenced with on-board RTK positioning. *Remote Sens.* **2018**, *10*, 311. [CrossRef]
21. Zahidi, I.; Yusuf, B.; Cope, M. Vegetative roughness estimation for hydraulic modelling: A review. *Res. Civ. Environ. Eng.* **2014**, *2*, 1–10.
22. Shakti, P.C.; Hirano, K.; Iizuka, S. Flood inundation mapping of the Hitachi region in the Kuji River basin, Japan, during the October 11–13, 2019 extreme rain event. *J. Disaster Res.* **2020**, *15*, 712–725. [CrossRef]
23. MLITT. National Land Data Information. Ministry of Land, Infrastructure, Transport and Tourism. Available online: https://nlftp.mlit.go.jp/ksj/gml/datalist/KsjTmplt-L03-b_r.html (accessed on 7 August 2022).
24. Moriguchi, S.; Matsugi, H.; Ochiai, T.; Yoshikawa, S.; Inagaki, H.; Ueno, S.; Suzuki, M.; Tobita, Y.; Chida, T.; Takahashi, K.; et al. Survey report on damage caused by 2019 Typhoon Hagibis in Marumori Town, Miyagi Prefecture, Japan. *Soils Found.* **2021**, *61*, 586–599. [CrossRef]
25. Tali, M.G.; Tavakolinia, J.; Heravi, A.M. Flood vulnerability assessment in northwestern areas of Tehran. *J. Disaster Res.* **2016**, *11*, 699–706. [CrossRef]
26. Tanaka, H.; Adityawan, M.B.; Mano, A. Morphological changes at the Nanakita River mouth after the Great East Japan Tsunami of 2011. *Coast. Eng.* **2014**, *86*, 14–26. [CrossRef]
27. Viet, N.T.; Tanaka, H.; Nakayama, D.; Yamaji, H. Effect of morphological changes and waves on salinity intrusion in the Nanakita River mouth. *Proc. Hydraul. Eng.* **2006**, *50*, 139–144. [CrossRef]
28. Pilailar, S.; Sakamaki, T.; Izumi, N.; Tanaka, H.; Nishimura, O. The characteristic change of fine particulate organic matter due to a flood in the Nanakita River. *Proc. Hydraul. Eng.* **2003**, *47*, 1033–1038. [CrossRef]
29. Japan Dam Foundation. Nanakita Dam [Miyagi Pref.]—Dams in Japan. Available online: http://damnet.or.jp/cgi-bin/binranA/enAll.cgi?db4=0302 (accessed on 7 August 2022).
30. Sato, S.; Kure, S.; Moriguchi, S.; Udo, K.; Imamura, F. Online information as real-time big data about heavy rain disasters and its limitations: Case study of Miyagi Prefecture, Japan, during Typhoons 17 and 18 in 2015. *J. Disaster Res.* **2017**, *12*, 335–346. [CrossRef]

31. Nastiti, K.D.; Kim, Y.; Jung, K.; An, H. The application of rainfall-runoff-inundation (RRI) model for inundation case in upper Citarum watershed, West Java-Indonesia. *Procedia Eng.* **2015**, *125*, 166–172. [CrossRef]
32. Bhagabati, S.; Kawasaki, A. Consideration of the rainfall-runoff-inundation (RRI) model for flood mapping in a deltaic area of Myanmar. *Hydrol. Res. Lett.* **2017**, *11*, 155–160. [CrossRef]
33. San, Z.M.L.T.; Zin, W.W.; Kawasaki, A.; Acierto, R.A.; Oo, T.Z. Developing flood inundation map using RRI and SOBEK models: A case study of the Bago River basin, Myanmar. *J. Disaster Res.* **2020**, *15*, 277–287. [CrossRef]
34. Sayama, T.; Ozawa, G.; Kawakami, T.; Nabesaka, S.; Fukami, K. Rainfall–runoff–inundation analysis of the 2010 Pakistan flood in the Kabul River basin. *Hydrol. Sci. J.* **2012**, *57*, 298–312. [CrossRef]
35. Ishizaki, H.; Matsuyama, H. Distribution of the annual precipitation ratio of radar/raingauge-analyzed precipitation to AMeDAS across Japan. *SOLA* **2018**, *14*, 192–196. [CrossRef]
36. Yamazaki, D.; Ikeshima, D.; Sosa, J.; Bates, P.D.; Allen, G.H.; Pavelsky, T.M. MERIT hydro: A high-resolution global hydrography map based on latest topography dataset. *Water Resour. Res.* **2019**, *55*, 5053–5073. [CrossRef]
37. Iwasa, Y.; Inoue, K. Mathematical simulations of channel and overland flood flows in view of flood disaster engineering. *J. Nat. Disaster Sci.* **1982**, *4*, 1–30.
38. Inoue, K.; Nakagawa, H.; Toda, K. Numerical analysis of overland flood flows by means of one-and two-dimensional models. In Proceedings of the 5th JSPS-VCC Seminar on Integrated Engineering, Engineering Achievement and Challenges, Johor Bahru, Malaysia, 12–14 November 1994; pp. 388–397.
39. Hashimoto, M.; Yoneyama, N.; Kawaike, K.; Deguchi, T.; Hossain, M.A.; Nakagawa, H. Flood and substance transportation analysis using satellite elevation data: A case study in Dhaka City, Bangladesh. *J. Disaster Res.* **2018**, *13*, 967–977. [CrossRef]
40. Japan Institute of Country-ology and Engineering. Manual of plans for river channel (in Japanese). In *Manual of Plans for River Channel*; Sankaidou: Tokyo, Japan, 2002.
41. Arcement, G.J.; Schneider, V.R. *Guide for Selecting Manning's Roughness Coefficients for Natural Channels and Flood Plains*; Water Supply Paper 2339; United States Federal Highway Administration: Washington, DC, USA, 1989; pp. 1–44. [CrossRef]
42. Weidner, U.; Förstner, W. Towards automatic building extraction from high-resolution digital elevation models. *ISPRS J. Photogramm. Remote Sens.* **1995**, *50*, 38–49. [CrossRef]
43. de Doncker, L.; Troch, P.; Verhoeven, R.; Bal, K.; Meire, P.; Quintelier, J. Determination of the Manning roughness coefficient influenced by vegetation in the river Aa and Biebrza river. *Environ. Fluid Mech.* **2009**, *9*, 549–567. [CrossRef]

water

MDPI

Article

An Analytical Solution to Predict the Distribution of Streamwise Flow Velocity in an Ecological River with Submerged Vegetation

Jiao Zhang [1], Zhangyi Mi [1], Wen Wang [1,*], Zhanbin Li [1], Huilin Wang [2], Qingjing Wang [3], Xunle Zhang [4] and Xinchun Du [4]

1 State Key Laboratory of Eco-Hydraulics in Northwest Arid Region, Xi'an University of Technology, Xi'an 710048, China
2 College of Water Conservancy and Civil Engineering, South China Agricultural University, Guangzhou 510642, China
3 Bureau of Hydrology, Changjiang Water Resources Commission, Wuhan 430010, China
4 Anhui Water Resources Development Co., Ltd., Hefei 230000, China
* Correspondence: wangwen1986@xaut.edu.cn

Abstract: Aquatic submerged vegetation is widespread in rivers. The transverse distribution of flow velocity in rivers is altered because of the vegetation. Based on the vegetation coverage, the cross-section of the ecological channels can be divided into the non-vegetated area and the vegetated area. In the vegetated area, we defined two depth-averaged velocities, which included the water depth-averaged velocity, and the vegetation height-averaged velocity. In this study, we optimized the ratio of these two depth-averaged velocities, and used this velocity ratio in the Navier–Stokes equation to predict the lateral distribution of longitudinal velocity in the open channel that was partially covered by submerged vegetation. Based on the Navier–Stokes equations, the term "vegetation resistance" was introduced in the vegetated area. The equations for the transverse eddy viscosity coefficient ξ, friction coefficient f, drag force coefficient C_d, and porosity α were used for both the non-vegetated area and the vegetated area, and the range of the depth-averaged secondary flow coefficient was investigated. An analytical solution for predicting the transverse distribution of the water depth-averaged streamwise velocity was obtained in channels that were partially covered by submerged vegetation, which was experimentally verified in previous studies. Additionally, the improved ratio proposed here was compared to previous ratios from other studies. Our findings showed that the ratio in this study could perform velocity prediction more effectively in the partially covered vegetated channel, with a maximum average relative error of 4.77%. The improved ratio model reduced the number of parameters, which introduced the diameter of the vegetation, the amount of vegetation per unit area, and the flow depth. This theoretical ratio lays the foundation for analyzing the flow structure of submerged vegetation.

Keywords: submerged vegetation; water depth-averaged streamwise velocity; Navier–Stokes equations; secondary flow coefficient; analytical solution; velocity ratio

Citation: Zhang, J.; Mi, Z.; Wang, W.; Li, Z.; Wang, H.; Wang, Q.; Zhang, X.; Du, X. An Analytical Solution to Predict the Distribution of Streamwise Flow Velocity in an Ecological River with Submerged Vegetation. *Water* **2022**, *14*, 3562. https://doi.org/10.3390/w14213562

Academic Editors: Hossein Afzalimehr and Jueyi Sui

Received: 13 September 2022
Accepted: 31 October 2022
Published: 5 November 2022

1. Introduction

Aquatic vegetation strongly influences ecosystem functions and improves water quality [1] by filtering nutrients, producing oxygen, and capturing suspended sediments. Aquatic vegetation also provides shelter to organisms in the river, providing them with nutrient-rich foods and spacious habitats, promoting the formation of natural food chains, and improving the ecology of the river [2]. Besides having ecological effects, aquatic vegetation also affects the flow structure owing to its drag force slowing the flow velocity. As a result of the large difference in velocity between vegetated and non-vegetated areas, shear force is generated at the junction of the two areas, and the turbulent flow structure becomes more complicated, ultimately affecting sediment transport and riverbed stability. Therefore, it is essential to study the effect of vegetation on flow velocity.

Several studies have investigated the flow structure in channels with vegetation, and have drawn various and meaningful conclusions. The type of vegetation can be divided into rigid vegetation and flexible vegetation according to its flexibility. Meanwhile, based on the relationship between vegetation height and flow depth, vegetation is divided into submerged and emergent vegetation. The submerged vegetation investigated in this study is similar to emergent vegetation, and the prediction of flow velocity distribution with emergent vegetation provides a reference for our research. Scholle and Aksel [3] presented a clear analytical solution of longitudinal velocity in an inclined channel with visco-capillary flow using the Navier–Stokes equations. White and Nepf [4] presented a method to predict the distribution of velocity and shear stress in shallow channels with a boundary of emergent vegetation. Using the vortex characteristics to predict momentum exchange, the model captured a two-layer structure that consisted of a rapidly varying shear layer across the vegetation interface, and a more gradual boundary layer in the main channel. Terrier [5] improved the Shiono and Knight method (SKM) [6] to account for the increase in turbulence activity due to the presence of vegetation. The researchers found that emergent vegetation significantly increased flow resistance, thus reducing velocity, decreasing boundary shear stress, and causing depth-averaged velocity profiles to converge more rapidly along the flume near the main channel–floodplain interface, compared to vegetation-free areas [7]. Regarding the influence of emergent vegetation on flow velocity, Liu et al. [8] found that near the upstream edge of a patch, lateral flow adjustments led to a decrease in velocity inside the patch, and an increase in velocity in the adjacent bare channel. They proposed a model based on exponential decay, to predict the longitudinal profiles of streamwise velocities upstream of, and inside a patch in the bare channel. Huai et al. [9] preliminarily estimated the secondary flow coefficient K, and proposed a two-dimensional analytical solution to predict the distribution of steady uniform flow velocity. They found that the K-value affected the accuracy of the transverse distribution of the depth-averaged velocity. Later, unlike the traditional emergent vegetation layout, Fu et al. [10] developed an improved analysis model, where the effects of boundary friction, vegetation resistance, lateral shear turbulence, and secondary flow were considered. Then, they predicted the transverse distribution of the water depth-averaged flow velocity in an open-channel flow using floating vegetation islands (FVIs), where the velocity distribution was similar to that of the submerged vegetation.

Concerning submerged vegetation, some researchers first began with rigid vegetation. Devi and Khatua [11] investigated the main channel–floodplain interface in the compound channel when studying the velocity distribution averaged in depth. Based on the friction coefficient calibration, the average absolute deviation of the water depth-averaged streamwise velocity prediction decreased. In order to determine the influence of flow characteristics with submerged vegetation, Sun et al. [12] evaluated the vertical velocity distribution of high and low vegetation on the basis of the height of submerged vegetation. The velocity of low vegetation at the top of the vegetation showed an approximately logarithmic distribution, but not the velocity of high vegetation. The flow velocity below the top of the vegetation was lower than the flow velocity without vegetation, while the flow velocity above the top was higher. This study showed that the transverse distribution of the velocity in the vegetation layer is different at different vegetation heights. Along with the impact of vegetation height on horizontal distribution, Liu et al. [13] also systematically studied the relationship between the secondary flow coefficient and the region. They also applied the model to the compound channel with a fully covered vegetated floodplain, and developed an analytical solution to predict the transverse distribution of the average flow velocity of the water depth. Unlike the artificial cylindrical vegetation previously studied, natural vegetation has a highly inconsistent shape. Yang et al. [14] measured the local flow velocities for different types of submerged vegetation (such as trees, shrubs, and grasses). In the absence of vegetation in the floodplain, all measured flow velocity distributions follow a logarithmic distribution. However, when vegetation is present in the floodplain, the vertical flow velocity has an S-shaped distribution [15,16]. Concerning transverse distribution of velocity with submerged vegetation, more general shapes of velocity profiles could be discussed based on a polynomial expansion, as used in Kowalski et al. [15] and in Koellermeier et al. [16]. Wang et al. [17] investigated the influence of vegetation on the water flow pattern along the vertical direction, and analyzed the dynamic characteristics of water flow passing over submerged vegetation of each subdivided vegetation segment, using the finite analytic method. They obtained the distribution of the water flow velocity within the flow depth range in the vegetation of an arbitrary shape. In addition to dividing the flow region in the transverse direction, Huai et al. [18] proposed a new three-layer model to predict the vertical velocity distribution in an open channel with submerged vegetation. The results showed that the velocity profile is composed of three hydrodynamic regimes (i.e., the upper non-vegetated layer, the outer layer, and the bottom layer within the vegetation). Multi-layer models have gained significant interest in the community of numerical methods for free-surface flows [19]. Based on the equation for the velocity distribution of cylindrical vegetation, some scholars studied the flow characteristics of flexible vegetation. Ghisalberti and Nepf [20] found that submerged flexible vegetation exhibits continuous eddy current oscillations, which can increase the flow velocity in the vegetation layer, and change transverse distribution as a result of the reduction in vegetation resistance. Compared with rigid cylindrical vegetation, the water flow pattern changes more with flexible vegetation. Wang et al. [21] proposed a new vegetation shape function to measure the upstream width. This function helped to obtain an analytical solution for the velocity profile from the momentum equation. The study of the water flow pattern with flexible vegetation led by Sukhodolov and Sukhodolova [22] found that submerged flexible vegetation converted the two-dimensional open-channel flow structure into a complex three-dimensional flow, making it more uneven, or even asymmetric, and improved the two-dimensional transverse velocity distribution. To summarize, the profile of the water depth-averaged streamwise velocity in a vegetated channel has been extensively investigated; moreover, more studies have examined the characteristics of the transverse velocity distribution in open channels that are partially covered by submerged vegetation.

As far as the flow velocity distribution model proposed for a channel with vegetation is concerned, the numerical simulation method has high accuracy. For example, Liu et al. [23] conducted a three-dimensional numerical simulation on the flow field with rigid emergent cylinders based on the hydrodynamic numerical model for free-surface flows, and studied the characteristics of velocity distribution. Yang et al. [24] used 'Fluent', a fluid simulation software, to simulate the fine-resolution of open-channel flow with rigid vegetation, and to determine the velocity distribution in channels with suspended and submerged vegetation in the vegetated area. However, numerical simulation with high precision requires high computer performance and more time. In determining the velocity distribution of flow with vegetation, using the analytical solution of velocity distribution is relatively simple. This method can quickly verify the initially developed model, and eliminate the error in the parameter settings. The analytical solution method may be applied in situations where the mathematical model can easily be expressed. Although the analytical solution method could not directly determine the change in the flow velocity, the results for the flow velocity distribution studied in this article were obtained.

In this study, we optimized the ratio of two depth-averaged velocities, and then applied this ratio in the Navier–Stokes equation to predict the lateral distribution of longitudinal velocity for an open channel partially covered by submerged vegetation. Using this method, the analytical solution for the water depth-averaged streamwise velocity was obtained.

2. Theoretical Analysis

The layout of the submerged vegetation in the open channel flume that we studied is shown in Figure 1. Based on the characteristics of distribution of vegetation, the flume was divided into two sub-areas along the cross-sectional direction, which included the non-vegetated area (I) and the vegetated area (II).

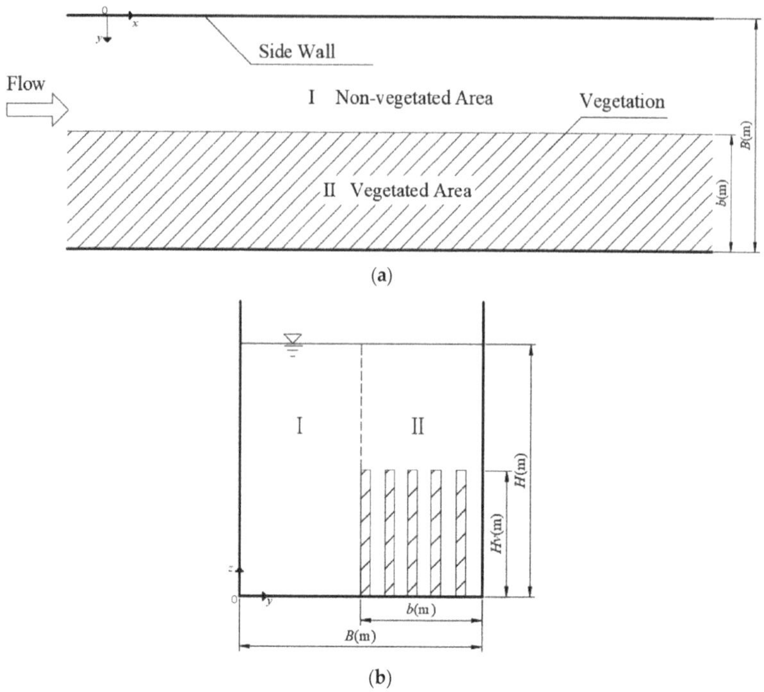

Figure 1. Layout of the vegetation in the flume. (**a**) Top view arrangement; (**b**) arrangement of the cross-section profile. Here, b represents the width of the vegetated area, B represents the width of the flume, H_v represents the height of the vegetation, and H represents the flow depth. I and II represent the non-vegetated area and the vegetated area, respectively.

For open-channel flow with vegetation, to determine the transverse distribution of the water depth-averaged streamwise velocity, a vegetation resistance term was introduced into the Navier–Stokes equation, in order to obtain the governing equation in the x direction, as follows [6]:

$$\rho\left[\frac{\partial(U^2)}{\partial x}+\frac{\partial(UV)}{\partial y}+\frac{\partial(UW)}{\partial z}\right]=\rho g S_0+\frac{\partial\tau_{xx}}{\partial x}+\frac{\partial\tau_{yx}}{\partial y}+\frac{\partial\tau_{zx}}{\partial z}-F_v, \quad (1)$$

where x represents the streamwise coordinate that is parallel to the slope of the bed, y represents the transverse coordinate parallel to the slope of the bed, and z represents the vertical coordinate. U, V, and W represent the time-averaged velocities in the x, y, and z directions. ρ represents the flow density, g represents the gravitational acceleration, and S_0 represents the channel bed slope. Along with the constitutive relations for the stresses, the τ_{xx}, τ_{yx}, and τ_{zx} represent forces acting on the surface of the control body in the derivation of the N–S equations [6]. F_v represents the drag force caused by vegetation per unit volume of a fluid, and can be expressed as follows [10]:

$$F_v=\frac{1}{2}\rho(C_d\beta A_v)U^2, \quad (2)$$

where C_d represents the drag force coefficient of vegetation, β represents the shape factor of vegetation, and A_v represents the projected area of vegetation per unit volume in the direction toward the downstream flow, $A_v = mD$; m represents the amount of vegetation per unit area, and D represents the stem diameter of the vegetation. Equations (1) and (2) were integrated over the flow depth H. As the water flow was uniform, the water surface in the lateral direction was parallel to the channel bed, indicating that the water depth was constant over the entire cross section [13]. $\partial(HU^2)/\partial x \approx 0$, $\partial(H\tau_{xx})/\partial x \approx 0$. Assuming that $W(H) = W(0) = 0$, Equation (1) can be simplified as follows [6]:

$$\rho\frac{\partial H(UV)_d}{\partial y} = \rho g H S_0 + \frac{\partial H\overline{\tau_{yx}}}{\partial y} - \tau_b - \int_0^H \frac{1}{2}\rho(C_d\beta A_v)U^2 dz, \tag{3}$$

where $(UV)_d$ represents the depth-averaged term, and $\tau_b = -\int_0^H \frac{\partial \tau_{zx}}{\partial z}dz$. The integral depth H for the vegetation resistance term can be divided into two parts: $0\sim H_v$ and $H_v\sim H$. H_v represents the height of the vegetation. No vegetation occurs in the range of $H_v\sim H$, and thus, the resistance is 0. The integral term used in Equation (3) can be presented as follows:

$$\begin{aligned}&\int_0^H \tfrac{1}{2}\rho(C_d\beta A_v)U^2 dz\\ &= \int_0^{H_v} \tfrac{1}{2}\rho(C_d\beta A_v)U^2 dz + \int_{H_v}^H \tfrac{1}{2}\rho(C_d\beta A_v)U^2 dz\\ &= \int_0^{H_v} \tfrac{1}{2}\rho(C_d\beta A_v)U^2 dz\\ &= \tfrac{1}{2}\rho(C_d\beta A_v)H_v {U_v}^2,\end{aligned} \tag{4}$$

where the depth-averaged streamwise velocity along the vegetation height is defined as follows:

$$U_v = \frac{1}{H_v}\int_0^{H_v} U dz. \tag{5}$$

In order to solve the governing equation, determining the relationship between the depth-averaged streamwise velocity along the vegetation height U_v and the water depth-averaged streamwise velocity U_d is necessary. Previous studies obtained different results for the relationship between U_v and U_d.

Stone and Shen [25] stated that for vegetation that covers almost half of the river, the relationship between these two velocities can be expressed as follows:

$$\frac{U_v}{U_d} = \sqrt{h^* k_v}, \tag{6}$$

$$k_v = \left(\frac{1 - Dm^{0.5}}{1 - h^* Dm^{0.5}}\right)^2. \tag{7}$$

Here, h^* is the relative height coefficient defined as min $[H_v, H]/H$. Therefore, the value of k_v is 1.0 for the emergent vegetation, and less than 1.0 for submerged vegetation.

Cheng [26] developed Equation (8), as follows:

$$\frac{U_v}{U_d} = \frac{\sqrt{\frac{2r_v}{C_{Dv}}}}{\left[\sqrt{\frac{\pi(1-\lambda)^3 D}{2C_{Dv}\lambda H_v}}\left(\frac{H_v}{H}\right)^{\frac{3}{2}} + 4.54\left(\frac{h_s}{D}\frac{1-\lambda}{\lambda}\right)^{\frac{1}{16}}\left(\frac{h_s}{H}\right)^{\frac{3}{2}}\right]\sqrt{H}}. \tag{8}$$

Here,

$$r_v = \pi(1-\lambda)D/(4\lambda), \tag{9}$$

$$\begin{aligned}C_{Dv} = &\frac{130}{\left[\pi(1-\lambda)D/(4\lambda)(gS_0/v^2)^{1/3}\right]^{0.85}}\\ &+0.8\left[1 - exp\left(-\frac{\pi(1-\lambda)D/(4\lambda)(gS_0/v^2)^{1/3}}{400}\right)\right],\end{aligned} \tag{10}$$

where $\lambda = m\pi D^2/4$, which represents the ratio of the vertical projected area of vegetation to the unit bed area; $h_s = H - H_v$. Regarding flow velocity ratios, in addition to the above-mentioned equations, another equation was presented by Huthoff et al. [27] as follows:

$$\frac{U_v}{U_d} = \frac{\sqrt{H/H_v}}{\sqrt{\frac{H_v}{H} + \frac{H-H_v}{H}\left(\frac{H-H_v}{l}\right)^{\frac{2}{3}\left(1-\left(\frac{H}{H_v}\right)^{-5}\right)}}}, \tag{11}$$

where l is the spacing between adjacent cylindrical vegetation zones.

Using the equations presented by Cheng [26] and Huthoff et al. [27] is a time-consuming process. However, the equation developed by Stone and Shen [25] is simpler than the other formulae. By introducing the diameter of the vegetation, the amount of vegetation per unit area, and the flow depth, the ratio of the depth-averaged streamwise velocity along the vegetation height to the water depth-averaged streamwise velocity for the emergent and submerged vegetation can be obtained. However, the limitation of this formula is that it is more suitable to be calculated on the basis of cylindrical vegetation. Since some parameters are removed, the accuracy of the formula needs to be verified by experiments.

We improved the model for the ratio of two depth-averaged velocities, based on the equation of Stone and Shen [25]. The relationship between the depth-averaged streamwise velocity along the vegetation height U_v and the water depth-averaged streamwise velocity U_d can be determined as follows:

$$\frac{U_v}{U_d} = \sqrt{h^* k_v \frac{H_v}{H}}. \tag{12}$$

By expressing U_v/U_d as φ, Equation (12) can be substituted into Equation (4) to obtain Equation (13), as follows:

$$\frac{1}{2}\rho(C_D\beta A_v)H_v {U_v}^2 = \frac{1}{2}\rho(C_D\beta A_v)H_v\varphi^2 {U_d}^2. \tag{13}$$

Equation (14) was obtained by substituting Equation (13) into Equation (3), as follows:

$$\rho\frac{\partial H(UV)_d}{\partial y} = \rho g H S_0 + \frac{\partial H\overline{\tau_{yx}}}{\partial y} - \tau_b - \frac{1}{2}\rho(C_D\beta A_v)H_v\varphi^2 {U_d}^2. \tag{14}$$

The vegetation porosity α [13], was defined as the volume ratio of fluid per unit volume, based on Equation (14) as follows:

$$\alpha\rho\frac{\partial H(UV)_d}{\partial y} = \alpha\rho g H S_0 + \alpha\frac{\partial H\overline{\tau_{yx}}}{\partial y} - \alpha\tau_b - \frac{1}{2}\rho(C_D\beta A_v)H_v\varphi^2 {U_d}^2. \tag{15}$$

Each term at the right side of Equation (15) shows the forces acting on an H-depth water column per unit of vegetated-bed area, and the left side of Equation (15) is the corresponding inertial force. After introducing the vegetation term, the volume of clear water in each term of the equation changes. The corresponding gravity component in the streamwise direction is $\alpha\rho g H S_0$. When vegetation blockage is considered, the gravity component and other terms in the equation can be obtained by multiplying the terms in Equation (15) by α. However, the drag force due to vegetation is still $\frac{1}{2}\rho(C_D\beta A_v)H_v\varphi^2 {U_d}^2$, indicating that the drag force term remains unchanged [13].

The depth-averaged transverse shear stress $\overline{\tau_{yx}}$ is expressed as the transverse gradient of the water depth-averaged streamwise velocity in Equation (15). The depth-averaged eddy viscosity coefficient $\overline{\varepsilon_{yx}}$ and the local shear velocity U_* are related to the flow depth H. Local shear velocity U_* is expressed as the relationship between the Darcy–Weisbach friction coefficient f and the water depth-averaged streamwise velocity. These parameters are expressed in Equations (16)–(19) [10]:

$$\overline{\tau_{yx}} = \rho\overline{\varepsilon_{yx}}\frac{\partial U_d}{\partial y}, \tag{16}$$

$$\overline{\varepsilon_{yx}} = \xi U_* H, \tag{17}$$

$$U_* = \sqrt{\frac{f}{8}}U_d, \tag{18}$$

$$\tau_b = \left(\frac{f}{8}\right)\rho {U_d}^2. \tag{19}$$

Here, ξ denotes the transverse eddy viscosity coefficient, and τ_b represents the comprehensive shear stress of the boundary. Equations (16)–(19) were substituted into (15) to obtain Equation (20), as follows:

$$\rho\alpha\frac{\partial H(UV)_d}{\partial y} = \rho\alpha g H S_0 + \alpha\frac{\partial}{\partial y}[\rho\xi H^2(\frac{f}{8})^{1/2}U_d\frac{\partial U_d}{\partial y}] - \alpha\rho(\frac{f}{8}){U_d}^2 - \frac{1}{2}\rho(C_d\beta A_v)H_v\varphi^2{U_d}^2. \tag{20}$$

In order to find the secondary flow term on the left side of Equation (20), the expression of $(UV)_d$ proposed by Liu et al. [13] is described as follows:

$$\frac{\partial H(UV)_d}{\partial y} = \frac{\partial H\overline{K}{U_d}^2}{\partial y}. \tag{21}$$

Here, $\overline{K}$ indicates the depth-averaged value of the secondary flow coefficient, which was obtained by integrating in the direction of flow depth.

For prediction of the distribution of vegetation, the solution of Equation (20) can be obtained as follows:

1. For the non-vegetated area, i.e., the area I in Figure 1b, the solutions of Equation (20) for the non-vegetated area and vegetated area are different. This is because, for the non-vegetated area, the drag force coefficient is 0 in Equation (20). Ignoring the vegetation resistance term "$\frac{1}{2}\rho(C_d\beta A_v)H_v\varphi^2{U_d}^2$", U_d is expressed as follows:

$${U_d}^{(1)} = \left[A_1 e^{\beta_1 y} + C_1 e^{\gamma_1 y} + \omega_1\right]^{\frac{1}{2}}, \tag{22}$$

where

$$\beta_1 = \frac{1}{\xi H}\left(\frac{8}{f_1}\right)^{\frac{1}{2}}(\overline{K_1} + \sqrt{\overline{K_1}^2 + \xi\frac{f_1}{4}\left(\frac{f_1}{8}\right)^{\frac{1}{2}}}),\ \gamma_1 = \frac{1}{\xi H}\left(\frac{8}{f_1}\right)^{\frac{1}{2}}(\overline{K_1} - \sqrt{\overline{K_1}^2 + \xi\frac{f_1}{4}\left(\frac{f_1}{8}\right)^{\frac{1}{2}}}),\ \omega_1 = \frac{8gHS_0}{f_1}. \tag{23}$$

2. For the vegetated area, i.e., area II in Figure 1b, U_d is expressed as follows:

$${U_d}^{(2)} = \left[A_2 e^{\beta_2 y} + C_2 e^{\gamma_2 y} + \omega_2\right]^{\frac{1}{2}}, \tag{24}$$

where

$$\beta_2 = \frac{1}{\xi H_f}\left(\frac{8}{f_2}\right)^{\frac{1}{2}}(\overline{K_2} + \sqrt{\overline{K_2}^2 + \xi(\frac{f_2}{4} + \frac{1}{\alpha}C_d\beta m D\varphi^2 H_v)\left(\frac{f_2}{8}\right)^{\frac{1}{2}}}, \tag{25}$$

$$\gamma_2 = \frac{1}{\xi H_f}\left(\frac{8}{f_2}\right)^{\frac{1}{2}}(\overline{K_2} - \sqrt{\overline{K_2}^2 + \xi(\frac{f_2}{4} + \frac{1}{\alpha}C_d\beta m D\varphi^2 H_v)\left(\frac{f_2}{8}\right)^{\frac{1}{2}}}),\ \omega_2 = \frac{gHS_0}{\frac{f_2}{8} + \frac{1}{2\alpha}C_d\beta m D\varphi^2 H_v}. \tag{26}$$

Here, A_1, C_1, A_2, and C_2 are unknown constants. The superscripts (1) and (2) indicate the non-vegetated area and vegetated area, respectively.

3. Boundary Conditions

In order to obtain the unknown constants A_1, C_1, A_2, and C_2 in Equations (22) and (24), four boundary conditions are required, which are described as follows:

(1) On the side wall, the no-slip boundary condition is present, and for the velocity near the side wall, i.e., when y = 0 and y = B, U_d = 0 (two boundary conditions).
(2) The velocity continuity condition exists at the junction between the non-vegetated area and the vegetated area, i.e., when y = B, $U_d(i) = U_d(i+1)$.
(3) The stress continuity condition is present when the water flow is uniform at the junction between the non-vegetated and vegetated areas. Thus, the flow depth transition is not abrupt at the junction between the areas. The stress continuity condition can be expressed as follows:

$$\left(\frac{\partial U_d}{\partial y}\right)^{(i)} = \left(\frac{\partial U_d}{\partial y}\right)^{(i+1)}. \tag{27}$$

Here, i = 1, 2, indicating the non-vegetated area and the vegetated area, respectively.

4. Parameter Determination

In order to obtain the analytical solutions of Equations (22) and (24), finding the model parameters ($\xi, f, \alpha, C_d, \overline{K_1}$, and $\overline{K_2}$) is necessary. The values of these parameters are generally different in the non-vegetated and vegetated areas. They can be calculated using the methods described below.

4.1. Transverse Eddy Viscosity Coefficient ξ

The method to calculate the transverse eddy viscosity coefficient was proposed by Abril and Knight [28], and Pasche and Rouvé [29], who presented as follows:

$$\xi_{non-vegetated} = \mathcal{K}/6, \tag{28}$$

$$\xi_{vegetated} = \left(-0.2 + 1.2{D_r}^{-1.44}\right), \tag{29}$$

$$D_r = H_v/H, \tag{30}$$

where $\xi_{non-vegetated}$ refers to the transverse eddy viscosity coefficient in the non-vegetated area, and $\xi_{vegetated}$ refers to the transverse eddy viscosity coefficient in the vegetated area. $\mathcal{K}$ denotes the Karman constant, and is usually 0.4. D_r is defined as the relative depth ratio, expressed as the ratio of vegetation depth to the water depth.

4.2. Darcy–Weisbach Friction Coefficient f

The friction coefficient was obtained by the formula that was developed by Rameshwaran and Shiono [30], as follows:

$$f = \left[-2\log\left(\frac{3.02v}{\sqrt{128H^3S_0}} + \frac{k_s}{\phi H}\right)\right]^{-2}, \tag{31}$$

where v represents the kinematic viscosity, and has a value of 1×10^{-6} m^2/s, and ϕ = {12.3, 1.2} in the non-vegetated and vegetated areas, respectively; k_s represents the equivalent roughness height, and can be calculated using the equation developed by Ackers [31], as follows:

$$k_s = (8.25n\sqrt{g})^6, \tag{32}$$

where n represents Manning's roughness coefficient. According to Naot et al. [32], n = 0.013 (concrete material), and according to Shi and Huai [33], n = 0.01 (glass material).

4.3. Porosity α

Porosity α is expressed as the volume ratio of fluid per unit volume in the vegetated area, and is calculated as follows:

$$\alpha = 1 - V_{vegetation}/V_{column}, \quad (33)$$

where $V_{vegetation}$ refers to the volume occupied by vegetation per unit of the water body. V_{column} represents the unit of water volume, and is expressed as follows:

$$V_{column} = 1 \times 1 \times H. \quad (34)$$

4.4. The Drag Force Coefficient C_d

The drag force coefficient C_d, described by Liu et al. [13], is related to the Reynolds number (*Re*), vegetation shape, and vegetation density. The drag force coefficient C_d decreases with an increase in the Reynolds number of simulated cylindrical vegetation, and increases with an increase in the volume fraction of vegetation [34]. James et al. [35] measured the drag force coefficient of cylindrical vegetation when $200 < Re < 10{,}000$, and found that C_d fluctuated around 1. The C_d value increases significantly with an increase in the number of leaves of cylindrical vegetation. The Reynolds numbers calculated for each case in this study are detailed in Table 1. The drag force coefficient C_d was considered to be 1 in this study.

Table 1. Summary of the experimental conditions.

Sources	Cases	H (m)	H_v (m)	D (m)	m (m^{-2})	β	$\overline{K_1}$	$\overline{K_2}$	Re
Naot et al. [32]	1	0.06	0.03	0.0036	278	0.51	−0.0009	0.15	9000
	2	0.06	0.03	0.0036	1111	0.43	−0.001	0.06	9000
	3	0.06	0.03	0.0036	4444	0.26	−0.001	0.06	9000
Shi and Huai [33]	4	0.31	0.25	0.008	400	1	−0.06	0.06	8550

4.5. Secondary Flow Coefficients $\overline{K_1}$ and $\overline{K_2}$

As a result of the differences in velocity between the vegetated and non-vegetated areas, a transverse eddy current, called the secondary flow, occurs at the junction between two areas. It is expressed as $\overline{K_1}$ and $\overline{K_2}$ for the non-vegetated area and the vegetated area, respectively. The $\overline{K}$-value is calculated, following the formulae provided by Liu et al. [13]. Generally, the values are negative in the main channel, and positive in the floodplain. In the Discussion, we also explain the effect of the $\overline{K}$-value to analytical solutions.

5. Experimental Data

In order to validate the applicability and accuracy of the analytical solutions, we used the experimental data that were obtained by Naot et al. [32] and Shi and Huai [33].

5.1. Experimental Data Obtained by Naot et al. [32]

The experiment was conducted by simulating an open channel. The channel was 0.36 m wide, with a slope S_0 of 0.0064. The cylindrical diameter D of the simulated rigid vegetation was 0.0036 m. The height of vegetation H_v was 0.03 m, and the flow depth H was 0.06 m. The non-dimensional vegetation density formula, defined as $N = mHD$, was proposed by Naot et al. [32] and used only in this case. The experiments were conducted with non-dimensional vegetation densities of $N = 0.06$, $N = 0.24$, and $N = 0.96$.

5.2. Experimental Data Obtained by Shi and Huai [33]

The study area had submerged vegetation that covered half of the river. The flume used in the experiment was 18 m long and B was 1 m wide, with an 8-meter-long section made from glass for observation; the vegetated area was 0.475 m wide. The vegetation was simulated using plexiglass rods. In the experiment, acoustic Doppler velocity (ADV) was used to measure a series of velocities at each point on each y-coordinate vertical line; then, the average was calculated as the water depth-averaged streamwise velocity of the y-coordinate. The open channel was equipped with an adjustable tailgate to adjust the water depth, leading to the water depth being parallel to the channel bed in both the longitudinal and lateral directions.

The experimental conditions established by Naot et al. [32] and Shi and Huai [33] are summarized in Table 1.

6. Comparison of Theoretical and Experimental Data

Equation (22) was used to predict the velocity distribution in the non-vegetated area, and Equation (24) was used to predict the velocity distribution in the vegetated area. A comparison was made between the experimental and analytical solutions for the water depth-averaged streamwise velocity under three cases, i.e., $N = 0.06$, $N = 0.24$, and $N = 0.96$, as proposed by Naot et al. [32], which are shown in Figures 2–4. The comparison between the experimental and analytical solutions provided by Shi and Huai [33], as calculated in case 4, is shown in Figure 5.

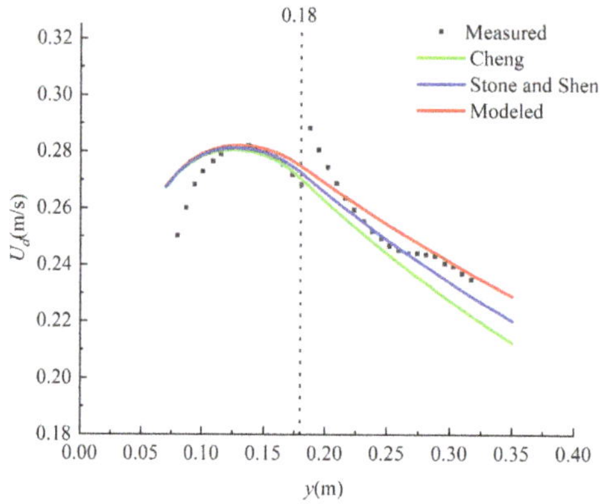

Figure 2. Water depth-averaged streamwise velocity between the experimental data and analytical solution in case 1. The green line occurs when $\varphi_1 = 0.8279$ (from Cheng [26]). The blue line occurs when $\varphi_2 = 0.6914$ (from Stone and Shen [25]). The red line occurs when $\varphi_3 = 0.4845$ (from the presented model ($\varphi = U_v/U_d$)). The dotted line indicates the boundary between the non-vegetated and vegetated areas.

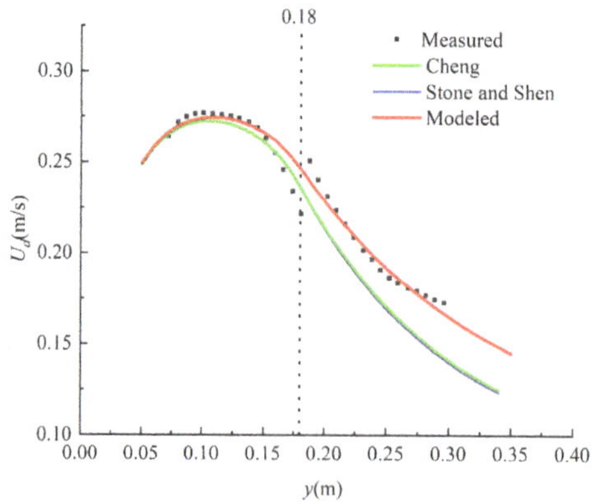

Figure 3. The water depth-averaged streamwise velocity between the experimental data and analytical solution in case 2. $\varphi_1 = 0.6522$, $\varphi_2 = 0.662$, and $\varphi_3 = 0.4681$.

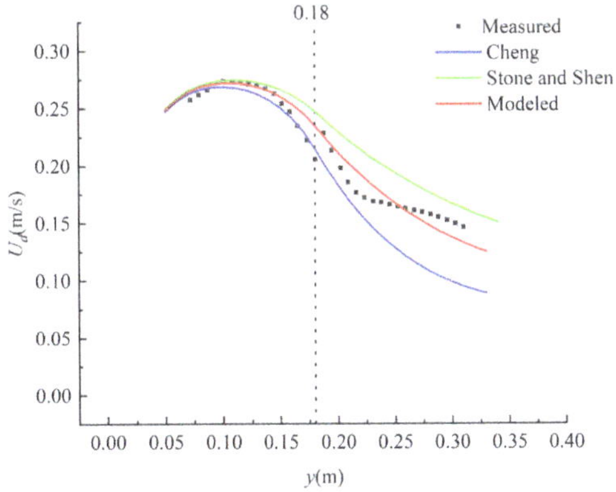

Figure 4. Water depth-averaged streamwise velocity between the experimental data and analytical solution in case 3. $\varphi_1 = 0.3044$, $\varphi_2 = 0.6107$, and $\varphi_3 = 0.4318$.

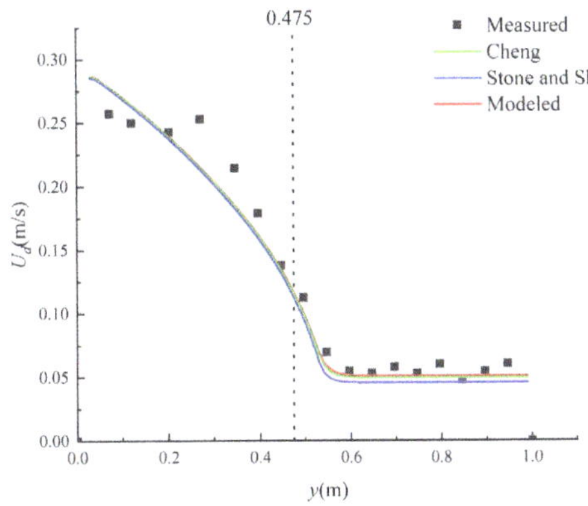

Figure 5. Water depth-averaged streamwise velocity between the experimental data and analytical solution in case 4; $\varphi_1 = 0.8013$, $\varphi_2 = 0.8685$, and $\varphi_3 = 0.7778$.

7. Discussion

The validation of the model using the experimental data obtained by Naot et al. [32] and Shi and Huai [33] showed that when vegetation density is high, the equations proposed by Cheng [26] and Stone and Shen [25] usually yielded relatively larger values. In cases 1 to 3, the analytical solution in the non-vegetated area is relatively consistent with the experimental data. In the vegetated area, there are some deviations between the analytical solution and experimental data, while in case 4, the analytical solution is relatively consistent with the experimental data. Optimizing of the model improved it for the ratio of two depth-averaged velocities, i.e., U_v/U_d. Table 1 shows that in four cases when the flow depth increased from 0.06 m to 0.31 m, the $\overline{K_1}$ values changed from -0.001 to -0.06, indicating that the absolute value of $\overline{K_1}$ increased with an increase in flow depth. In order to quantitatively describe the difference between the results of the model and the experimental data, we performed an error analysis from two perspectives: the average values of the absolute error $\overline{\varepsilon}$, and the relative error $\overline{\varepsilon'}$. The absolute error ε is expressed as follows:

$$\varepsilon = \left| U_{d\_measured} - U_{d\_calculated} \right|, \tag{35}$$

where the subscripts "measured" and "calculated" represent the measured values and the analytical solutions, respectively. In order to obtain the average error, it is necessary to calculate the average value of the error for each point along the y direction on the cross section. The average value of the absolute error $\bar{\varepsilon}$ can be expressed as follows:

$$\bar{\varepsilon} = \frac{1}{N_m}\sum_{i=1}^{N_m}|\varepsilon_i|. \tag{36}$$

Here, N_m represents the number of experimental measurement points. The relative error ε' is expressed as follows:

$$\varepsilon' = \frac{\varepsilon}{U_{d\_measured}}. \tag{37}$$

The average value of the relative error $\overline{\varepsilon'}$ is expressed as follows:

$$\overline{\varepsilon'} = \frac{1}{N_m}\sum_{i=1}^{N_m}|\varepsilon_i'|. \tag{38}$$

The average absolute error and relative error values in cases 1–4 are shown in Table 2.

Table 2. Error statistics of the transverse distribution of the water depth-averaged streamwise velocity calculated by the models.

Sources	The Average Value of Error	Cases			
		Case 1	Case 2	Case 3	Case 4
Cheng [26]	$\bar{\varepsilon}(m/s)$	0.0072	0.0125	0.0191	0.0045
	$\overline{\varepsilon'}(\%)$	2.8	6.15	10.3	5.47
Stone and Shen [25]	$\bar{\varepsilon}(m/s)$	0.0047	0.0131	0.02	0.0059
	$\overline{\varepsilon'}(\%)$	1.81	6.47	11.7	8.04
Present Model	$\bar{\varepsilon}(m/s)$	0.0048	0.0041	0.0080	0.004
	$\overline{\varepsilon'}(\%)$	1.85	1.84	4.34	4.77

The average values of the absolute error and the relative error for the presented model were the smallest (Table 2). For cases 1 to 4, the average absolute error $\bar{\varepsilon}$ in this model was within 0.008. The maximum value of $\bar{\varepsilon}$ was 0.008 (in case 3). The minimum value of $\bar{\varepsilon}$ was 0.004 (in case 4). However, the average value of the relative error $\overline{\varepsilon'}$ was less than 5%. The maximum value of $\overline{\varepsilon'}$ was 4.77% (in case 4), and the minimum value of $\overline{\varepsilon'}$ was 1.84% (in case 2).

The variation in the values of the secondary flow coefficients $\overline{K_1}$ and $\overline{K_2}$, used for the calculation, are shown in Table 3. In four cases, when the flow depth increased from 0.06 m to 0.31 m, the $\overline{K_1}$ values changed from −0.001 to −0.06, indicating a positive relationship between the absolute value of $\overline{K_1}$ and the flow depth; meanwhile, the difference in the absolute value of $\overline{K_2}$ was small when the flow depth increased in case 4, compared to the value of $\overline{K_2}$ in cases 1 to 3. When the depth-averaged velocity was calculated, we found that the value of the secondary flow coefficient had a greater effect on the results of analytical solution. Liu et al. [13] proposed that the results of analytical solutions were unsatisfactory, especially in the non-vegetated main channel, when the effect of secondary flow coefficient was ignored or it was considered to be constant. In order to evaluate the importance of the values of $\overline{K_1}$ and $\overline{K_2}$ in the calculation model, we considered four cases to modify the sign of one of the coefficients and ignore the coefficient $\overline{K_1} = \overline{K_2} = 0$. We selected cases 2 and 4 for the specific modification methods (Table 3). We found that the secondary flow coefficient greatly influenced the model results, as shown in Figure 6. In case 2, modifying $\overline{K_2}$ and ignoring the secondary flow coefficient greatly affected the results of analytical solution. In case 4, however, modifying $\overline{K_1}$ and ignoring the secondary flow coefficient strongly influenced the results of the analytical solution. Therefore, calibrating the secondary flow coefficient in different regions and different cases might affect the calculations in other regions. Additionally, for the floodplain, the intensity and range of secondary flow coefficient increased with an increase in the elevation of the floodplain [36]; thus, it should be considered in the calculation.

Table 3. Summary of the modification of the $\overline{K}$-value.

Sources	$\overline{K}$			
	Present Model	Modify $\overline{K_1}$	Modify $\overline{K_2}$	Ignore $\overline{K_1}$, $\overline{K_2}$
Case 2	$\overline{K_1} = -0.001$ $\overline{K_2} = 0.06$	$\overline{K_1} = 0.001$ $\overline{K_2} = 0.06$	$\overline{K_1} = -0.001$ $\overline{K_2} = -0.06$	$\overline{K_1} = \overline{K_2} = 0$
Case 4	$\overline{K_1} = -0.06$ $\overline{K_2} = 0.06$	$\overline{K_1} = 0.06$ $\overline{K_2} = 0.06$	$\overline{K_1} = -0.06$ $\overline{K_2} = -0.06$	$\overline{K_1} = \overline{K_2} = 0$

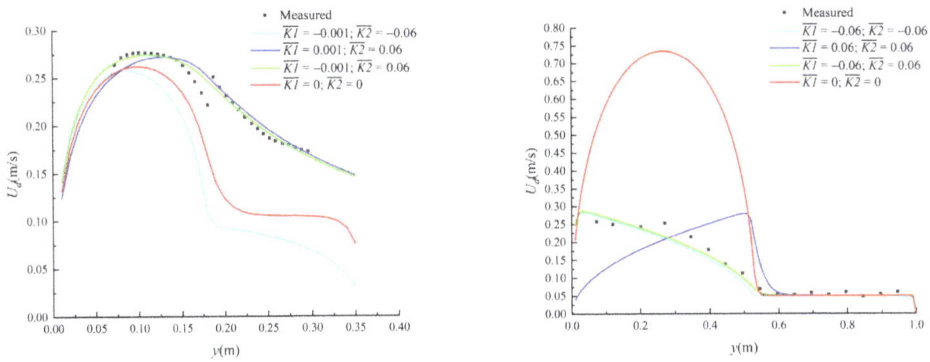

(**a**) (**b**)

Figure 6. Effect of the secondary flow coefficient, $\overline{K}$, on the results of the model for the prediction of the transverse distribution of water depth-averaged streamwise velocity with submerged vegetation. (**a**) For case 2, in the presented model, $\overline{K_1} = -0.001$ and $\overline{K_2} = 0.06$. (**b**) For case 4, in the presented model, $\overline{K_1} = -0.06$ and $\overline{K_2} = 0.06$.

We found that the experimental data obtained by Naot et al. [32] showed an abrupt change at the junction between the non-vegetated area and the vegetated area, as shown in Figures 2–4. The reason for this may be because the flow velocity data in the vegetation layer were not measured, and only the flow velocity data above the vegetation layer were measured. As a result, the velocity inside the vegetation layer is smaller than that outside the vegetation layer, and the depth-averaged streamwise velocity above the vegetation is larger than the water depth-averaged velocity; this leads to a sudden increase at the junction ($y = 0.18$ m). When the velocity continuity condition was adopted in the present analytical solutions, there was no sudden increase at the junction. The study shows that some deviations between the analytical solution and the experimental data can be easily found in cases 1 to 3. In these three cases, the depth-averaged streamwise velocity above the vegetation was adopted for the vegetated area, and the water depth-averaged streamwise velocity was adopted for the non-vegetated area and the junction (Figure 1b) [37]. For the analysis of the measurement of the experimental data obtained by Shi and Huai [33], the measurement points were arranged in the vegetation layer and above the vegetation layer. Therefore, at the junction between the regions, the water depth-averaged streamwise velocity did not show an abrupt change.

8. Conclusions

This paper optimizes the ratio of the vegetation height-averaged velocity to the water depth-averaged velocity, which can be applied to the N–S equation. The N–S equation with optimized velocity ratio can predict the lateral distribution of the water depth-averaged streamwise velocity in the open channel partially covered by the submerged vegetation. In both the non-vegetated area and submerged vegetation area, different parameters, including the transverse eddy viscosity coefficient ξ, friction coefficient f, porosity α, and the drag force coefficient C_d, were introduced to determine the analytical solution. Additionally, we discussed the methods for calculating the parameters in different zones. Close inspection of the range of the secondary flow coefficient showed that the secondary flow coefficient $\overline{K}$ could not be ignored in different regions. $\overline{K}$ could be determined by the flow depth and the depth of the vegetation layer. According to the error analysis for the velocity data from the analytical solution and experiments, the average relative error is smaller than those from previous studies, revealing that a relative satisfactory prediction of the transverse distribution of water depth-averaged streamwise velocity in the channel flow with submerged vegetation was obtained using the present model. Additionally, when the flow is not assumed to be uniform, the model parameters may be modified, although further experiments are required to verify this. Future studies may put emphasis on the secondary flow coefficient, and extend the model into flows with flexible vegetation.

Author Contributions: All authors made contributions to the conception and design of the study. Material preparation and data collection and analysis were performed by J.Z., Z.M., W.W., Z.L., H.W., Q.W., X.Z. and X.D. The first draft of the paper was completed by J.Z. and Z.M., and all authors commented on previous versions of the manuscript. All authors have read and agreed to the published version of the manuscript.

Funding: This study received funding by the National Natural Science Foundation of China [grant numbers U2040208; 52109100], the Postdoctoral Research Foundation of China [grant number 2021M702643], and by AnHui Water Resources Decelopment Co.,Ltd. [grant number KY-2021-13].

Institutional Review Board Statement: Not applicable.

Informed Consent Statement: Not applicable.

Conflicts of Interest: The authors have no relevant financial or non-financial interests to disclose.

Notation

The following main parameters were used in this article:

A_v	projected vegetated area per unit volume in the direction of downstream flow
B	width of the flume
b	width of the vegetation layer
C_d	drag force coefficient for vegetation
D	vegetation stem diameter
D_r	relative depth ratio
F_v	drag force
f	Darcy–Weisbach friction coefficient
g	gravitational acceleration
H	flow depth
H_v	height of vegetation
$\overline{K}$	secondary flow coefficient
m	number of vegetation per unit of area
N	non-dimensional vegetation density
S_0	channel bed slope
U_v	depth-averaged streamwise velocity along the vegetation height
U_d	water depth-averaged streamwise velocity
ρ	flow density
α	porosity
β	shape factor of the vegetation
ξ	transverse eddy viscosity coefficient
φ	U_v/U_d
$\overline{\varepsilon}$	average value of absolute error
$\overline{\varepsilon'}$	average value of relative error

References

1. Harvey, J.W.; Noe, G.B.; Larsen, L.G.; Nowacki, D.J.; Mcphillips, L.E. Field flume reveals aquatic vegetation's role in sediment and particulate phosphorus transport in a shallow aquatic ecosystem. *Geomorphology* **2011**, *126*, 297–313. [CrossRef]
2. Cullen-Unsworth, L.; Unsworth, R. Seagrass meadows, ecosystem services, and sustainability. *Environ. Sci. Policy Sustain. Dev.* **2013**, *55*, 14–28. [CrossRef]
3. Scholle, M.; Aksel, N. An exact solution of visco-capillary flow in an inclined channel. *Z. Angew. Math. Phys.* **2001**, *52*, 749–769. [CrossRef]
4. White, B.L.; Nepf, H.M. A vortex-based model of velocity and shear stress in a partially vegetated shallow channel. *Water Resour. Res.* **2008**, *44*, W01412. [CrossRef]
5. Terrier, B. Flow Characteristics in Straight Compound Channels with Vegetation along the Main Channel. Ph.D. Thesis, Loughborough University, Loughborough, UK, 2010.
6. Shiono, K.; Knight, D.W. Turbulent open-channel flows with variable depth across the channel. *J. Fluid Mech.* **1991**, *222*, 617–646. [CrossRef]
7. Liu, C.; Shan, Y. Impact of an emergent model vegetation patch on flow adjustment and velocity. In *Proceedings of the Institution of Civil Engineers—Water Management*; Thomas Telford Ltd.: London, UK, 2022; Volume 175, pp. 55–66. [CrossRef]
8. Liu, C.; Shan, Y.; Sun, W.; Yan, C.; Yang, K. An open channel with an emergent vegetation patch: Predicting the longitudinal profiles of velocities based on exponential decay. *J. Hydrol.* **2020**, *582*, 124429. [CrossRef]
9. Huai, W.; Gao, M.; Zeng, Y.; Li, D. Two-dimensional analytical solution for compound channel flows with vegetated floodplains. *Appl. Math. Mech.* **2009**, *30*, 1049–1056. [CrossRef]
10. Fu, X.; Wang, F.; Liu, M.; Huai, W. Transverse distribution of the streamwise velocity for the open-channel flow with floating vegetated islands. *Environ. Sci. Pollut. Res. Int.* **2021**, *28*, 51265–51277. [CrossRef]
11. Devi, K.; Khatua, K.K. Prediction of depth averaged velocity and boundary shear distribution of a compound channel based on the mixing layer theory. *Flow Meas. Instrum.* **2016**, *50*, 147–157. [CrossRef]
12. Sun, Z.; Zheng, J.; Zhu, L.; Chong, L.; Liu, J.; Luo, J. Influence of submerged vegetation on flow structure and sediment deposition. *J. Zhejiang Univ. (Eng. Sci.)* **2021**, *55*, 71–80. [CrossRef]
13. Liu, C.; Luo, X.; Liu, X.; Yang, K. Modeling depth-averaged velocity and bed shear stress in compound channels with emergent and submerged vegetation. *Adv. Water Resour.* **2013**, *60*, 148–159. [CrossRef]
14. Yang, K.; Cao, S.; Knight, D.W. Flow patterns in compound channels with vegetated floodplains. *J. Hydraul. Eng.* **2007**, *133*, 148–159. [CrossRef]
15. Kowalski, J.; Torrilhon, M. Moment approximations and model cascades for shallow flow. *Commun. Comput. Phys.* **2019**, *25*, 669–702. [CrossRef]
16. Koellermeier, J.; Rominger, M. Analysis and numerical simulation of hyperbolic shallow water moment equations. *Commun. Comput. Phys.* **2020**, *28*, 1038–1084. [CrossRef]
17. Wang, W.; Liu, Z.; Chen, Y.; Zhu, D. Vertical profile of horizontal velocity in the flow with submerged vegetation. *J. Sichuan Univ. (Eng. Sci. Ed.)* **2012**, *44*, 253–257. [CrossRef]
18. Huai, W.; Zeng, Y.; Xu, Z.; Yang, Z. Three-layer model for vertical velocity distribution in open channel flow with submerged rigid vegetation. *Adv. Water Resour.* **2009**, *32*, 487–492. [CrossRef]
19. Castro, M.J.; Macias, J.; Pares, C. A q-scheme for a class of systems of coupled conservation laws with source term. application to a two-layer 1-d shallow water system. *ESAIM: Math. Model. Numer. Anal.* **2001**, *35*, 107–127. [CrossRef]
20. Ghisalberti, M.; Nepf, H. Shallow flows over a permeable medium: The hydrodynamics of submerged aquatic canopies. *Transp. Porous Media* **2009**, *78*, 309. [CrossRef]
21. Wang, W.; Huai, W.; Li, S.; Wang, P.; Wang, Y.; Zhang, J. Analytical solutions of velocity profile in flow through submerged vegetation with variable frontal width. *J. Hydrol.* **2019**, *578*, 124088. [CrossRef]
22. Sukhodolov, A.N.; Sukhodolova, T.A. Case study: Effect of submerged aquatic plants on turbulence structure in a Lowland River. *J. Hydraul. Eng.* **2010**, *136*, 434–446. [CrossRef]
23. Liu, H.; Zhang, J.; Hu, T. Analysis of local flow field characteristics of non-submerged cylinder group. *Shanghai Jiao Tong Univ.* **2016**, *31*, 161–170.
24. Yang, Y.; Ma, Y.; Zhan, Z.; Fang, S.L.; Zhang, M. Fine numerical simulation of three-dimensional hydrodynamics in vegetation area under submerged and floating vegetation. In *Proceedings of the 31st National Symposium on Hydrodynamics*; Ocean Press: Lancing, UK, 2020; Volume II.
25. Stone, B.M.; Shen, H. Hydraulic resistance of flow in channels with cylindrical roughness. *J. Hydraul. Eng.* **2002**, *128*, 500–506. [CrossRef]
26. Cheng, N.S. Representative roughness height of submerged vegetation. *Water Resour. Res.* **2011**, *47*, W08517. [CrossRef]
27. Huthoff, F.; Augustijn, D.C.M.; Hulscher, S.J.M.H. Analytical solution of the depth-averaged flow velocity in case of submerged rigid cylindrical vegetation. *Water Resour. Res.* **2007**, *43*. [CrossRef]
28. Abril, J.B.; Knight, D.W. Stage-discharge prediction for rivers in flood applying a depth-averaged model. *J. Hydraul. Res.* **2004**, *42*, 616–629. [CrossRef]
29. Pasche, E.; Rouvé, G. Overbank flow with vegetatively roughened flood plains. *J. Hydraul. Eng.* **1985**, *111*, 1262–1278. [CrossRef]

30. Rameshwaran, P.; Shiono, K. Quasi two-dimensional model for straight overbank flows through emergent. *J. Hydraul. Res.* **2007**, *45*, 302–315. [CrossRef]
31. Ackers, P. Hydraulic design of straight compound channels. *Detail. Dev. Des. Method* **1991**, *2*, 1–139.
32. Naot, D. Hydrodynamic behavior of partly vegetated open channels. *J. Hydraul. Eng.* **1996**, *122*, 625–633. [CrossRef]
33. Shi, H.; Huai, W. Two-Dimensional Analytical Solution for a Compound Open Channel Flow with Submerged Vegetation. *Sciencepaper Online*. 2016. Available online: http://www.paper.edu.cn/releasepaper/content/201612-315 (accessed on 12 September 2022).
34. Tanino, Y.; Nepf, H.M. Laboratory investigation of mean drag in a random array of rigid, emergent cylinders. *J. Hydraul. Eng.* **2008**, *134*, 34–41. [CrossRef]
35. James, C.S.; Birkhead, A.L.; Jordanova, A.A.; O'Sullivan, J.J. Flow resistance of emergent vegetation. *J. Hydraul. Res.* **2004**, *42*, 390–398. [CrossRef]
36. Wang, C.; Zhang, H. Hydrodynamic and mixing characteristics of a river confluence with floodplain. *J. Hohai Univ. (Nat. Sci.)* **2022**, 1–13. Available online: http://kns.cnki.net/kcms/detail/32.1117.TV.20220628.1720.008.html (accessed on 12 September 2022).
37. Zhang, M. Study on Flow Characteristics of Compound Channel with Vegetation Floodplain. Ph.D. Thesis, Tsinghua University, Beijing, China, 2011.

water

MDPI

Article

A Comparative Study on 2D CFD Simulation of Flow Structure in an Open Channel with an Emerged Vegetation Patch Based on Different RANS Turbulence Models

Songli Yu [1], Huichao Dai [2], Yanwei Zhai [2], Mengyang Liu [3] and Wenxin Huai [1,*]

1 State Key Laboratory of Water Resources and Hydropower Engineering Science, Wuhan University, Wuhan 430072, China
2 Science and Technology Research Institute, China Three Gorges Corporation, Beijing 100038, China
3 State Key Laboratory of Hydrology-Water Resources and Hydraulic Engineering, Hohai University, Nanjing 210098, China
* Correspondence: wxhuai@whu.edu.cn

Abstract: Aquatic plants widely exist in rivers, which can affect the flow structure in rivers and have an important impact on the evolution of river morphology. The emerged vegetation is an important member of aquatic vegetation in the river, so studying the flow structure around the emerged vegetation patches is of great significance. Computational fluid dynamics (CFD) simulation provides support for the related research works. Applying the appropriate turbulence model is crucial to achieving realistic numerical simulation results. In this study, two-dimensional numerical simulations were carried out and compared with experimental data by six different Reynolds-Averaged Navier–Stokes (RANS) turbulence models, i.e., Standard k-ε model, Renormalization group (RNG) k-ε model, Realizable k-ε model, Standard k-ω model, Shear-stress transport (SST) k-ω Model, and the Reynolds stress model (RSM). CFD is an effective research method, and the results showed that there are different simulation performances with different turbulence models. The shear stress transport k-ω model achieves the most consistent numerical simulation results with the experimental data for the longitudinal mean flow velocity distribution at the centerline, and the Reynolds stress model provides the least consistent numerical simulation with the experimental data. Then the performance of the six models in simulating the flow field characteristics and longitudinal outflow after vegetation patch was compared.

Keywords: emerged vegetation; Computational fluid dynamics; turbulence model; 2D numerical simulation

Citation: Yu, S.; Dai, H.; Zhai, Y.; Liu, M.; Huai, W. A Comparative Study on 2D CFD Simulation of Flow Structure in an Open Channel with an Emerged Vegetation Patch Based on Different RANS Turbulence Models. *Water* **2022**, *14*, 2873. https://doi.org/10.3390/w14182873

Academic Editors: Jueyi Sui and Hossein Afzalimehr

Received: 17 August 2022
Accepted: 12 September 2022
Published: 15 September 2022

1. Introduction

Aquatic plants are an important part of the ecological environment and are widely distributed in rivers, lakes, and offshore areas. In natural rivers, aquatic vegetation is generally classified as emerged and submerged, rigid and flexible. Positive and negative feedback widely exists between various aquatic vegetation and river morphology [1]. Vegetation can affect sediment transport and flow structure. The density and arrangement of vegetation affect the water flow to some extent, and the increase in plant density also increases flow resistance [2]. The significance of the interaction between aquatic plants and water flow in natural rivers must be studied. Zong and Nepf represented vegetation patches and plants with 2D circular porous obstruction and rigid cylinders, respectively, because rigid cylinders can approximately simulate stems [3]. They also experimentally investigated the turbulent wake behind a single 2D emerged vegetation patch, described the length and stability of this wake and investigated the effect of patch diameter and porosity on this wake [3]. This type of simulation has also been used in many studies. Liu et al. studied the resistance estimation of rigid submerged vegetation using similar cylindrical arrays [4].

Caroppl et al. measured the characteristics of wake flows after a real riparian vegetation patch and found that the presence of leaves is a key factor in the obstructing effect of vegetation patches on water flow [5]. Huai and Zhang et al. conducted physical experiments to study the flow structure near flexible submerged aquatic vegetation with leaves [6]. By contrast, cylindrical arrays are easy to implement and are effective in simulating vegetation patches. Yu and Shan et al. conducted experiments using rigid cylinders as representative of vegetation to investigate the wake structure behind an individual vegetation patch and found that variations in the density and length of vegetation patches along the streamwise direction affect the steady wake flow [7]. White and Nepf also used rigid cylindrical arrays to represent aquatic plants and performed laboratory experiments to study and characterize the flow structure and characteristics around the cylindrical arrays [8]. Related numerical simulation studies also simulated vegetation communities as rigid cylinder arrays. Liu and Huai et al. studied the flow structure and characteristics around submerged vegetation patches through three-dimensional CFD numerical simulation [9]. Because of the important influence of aquatic vegetation on rivers, further research is necessary to study the interaction between aquatic plant patches and water flow, which is also the starting point of the current work.

Computational fluid dynamics (CFD), a branch of fluid dynamics, is an available and reliable tool commonly used in various research fields. CFD provides an operational platform for engineers to simulate actual working conditions and has been widely used to study the interaction between water flow and aquatic plants. RANS, large eddy simulation (LES), and direct numerical simulation are currently the main methods for numerical fluid simulation. Common turbulent flow models include the Spalart–Allmaras model, standard k-ε model, RNG k-ε model, realizable k-ε model, standard k-ω model, SST k-ω model, Reynolds stress model (RSM), and LES, etc. An important research step is to choose a proper turbulence model to describe the turbulence. Liu and Huai et al. conducted a numerical simulation on the resistance characteristics of rigid submerged vegetation, in which LES was used to simulate the flow around the vegetation patches [10]. Liu and Chen used an improved RNG k-ε turbulence model to simulate the flow in open waters and vegetation waters. The predicted results are in good agreement with the experimental data, proving that the improved RNG k-ε turbulence model has good performance in the flow field near vegetation [11]. Liu and Huai et al. used 3D LES to numerically study hydrodynamics in open channels with an array of square vegetation patches that are discontinuously distributed along the river bank. They found that LES performs well in predicting the variation of turbulence structure with different densities and distances of vegetation array [12]. Anjum and Ghani et al. developed a 3D geometric model to study the internal flow structure of a two-layer vegetated patch, which was solved with the 3D RSM to obtain the distribution of mean velocity and Reynolds stress at different flow rates [13]. Anjum and Tanaka used RSM for 3D numerical simulations to investigate the turbulent flow characteristics of water flow in a channel arranged with double-layered vegetation, submerged vegetation, and emerged vegetation with the same vegetation density [14]. Meanwhile, different works have used different turbulent flow models for the simulation of turbulent flow behind vegetation patches in the flow channel. In the case of complex numerical fluid dynamics, the accuracy and efficiency of simulation vary among different turbulent flow models. A comparison of basic turbulence models is currently needed to provide guidance for future research and support the application of CFD in the study of vegetation–water flow interaction.

Submerged vegetation and emerged vegetation are common types of aquatic vegetation in river channels. 2D models are suitable for the numerical simulation of rigid emerged vegetation. Meanwhile, 3D models are usually adopted for the numerical simulation of submerged vegetation patches. For emerged vegetation in rivers, its interaction with water flow has been investigated for vegetation patches of different shapes and distributions. Qu and Yu conducted a 2D numerical simulation of an isolated emerged vegetation patch in a channel. By using a simple formula of flow velocity distribution, this highly simplified method captures the key features of stable wake, wake recovery, and von–Kármán vortex street [15]. Zhan and Hu et al. introduced a nonconstant inertial resistance coefficient and used a porous media approach to numerically simulate the emerged vegetation in a 2D channel. They found that the improved porous media model can reasonably predict water flow [16]. Yamasaki and Lima et al. investigated the interaction between the emerged vegetation patches and water flow by 2D CFD and simulated the evolutionary processes of the patch erosion and growth of emerged vegetation in the channel [17]. Zhu and Yang et al. investigated the evolution trends of vegetation patches using a 2D shallow water equation and simulated the evolutionary behavior of vegetation patches in the river channel by setting different initial conditions [18]. Relevant physical experiments were also conducted. Li and Huai et al. performed laboratory experiments to investigate the hydrodynamics and turbulent structure in a channel with multiple emerged vegetation patches distributed on one side by simplifying the vegetation to rigid groups of fine cylinders and setting different vegetation densities, diameters of vegetation patches, and distances between adjacent vegetation patches [19].

The Reynolds average models have a wide range of applications and require relatively fewer computational costs. The application of RANS-based turbulent flow models in the simulation of turbulence in river channels was investigated. Farhadi and Mayrhofer et al. selected the standard k-ε model and two k-ω models to simulate turbulent structures in a river section of a runway channel. The comparison with the experimental data revealed that all three models underestimated the intensity of turbulence in the river channel, although they were able to achieve good predictions of the mean velocity [20]. Shaheed et al. investigated the comparative simulation performance of the standard k-ε model and realizable k-ε model for flow characteristics in the bends and confluences of an open channel [21]. Only a few comparative studies have been conducted on the performance of RANS-based turbulence models in the numerical simulation of wake after rigid emerged vegetation patches. For the numerical simulation of the interaction between the emerged aquatic vegetation and water flow in an open channel, the applied RANS model varied in different reports. Using the k-ε model, Lina and Nepf conducted a 2D simulation study of the flow field near two adjacent circular vegetation patches of equal diameter and investigated different distributions of two adjacent vegetation patches on the wake and interaction [22]. Brito and Fernandes et al. used the porous media model to carry out the 3D RANS numerical simulation of water flow with submerged vegetated floodplains and obtained the correct simulation [23]. In the current work, a simple numerical simulation experiment of a straight channel arranged with an isolated emerged vegetation patch was performed to investigate the applicability of several existing available turbulent flow models in the simulation of the interaction between the emerged vegetation and water flow. The results of numerical simulations were analyzed and compared with those published by Zong and Nepf [3]. The implementation of the numerical simulation is described in the next section.

2. Materials and Methods

2.1. Numerical Models

The Reynolds–Average model decomposes the instantaneous variables in the Navier–Stokes equations into two parts, the average and the fluctuating. Considering the incompressible turbulent flow without temperature change, the governing equations based on RANS are [24–26]:

$$\frac{\partial \overline{u}_i}{\partial x_i} = 0 \tag{1}$$

$$\rho\frac{\partial \overline{u}_i}{\partial t} + \frac{\partial}{\partial x_j}\left(\rho\overline{u}_i\overline{u}_j + \rho\overline{u_i'u_j'}\right) = \overline{f}_i - \frac{\partial \overline{p}}{\partial x_i} + \frac{\partial \overline{\tau}_{ij}}{\partial x_j} \tag{2}$$

where $\overline{u}$ and u' represent the mean value and fluctuating value of velocity, respectively; ρ represents the density; $\overline{p}$ represents the mean pressure; and $\overline{\tau}_{ij}$ represents the mean viscous stress tensor:

$$\overline{\tau}_{ij} = \mu\left(\frac{\partial \overline{u}_i}{\partial x_j} + \frac{\partial \overline{u}_j}{\partial x_i}\right) \tag{3}$$

The models selected in this paper include the standard k-ε model, RNG k-ε model, realizable k-ε model, standard k-ω model, shear-stress transport (SST) k-ω model, and Reynolds stress model (RSM).

2.1.1. Standard k-ε Model

The standard k-ε model is the most classical two-equation model proposed by Launder and Spalding, which is based on transport equations for the turbulence kinetic energy (k) and the turbulence dissipation rate (ε) [27]. The model is a widely used semi-empirical model, which assumes that the flow is completely turbulent, ignoring the molecular viscosity [28]. For some unconstrained flow, strong swirl, etc., the k-ε model performs poorly [28].

2.1.2. RNG k-ε Model

The RNG k-ε model is a turbulent model derived from the renormalization group theory by Yakhot and Orszag of Princeton University [29]. The RNG k-ε model has a similar form to the standard k-ε model. It improves the standard k-ε model by adding additional source terms to the turbulent kinetic energy dissipation ε transport equation and modifies the large-scale motion and viscosity to remove the influence of small-scale motion [30]. The RNG k-ε model is more suitable for the numerical simulation of flow in a rotating cavity because of its higher accuracy [31].

2.1.3. Realizable k-ε Model

Compared with the standard k-ε model, the realizable k-ε model replaces the original dissipation rate equation and adopts a new realizable eddy viscosity formula. In some comparative test cases, the stability and prediction accuracy of the realizable k-ε model are better than those of the standard k-ε model [32,33]. The realizable k-ε model is more accurate than the standard k-ε model in predicting the dissipation rate of flat and round jets and better in predicting the characteristics of the boundary layer with a large pressure gradient [31].

2.1.4. Standard k-ω Model

The standard k-ω model is based on the Boussinesq assumption and Reynolds average rule. Launder and Spalding (1972) improved and optimized the Kolmogorov model. Wilcox (2006) improved the previous version of the k-ω model and made the eddy viscosity a function of k and ω, thus allowing the model to be applied in extensive situations [34–36]. Although the standard k-ω model can well describe the near-wall flow, the boundary conditions and the initial turbulence level can affect the simulation results of some flows very sensitively [31].

2.1.5. SST k-ω Model

The SST k-ω model combines the accurate Wilcox k-ω model in the near wall region and the stable k-ε model in the far field region by introducing functions; in 1993, Menter proposed redefining the eddy viscosity in this model to solve the problem of excessive prediction of eddy viscosity [37]. The good accuracy of the shear pressure transport k-ω model is mainly attributed to its efficient near-wall formulation [38].

2.1.6. RSM

The RSM directly solves the transport equation of each component of the Reynolds stress tensor without applying the isotropic eddy viscosity assumption in the Boussinesq hypothesis [39,40]. Five additional transport equations must be solved in the 2D flow problem. RSM can accurately predict complex flow, making it a precise turbulence model; however, the calculation cost is high, and RSM cannot always obtain more accurate results compared with some simple models [41]. It should be noted that the RSM model used here is based on the ε equation. In some numerical simulation cases of swirl flow, RSM produces the worst results [42].

2.2. Simulation Setup

The arrangement of the numerical simulation in this study was adopted from the work of Zong and Nepf [3]. In the experiments, rigid cylinders were used to represent the aquatic vegetation. The water depth in the channel was smaller than the length of the cylinders, allowing the study of the interaction between an emerged vegetation patch and water flow. Hence, 2D numerical simulation is applicable in this case. In the experiments, the total length of the experimental section was 13 m, the width was 1.2 m, and the channel was kept horizontal. In the numerical simulation, the total length of the calculation domain was 7.01 m ($\approx$31.86 D), and the width was 1.2 m ($\approx$5.45 D), where D represents the diameter of the vegetation patch. Since the measurements in the experiments of Zong and Nepf [3]. were conducted only for part of the experimental segment, the computational domain length of the numerical simulations was shorter than that of the experimental segment to save computational resources. This did not affect the subsequent comparison and analysis. In the experiments, an emerged vegetation patch was simulated by a rigid cylinder group with a diameter of d = 0.6 cm. In the numerical simulation, the vegetation patch was simulated by a hollow circular hole group with a diameter of d = 0.6 cm. The diameter of the rigid emerged vegetation patch was D = 22 cm in the experiments and numerical simulation. In the experiments, the vegetation patches were located in the center of the channel, and the center of the vegetation patches coincided with the centerline of the channel. The waterward front edge was 3 m away from the water inlet. In the numerical simulation, given that the total length of the calculation domain was smaller than the total length of the channel test section in the laboratory, the upstream of the vegetation patch was set as 1.54 m (=7 D) away from the inlet at the center line. The center of the upstream edge of the vegetation patch was the origin of the x-axis, and the x-axis pointed to the flow direction, while the origin of the y-axis was the same as that of the x-axis. Φ represents the average solid volume fraction, a represents the frontal area in unit volume, and n represents the number of cylinders in the unit area. Their relationships were $\Phi = \pi ad/4$ and $a = nd$. In the experiment, the vegetation density of the vegetation patch was $\Phi = 0.1$, and the upstream inflow velocity was $U_\infty = 9.8 \pm 0.1$ cm/s. In the numerical simulation, the inflow velocity was defined as $U_\infty = 9.8$ cm/s. Reynolds number based on vegetation stem scale d was defined as [2]:

$$Re_d = \frac{U_\infty d}{\nu} \quad (4)$$

where ν represents the kinematic viscosity coefficient of the water. Figure 1 shows the schematic of the numerical simulation model.

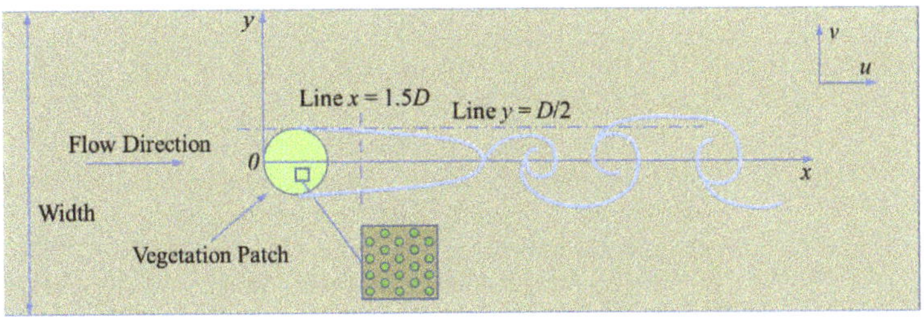

Figure 1. Top view of the numerical simulation arrangement. An emerged vegetation patch is arranged at the central line of a rectangular channel.

For numerical simulation using the six turbulence models, the same set of grids was shared. The grid was hybrid, and the grid of vegetation patch area and wake area was refined. The total number of grid units was 349 k, where k represents a thousand. Figure 2 shows partial details of the grid.

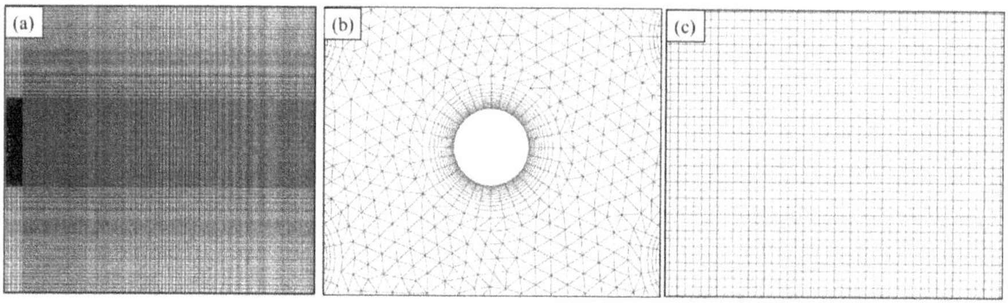

Figure 2. Computational grid for the cases of D = 22 cm and ϕ = 0.1: (**a**) a close view of the wake area grid, (**b**) a detailed view of the grid near the circular hole, and (**c**) a feature view of the grid near the wake behind the vegetation patch.

SIMPLEC was used for pressure–velocity coupling. The steps of the SIMPLEC algorithm are the same as those of SIMPLE, and the correction makes SIMPLEC more economical than SIMPLER and SIMPLE [43]. The same settings were used for the boundary conditions of all numerical simulation cases. The three main types of boundary conditions are inlet, outlet, and wall. The inlet boundary adopted the velocity inlet condition, and the flow velocity was set to 9.8 cm/s. The outlet boundary adopted the pressure outlet condition, and the pressure was set to 0. The wall boundary adopted the non-slip boundary condition, including the boundary on both sides of the calculation domain and the circular holes in the vegetation area. For the non-slip wall boundary in all the cases, the enhanced wall treatment was adopted. The information of all cases is listed in Table 1.

Table 1. Information of numerical simulation cases. Geometric and hydrodynamic parameters are given.

Case	Φ	a (cm^{-1})	D (cm)	D (cm)	U_∞ (cm/s)	Re_d	Turbulence Model	Time Step (s)	Grid Cells (k)
I	0.1	0.2	22	0.6	9.8	588	Standard k-ε	0.01	349
II	0.1	0.2	22	0.6	9.8	588	RNG k-ε	0.01	349
III	0.1	0.2	22	0.6	9.8	588	Realizable k-ε	0.01	349
IV	0.1	0.2	22	0.6	9.8	588	Standard k-ω	0.01	349
V	0.1	0.2	22	0.6	9.8	588	SST k-ω	0.01	349
VI	0.1	0.2	22	0.6	9.8	588	RSM	0.01	349
VII	0.1	0.2	22	0.6	9.8	588	SST k-ω	0.01	258
VIII	0.1	0.2	22	0.6	9.8	588	SST k-ω	0.01	489
IX	0.1	0.2	22	0.6	9.8	588	SST k-ω	0.05	349
X	0.1	0.2	22	0.6	9.8	588	SST k-ω	0.005	349

2.3. Model Validation

Transient numerical simulations were used for all of the current cases. Therefore, determining the independence between the number of grid units used in the simulation and the experimental results and proving the independence between the time step and simulation results are necessary. With reasonable grid and time step independence verification, a small number of computational grid cells and a relatively large time step can be selected to obtain the correct results, which will make the computation faster and save computational resources. Table 2 shows the information on the numerical simulation cases used for model validation.

Table 2. Details of the cases for model validation.

Model Validation		Turbulence Model	Time Size (s)	Grid Cells(k)
Grid independence validation	Case VII	SST k-ω	0.01	258
	Case V	SST k-ω	0.01	349
	Case VIII	SST k-ω	0.01	489
Time step independence validation	Case IX	SST k-ω	0.05	349
	Case V	SST k-ω	0.01	349
	Case X	SST k-ω	0.005	349

2.3.1. Grid Independence Validation

Numerical cases V, VII, and VIII were selected to verify the mesh independence. The number of grid cells in case VII was 258 k. The grids of cases V and VIII are globally encrypted by the case VII grid. The number of grid cells of case V was 349 k, which is 1.35 times that in case VII. The number of grid cells in case VIII was 489 k, which is 1.40 times that in case V. The longitudinal instantaneous velocity $\overline{u}$ at the wake centerline of the simulation results of cases V and VIII was verified to be consistently distributed. Meanwhile, the simulation results of cases VII and VIII showed poor consistency in some regions. This finding proves that the number of grid cells has a great influence on the simulation, but the simulation results are extremely close when the number of grid cells exceeds 349 k. When the number of grid cells in case V continued to increase to a certain extent, the simulation results stabilized, and the grid independence of the case V grid was consequently verified. Only the case V grid was applied in the next section to save computing resources. Figure 3 depicts the longitudinal mean velocity $\overline{u}$ at the centerline of cases V, VII, and VIII. Table 3 shows several numerical simulation data of three cases for grid independence validation. The longitudinal flow velocity u and the transverse coordinate x are dimensionless and expressed as $\overline{u}/U_\infty$ and x/D, respectively.

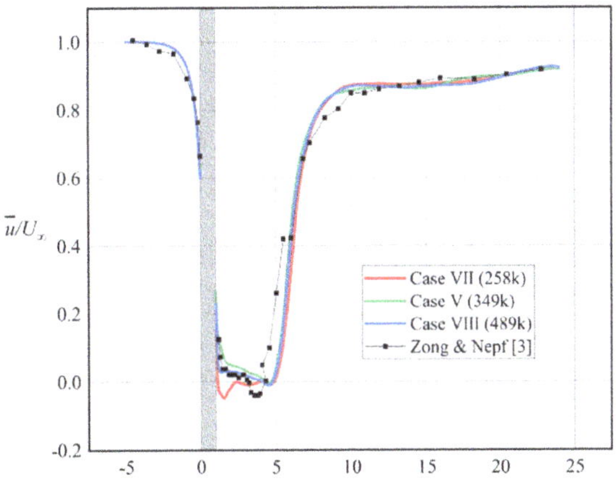

Figure 3. The simulation results of case V, case VII, and case VIII are compared with those of Zong and Nepf [3]. The diameter of the vegetation patch is 22 cm, and the data in the figure are the centerline (y = 0) longitudinal mean velocity ($\overline{u}$) distribution. The gray part indicates the internal area of vegetation patches.

Table 3. Some results of cases V, VII, and VIII for grid independence validation, where L_1 is defined in Section 3.

Case	Global Maximum Mean Velocity (cm/s)	Maximum Longitudinal Mean Velocity (cm/s)	L_1 (D)
VII	13.08022	13.07509	4.6
V	13.07974	13.07616	4.5
VIII	13.07767	13.07593	4.5

2.3.2. Time Step Independence Validation

Numerical simulation cases V, IX, and X were selected to verify the independence of time step size. The grid numbers of the three cases are all 349 k. The time step sizes of the three cases were $\Delta t = 0.01$ s, $\Delta t = 0.05$ s, and $\Delta t = 0.005$ s, and time step independence validation was carried out. Figure 4 illustrates the comparison of the simulation results of cases V, IX, and X for the distribution of mean longitudinal velocity $\overline{u}$ at the centerline. The simulation results of cases V and X were in good agreement. A relatively large deviation was observed between the distribution of the longitudinal mean velocity $\overline{u}$ in cases V and IX, and the mean velocity distribution characteristics of case IX at the centerline were not consistent with the expectations. For correct results, time steps that have a minimal influence on the numerical simulation results and satisfy the stability requirements should be prioritized to save computational resources. Therefore, in the current numerical simulation cases, the time step was set as $\Delta t = 0.01$ s. Table 4 shows some of the numerical simulation data for time step independence validation.

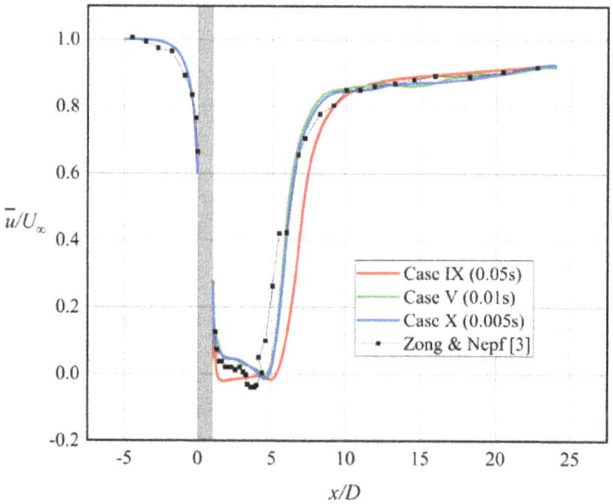

Figure 4. The simulation results of case V, case VIII, and case X are compared with those of Zong and Nepf [3]. The data in the figure is the centerline ($y = 0$) longitudinal mean velocity ($\overline{u}$) distribution. The gray part indicates the internal area of vegetation patches.

Table 4. Some results of cases V, IX, and X for time step independence validation, where L_1 is defined in Section 3.

Case	Global Maximum Mean Velocity (cm/s)	Maximum Longitudinal Mean Velocity (cm/s)	L_1 (D)
IX	13.05878	13.05489	4.9
V	13.07974	13.07616	4.5
X	13.08012	13.08474	4.5

3. Results and Discussion

Analysis was conducted on the performance of the six commonly used turbulence models in simulating and predicting the interaction between vegetation patches and water flow. Figure 5 shows the wake behind the vegetation patch, the shear flow on both sides of the wake, and the special flow velocity distribution due to the von-Kármán vortex street behind the stable wake section. The smooth streamline diagram of flow after an emerged vegetation patch arranged in the channel is provided in Figure 5.

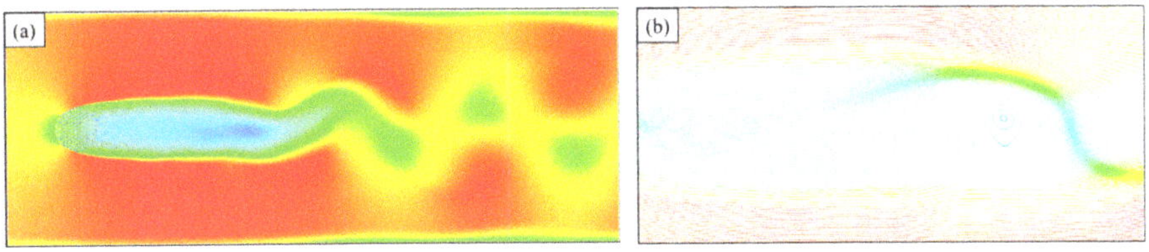

Figure 5. (**a**) Contour of longitudinal velocity u distribution of water flow obstructed by a vegetation patch, and (**b**) streamlines colored by velocity u near the stable wake segment; both are the numerical simulation results of SST k-ω turbulence model.

Represented by L, the total wake segment length spans from the downstream edge of the vegetation patch to the location at the centerline, where the velocity recovery rate is reduced to 0.1, i.e.,

$$\frac{\partial(\overline{u}/U_\infty)}{\partial(x/D)} < 0.1 \quad (5)$$

The wake segment is composed of a stable wake segment and a wake recovery segment. The stable wake segment is an area between the downstream edge of the vegetation patch and the formation location of the vortex street, and its length is represented by L_1. The length of the wake recovery segment is represented by L_2, $L_2 = L - L_1$ [3].

In the preliminary analysis, all six turbulence models were able to make predictions of the interaction between an emerged vegetation patch and the water flow. The simulation results of the six turbulence models were all consistent with the physical characteristics derived from the experiments of Zong and Nepf [3]. However, differences in results were observed for each numerical simulation, and all of the model results exhibited different degrees of deviation from the experimental results. Among them, the RSM model showed the most unsatisfactory performance in predicting the vegetation patch wake.

3.1. Comparative Analysis of Mean Velocity Profile and Wake Segment Simulation

Numerical simulation data of the six turbulence models were collected. The data at the lines $y = 0$ and $y = 0.11$ m shown in Figure 1 were statistically analyzed. Figures 6 and 7 depict the time-averaged velocity distribution of each example at the two lines for each simulation case in comparison with the experimental data of Zong and Nepf [3].

Figure 6 depicts the time-averaged longitudinal flow velocity distribution of the six turbulence models at the wake centerline. On the basis of the experimental results of Zong and Nepf [3], the results of six numerical simulation examples were consistent with the experimental physical characteristics of laboratory experiments. All six turbulence models were able to simulate reasonable longitudinal mean flow velocity distributions at the centerline, including the stable wake segment and wake recovery segment. The values are shown in Figure 8. Figure 6 illustrates that the predictions of the numerical simulations deviated from the experimental data. Case V simulated by the SST k-ω model had the simulation results with the least overall deviation from the experimental data. The other five turbulence models had great deviations at different locations. In Figure 6, L_1 and L_2, respectively, mark the distribution of the mean flow velocity $\overline{u}$ in the steady wake segment and the wake recovery segment from the experimental data of Zong and Nepf [3]. The flow velocities in the wake stable segment L_1 simulated by the three turbulence models of the

k-ε series were remarkably larger than those of the experimental data. In addition to the SST k-ω model, the other five turbulence models predicted that the wake recovery segment was highly downstream. Only the realizable k-ε model and RSM model poorly predicted the distribution and magnitude of the mean flow velocity $\overline{u}$ after the wake. Among the six turbulence models, the worst prediction results were provided by the RSM model. The simulation results of the RSM model were unsatisfactory in the wake segment, and the flow velocity prediction after the wake section highly differed from that in the experimental data. The relative velocity of the experimental data at x = 23D is about 0.9, while about 0.8 for RSM. Meanwhile, the numerical simulation results of the standard k-ε model were better than those of the RNG k-ε model and realizable k-ε model. This finding is unexpected because the realizable k-ε model shows better performance than the standard k-ε model in most previous numerical simulation examples.

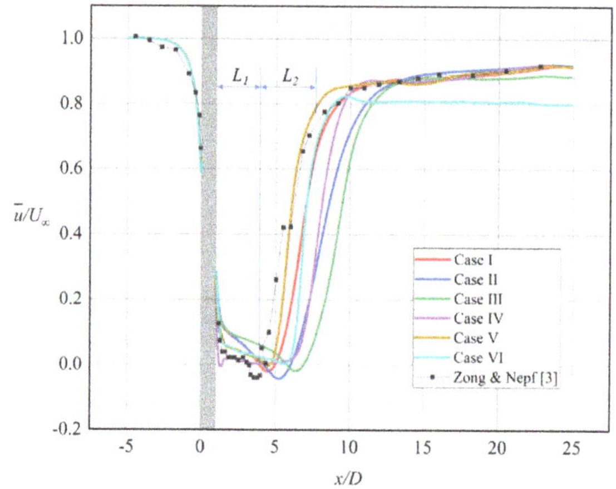

Figure 6. The longitudinal flow profiles simulated by six turbulence models are illustrated in comparison with the experimental results of Zong and Nepf (Φ = 0.1, D = 22 cm) [3], where the longitudinal mean velocity ($\overline{u}$) is measured at the centerline, y = 0. The gray part indicates the internal area of vegetation patches.

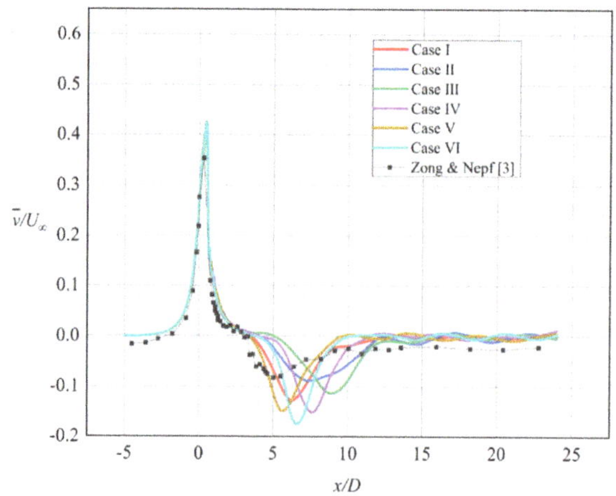

Figure 7. Transverse velocity $\overline{v}$ is measured at line $y = D/2$; the black dot plot in the figure is the experimental data of Zong and Nepf (Φ = 0.1, D = 22 cm) [3].

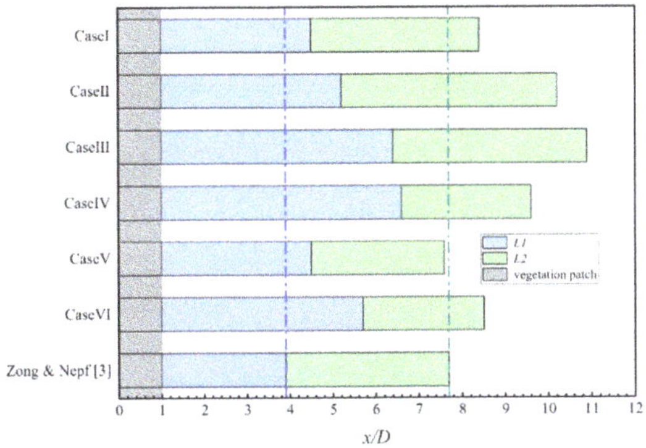

Figure 8. The lengths of the wake segments obtained from the cases and experiments of Zong and Nepf are illustrated [3]; blue indicates the stable wake segment, and green indicates the wake recovery segment; the blue and green dashed line indicates the L_1 and L_2 of experiments. The gray part indicates the internal area of vegetation patches.

Figure 7 depicts the numerical simulations results of the transverse mean flow velocity $\overline{v}$ profiles at the line $y = D/2$. Although the six turbulence models were able to correctly represent the physical characteristics consistent with the experiments of Zong and Nepf [3], they showed varying degrees of deviation in specific values. The minimum position of the transverse mean velocity $\overline{v}$ at the line $y = D/2$ was more downstream than that in the experimental data. Among the six turbulence models, the SST k-ω model simulated the most consistent results with the experimental data but underestimated the minimum value of $\overline{v}$. The minimum value of $\overline{v}$ predicted by the RNG k-ε model was consistent with the experimental data of Zong and Nepf [3], but the location of the minimum value was more downstream. A high recovery rate of the minimum value is associated with a fast shear flow on both sides of the vegetation patch and a short wake recovery segment. On this basis, the predicted length of wake recovery segment by the RSM model was too small.

Figure 8 summarizes the numerical simulation results of the wake segment length for the six numerical cases. All of the turbulence models predicted greater length L for the emerged vegetation patch compared with that in Zong and Nepf experiments [3] (about 10% to 47.8% excess of the experimental data), except for the prediction results of the SST k-ω model that were in good agreement with the experimental data. Among them, the realizable k-ε model predicted the highest L value compared to the experimental data, about 47.8% of the experimental data. For the stable wake segment length L_1, the six turbulence models predicted longer values than that obtained from the experiments (about 20.7% to 93.1% excess of the experimental data). For the wake recovery segment length L_2, the simulation results of the standard k-ε model were the most consistent with the experimental data (the difference is less than 3% of the experimental data). The predicted L_2 values by the RNG k-ε model and realizable k-ε model were larger than the experimental values, and the predictions of the standard k-ω model, SST k-ω model, and RSM model were smaller.

3.2. Simulation Results of Flow Field after Vegetation Patch

Figure 9 shows the transient velocity distribution contours simulated by the six turbulence models and colored by the longitudinal flow velocity u. The obstructive effect of an emerged vegetation patch led to the formation of a low flow velocity segment (the wake) behind the vegetation patch, and the flow velocity on both sides of the vegetation patch was increased by the squeezing effect, generating shear layers. The shear layers gradually developed and widened with the flow until the shear layers on both sides interacted, forming Kármán vortex street after the wake region. The distribution of the Kármán vortices street in the contours for each case is marked in Figure 9. These positions were confirmed based on the flow velocity vector diagram and the x values for which $v(x)$ takes the extremum. The comparison indicates that all of the applied turbulence models were able to make predictions for the two vortex sheets behind the wake. In the calculation domain, all turbulence models did not capture the breakdown of the main vortex street, except the standard *k-ω* model. The simulation by the standard *k-ω* model demonstrated the instability and intermixing of the two vortex sheets behind the wake. Compared with that predicted by the standard *k-ω* model, the main vortex street lengths after the wake predicted by the standard *k-ε* model, RNG *k-ε* model, realizable *k-ε* model, SST *k-ω* model, and RSM model simulations were larger and more stable. The distances between neighboring vortices at the same vortex sheet of the main vortex street are represented by a_1 and a_2. Here a_1 and a_2 are consecutive, and a_1 is closer to the upstream. In other words, a_1 and a_2 include three consecutive vortices on the same vortex sheet.

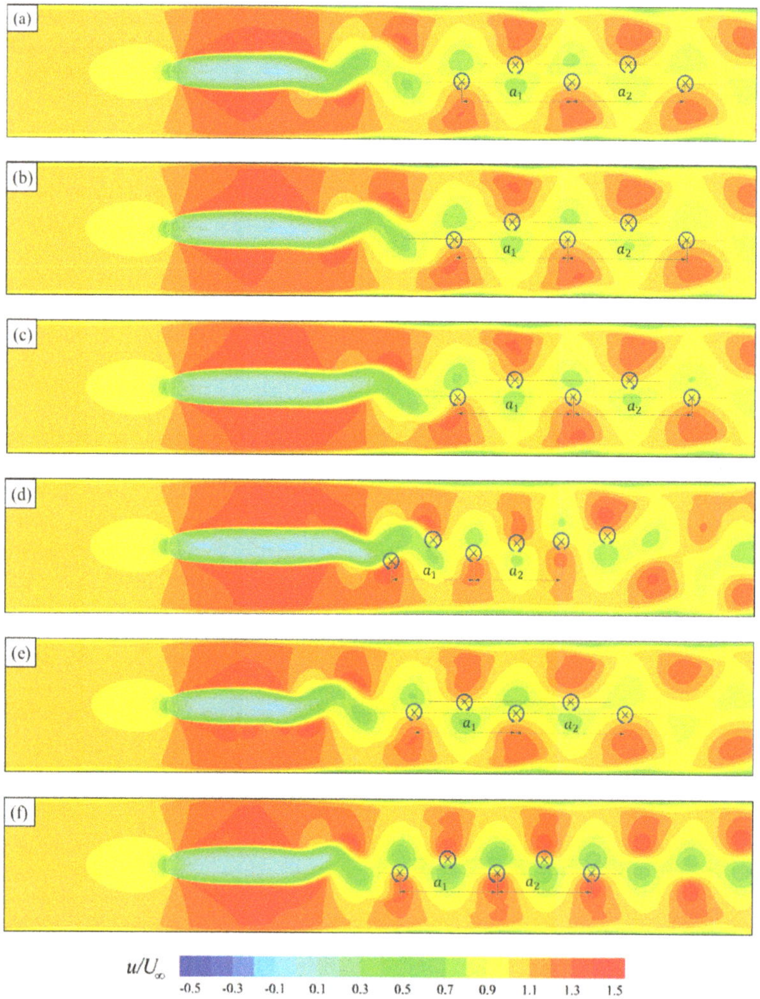

Figure 9. Contours of instantaneous velocity distribution for Φ = 0.1 and *D* = 22 cm colored by velocity *u*. (**a**) case I, (**b**) case II, (**c**) case III, (**d**) case IV, (**e**) case V, and (**f**) case VI. Arrows indicate the direction of the vortex rotation.

a_1 and a_2 were averaged to represent the wavelength of the main vortex street for each turbulence model prediction result.

$$a = \frac{a_1 + a_2}{2} \tag{6}$$

Figure 10 illustrates the comparison of the numerical simulation results of six turbulence models for the wavelength of the main vortex street. Figure 10 illustrates the wavelength of the main vortex street of the numerical simulation results presented in Figure 9.

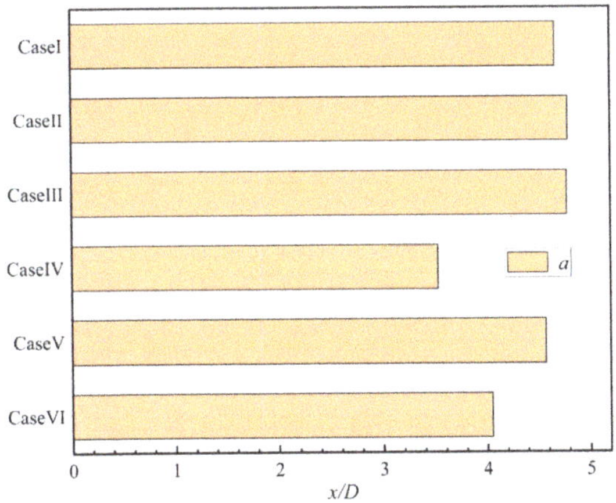

Figure 10. Wavelength comparison of main vortex street for six cases.

The standard k-ε model, RNG k-ε model, and realizable k-ε model produced relatively consistent prediction results of the main vortex street wavelength after the wake. The difference between the prediction results of the SST k-ω model and the standard k-ε model was extremely small because the calculation of the far-field region of the SST k-ω model was consistent with that of the standard k-ε model. According to the prediction results of the standard k-ω model, its predicted main vortex street wavelength was significantly small. When the wavelength decreases, the velocity fluctuation frequency of the main vortex street increases, corresponding to the smaller vortex shedding frequency behind the vegetation patch.

3.3. Comparative Analysis of Longitudinal Outflow Intensity

The emerged vegetation patch are porous areas, and some of the water flow can form outflows on the sides and back of the patch through the gaps between the plants. The longitudinal outflow in an emerged vegetation patch will form a stable wake segment, a wake recovery section, and a Kármán vortex street behind the vegetation patch. The longitudinal outflow intensity of an emerged vegetation patch is closely related to the diameter D of the vegetation patch and the vegetation density Φ [9]. Therefore, the longitudinal outflow of the simulation results of the six turbulence models can be compared by comparing the transverse distribution of the longitudinal mean flow velocity $\overline{u}$ behind the vegetation patch. It should be noted that the instantaneous longitudinal velocity distributions in Figure 11 are not at the same time, and there is no need to compare the instantaneous longitudinal velocity distributions of the six turbulence models. Figure 11 depicts the longitudinal outflow from the numerical simulation results under the six turbulence models. The distributions of flow velocity u and $\overline{u}$ at line $x = 1.5D$ in the stable wake segment were collected, and the position of the transverse line is marked in Figure 1. The dimensionless distributions of the mean longitudinal velocity $\overline{u}$ and the transient longitudinal velocity u simulated by the six turbulence models were compared, as shown in Figure 11.

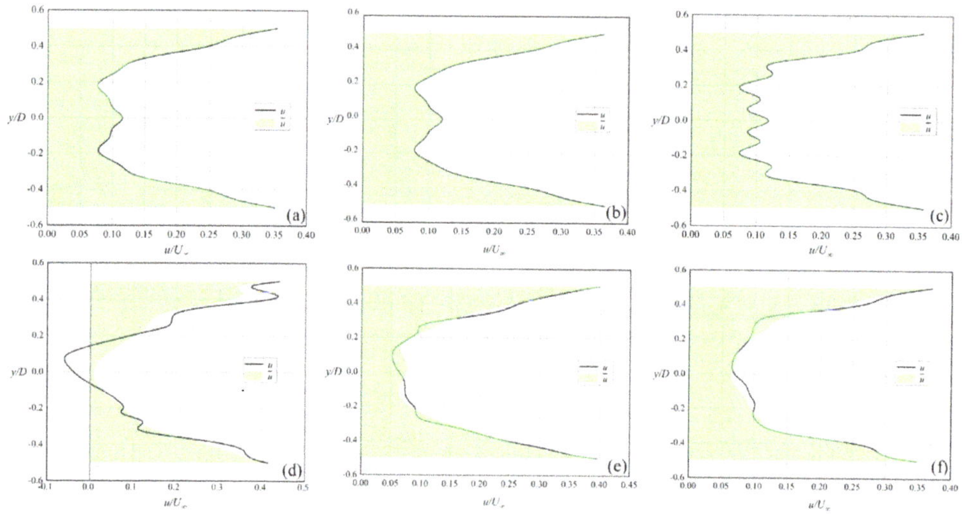

Figure 11. The black curve indicates the transient longitudinal flow velocity u distribution, and the green area indicates the cumulative plot of the mean longitudinal flow velocity $\overline{u}$ to the line $\overline{u}/U_\infty = 0$. (**a**) case I, (**b**) case II, (**c**) case III, (**d**) case IV, (**e**) case V, and (**f**) case VI.

Given that the shear flow on both sides of the emerged vegetation patch gradually developed along the flow direction, the mean flow velocity $\overline{u}$ and transient flow velocity u located near $y = \pm 0.5D$ were larger than the flow velocity near the centerline but were not affected by the development of shear flow. Figure 11a,b shows the good consistency between the simulations of the mean longitudinal outflow at $x = 1.5D$ by the standard k-ε model and the RNG k-ε model. The large overlap between the distributions of the mean flow velocity $\overline{u}$ and the transient flow velocity u in Figure 11a–c indicated the stability in time of the steady wake segment, simulated by the standard k-ε model, RNG k-ε model, and realizable k-ε model. The multiple maximum and minimum values of longitudinal flow velocities in Figure 11c revealed that the turbulence in the stable wake segment simulated by the realizable k-ε model was extremely weak, and the interaction between the longitudinal outflows behind the porous region occurred in the further downstream. In Figure 11d, the distribution of transient velocity u had negative values, indicating that the turbulence in the stable wake segment predicted by the standard k-ω model was intensive, and reflux can be observed in the segment. The distributions of the mean flow velocity $\overline{u}$ illustrated in Figure 11d–f did not match with the distribution of the transient flow velocity u. This finding indicates that the longitudinal outflow fluctuated in time at line $x = 1.5D$ simulated by the standard k-ω model, SST k-ω model, and RSM model. Therefore, different degrees of turbulence existed in the stable wake segment predicted by these three turbulence models.

3.4. Computational expenses

In the application of numerical simulation, an important concern is the calculation cost. Usually, fast and economical methods will be more widely adopted. For this reason, the calculation demands of the six models relative to the standard k-e model simulation are given in Table 5. It should be noted that all of the simulations were intentionally conducted on the same computer.

Table 5. Computational expenses of the six turbulence models.

Case	Turbulence Model	Grid Cells (k)	Time Step (s)	Relative Iteration Time
I	Standard k-ε	349	0.1	1.00
II	RNG k-ε	349	0.1	1.07
III	Realizable k-ε	349	0.1	1.06
IV	Standard k-ω	349	0.1	1.46
V	SST k-ω	349	0.1	0.87
VI	RSM	349	0.1	1.21

It is worth mentioning that the SST k-ω model obtained the most consistent fluency characteristics with the experimental data of Zong and Nepf [3] in Section 3.1. In Table 5, the relative iteration time of SST k-ω is also the smallest, which means the lowest computational expense. In other words, the SST k-ω model is relatively accurate and efficient under the current research background and conditions.

4. Conclusions

This work mainly investigates the interaction between an emerged vegetation patch and water flow in a straight channel by 2D CFD. A comparative analysis is also conducted on the simulation results of different turbulence models. The interaction between vegetation patches and water flow is extremely sensitive to the diameter and density of the vegetation patch, so the current results are only strictly valid for the current situation [2]. All six turbulence models currently selected are able to make reasonable simulations and predictions of the flow field with different degrees of accuracy.

In the computational domain, all six turbulence models are able to simulate the shear flow on both sides of the emerged vegetation patch, the wake, and the Kármán vortex street behind the patch. The standard k-ε model is able to simulate reasonable and accurate simulation predictions for the current range of flow field characteristics. Moreover, it is better than the RNG k-ε model and realizable k-ε model in terms of consistency with the experimental data of Zong and Nepf [3]. The SST k-ω model has the best agreement with the experimental data in stimulating the interaction between an emerged vegetation patch and water flow in a straight channel, including the mean longitudinal velocity distribution at the central line, the transverse mean velocity distribution at the line $y = D/2$, length of the wake segment. The standard k-ω model and the RSM do not perform well. From the simulation results, the turbulence intensity of RSM, the standard k-ω model, and the SST k-ω model is stronger than that of the standard k-ε model, RNG k-ε model, and realizable k-ε model. For the prediction of turbulent structure around rigid emerged vegetation by 2D numerical simulation, the SST k-ω model can achieve better results than the standard k-ε model, RNG k-ε model, realizable k-ε model, standard k-ω model, and Reynolds stress model (RSM). In addition, the SST k-ω model has the lowest computational expense in current research.

Author Contributions: Conceptualization, W.H. and S.Y.; methodology, W.H., M.L. and S.Y.; validation, H.D. and S.Y.; visualization, Y.Z. and S.Y.; supervision, W.H.; writing—original draft preparation, S.Y.; writing—review and editing, W.H., M.L. and S.Y. All authors have read and agreed to the published version of the manuscript.

Funding: This work was supported by the National Natural Science Foundation of China [grant numbers 52020105006, 12272281, 11872285].

Data Availability Statement: Data are contained within the article.

Conflicts of Interest: The authors declare no conflict of interest.

References

1. Huai, W.; Li, S.; Katul, G.G.; Liu, M.; Yang, Z. Flow dynamics and sediment transport in vegetated rivers: A review. *J. Hydrodyn.* **2021**, *33*, 400–420. [CrossRef]
2. Wang, C.; Zheng, S.; Wang, P.; Hou, J. Interactions between vegetation, water flow and sediment transport: A review. *J. Hydrodyn.* **2015**, *27*, 24–37. [CrossRef]
3. Zong, L.; Nepf, H. Vortex development behind a finite porous obstruction in a channel. *J. Fluid Mech.* **2012**, *691*, 368–391. [CrossRef]
4. Liu, M.; Huai, W.; Yang, Z.; Zeng, Y. A genetic programming-based model for drag coefficient of emergent vegetation in open channel flows. *Adv. Water Resour.* **2020**, *140*, 103582. [CrossRef]
5. Caroppi, G.; Västilä, K.; Järvelä, J.; Lee, C.; Ji, U.; Kim, H.S.; Kim, S. Flow and wake characteristics associated with riparian vegetation patches: Results from field-scale experiments. *Hydrol. Process.* **2022**, *36*, e14506. [CrossRef]
6. Huai, W.; Zhang, J.; Katul, G.G.; Cheng, Y.; Tang, X.; Wang, W. The structure of turbulent flow through submerged flexible vegetation. *J. Hydrodyn.* **2019**, *31*, 274–292. [CrossRef]
7. Yu, Z.; Shan, Y.; Liu, C.; Liu, X. Wake flow and vortex structures behind emergent vegetation patches elongated in the longitudinal direction. *J. Hydrodyn.* **2021**, *33*, 1148–1161. [CrossRef]
8. White, B.L.; Nepf, H.M. Shear instability and coherent structures in shallow flow adjacent to a porous layer. *J. Fluid Mech.* **2007**, *593*, 1–32. [CrossRef]
9. Liu, M.; Huai, W.; Ji, B. Characteristics of the flow structures through and around a submerged canopy patch. *Phys. Fluids* **2021**, *33*, 035144. [CrossRef]
10. Liu, M.; Huai, W.; Ji, B.; Han, P. Numerical study on the drag characteristics of rigid submerged vegetation patches. *Phys. Fluids* **2021**, *33*, 085123. [CrossRef]
11. Liu, Z.; Chen, Y.; Wu, Y.; Wang, W.; Li, L. Simulation of exchange flow between open water and floating vegetation using a modified RNG k-ε turbulence model. *Environ. Fluid Mech.* **2017**, *17*, 355–372. [CrossRef]
12. Liu, M.; Yang, Z.; Ji, B.; Huai, W.; Tang, H. Flow dynamics in lateral vegetation cavities constructed by an array of emergent vegetation patches along the open-channel bank. *Phys. Fluids* **2022**, *34*, 035122. [CrossRef]
13. Anjum, N.; Ghani, U.; Pasha, G.A.; Latif, A.; Sultan, T.; Ali, S. To Investigate the Flow Structure of Discontinuous Vegetation Patches of Two Vertically Different Layers in an Open Channel. *Water* **2018**, *10*, 75. [CrossRef]
14. Anjum, N.; Tanaka, N. Investigating the turbulent flow behaviour through partially distributed discontinuous rigid vegetation in an open channel. *River Res. Appl.* **2020**, *36*, 1701–1716. [CrossRef]
15. Qu, J.; Yu, J. A numerical modelling of flows in an open channel with emergent vegetation. *J. Hydraul. Res.* **2021**, *59*, 250–262. [CrossRef]
16. Zhan, J.; Hu, W.; Cai, W.; Ye-Jun, Y.; Li, C. Numerical simulation of flow through circular array of cylinders using porous media approach with non-constant local inertial resistance coefficient. *J. Hydrodyn.* **2017**, *29*, 168–171. [CrossRef]
17. Yamasaki, T.N.; de Lima, H.S.; Silva, D.F.; Preza, C.G.D.A.; Janzen, J.G.; Nepf, H.M. From patch to channel scale: The evolution of emergent vegetation in a channel. *Adv. Water Resour.* **2019**, *129*, 131–145. [CrossRef]
18. Zhu, Z.; Yang, Z.; Huai, W.; Wang, H.; Li, D.; Fan, Y. Growth-decay model of vegetation based on hydrodynamics and simulation on vegetation evolution in the channel. *Ecol. Indic.* **2020**, *119*, 106857. [CrossRef]
19. Li, D.; Huai, W.; Liu, M. Investigation of the flow characteristics with one-line emergent canopy patches in open channel. *J. Hydrol.* **2020**, *590*, 125248. [CrossRef]
20. Farhadi, A.; Mayrhofer, A.; Tritthart, M.; Glas, M.; Habersack, H. Accuracy and comparison of standard k-ε with two variants of k-ω turbulence models in fluvial applications. *Eng. Appl. Com Fluid* **2018**, *12*, 216–235.
21. Shaheed, R.; Mohammadian, A.; Gildeh, H.K. A comparison of standard k–ε and realizable k–ε turbulence models in curved and confluent channels. *Environ. Fluid Mech.* **2019**, *19*, 543–568. [CrossRef]
22. de Lima, H.S.; Janzen, J.G.; Nepf, H.M. Flow patterns around two neighboring patches of emergent vegetation and possible implications for deposition and vegetation growth. *Environ. Fluid Mech.* **2015**, *15*, 881–898. [CrossRef]
23. Brito, M.; Fernandes, J.; Leal, J.B. Porous media approach for RANS simulation of compound open-channel flows with submerged vegetated floodplains. *Environ. Fluid Mech.* **2016**, *16*, 1247–1266. [CrossRef]
24. Versteeg, H.; Malalasekera, W. *An Introduction to Computational Fluid Dynamics: The Finite Volume Method*, 2nd ed.; Pearson: London, UK, 2007.
25. Ferziger, J.H.; Peric, M.; Leonard, A. Computational Methods for Fluid Dynamics. *Phys. Today* **1997**, *50*, 80–84. [CrossRef]
26. Xiao, H.; Cinnella, P. Quantification of model uncertainty in RANS simulations: A review. *Prog. Aeros. Sci.* **2019**, *108*, 1–31. [CrossRef]
27. Launder, B.E.; Spalding, D.B. *Lectures in Mathematical Model of Turbulence*; Academic Press: Cambridge, MA, USA, 1972.
28. Ben Ramoul, L.; Korichi, A.; Popa, C.; Zaidi, H.; Polidori, G. Numerical study of flow characteristics and pollutant dispersion using three RANS turbulence closure models. *Environ. Fluid Mech.* **2019**, *19*, 379–400. [CrossRef]
29. Yakhot, V.; Orszag, S.A. Renormalization group analysis of turbulence. I. Basic theory. *J. Sci. Comput.* **1986**, *1*, 3–51. [CrossRef]
30. Koutsourakis, N.; Bartzis, J.G.; Markatos, N.C. Evaluation of Reynolds stress, k-ε and RNG k-ε turbulence models in street canyon flows using various experimental datasets. *Environ. Fluid Mech.* **2012**, *12*, 379–403. [CrossRef]

31. Bulat, M.; Bulat, P. Comparison of Turbulence Models in the Calculation of Supersonic Separated Flows. *World Appl. Sci. J.* **2013**, *27*, 1263–1266.
32. Shih, T.H.; Liou, W.W.; Shabbir, A.; Yang, Z.; Jiang, Z. A new k-e eddy viscosity model for high Reynolds number turbulent flows. *Comput. Fluids* **1995**, *24*, 227–238. [CrossRef]
33. Duan, Y.; Jackson, C.; Eaton, M.D.; Bluck, M.J. An assessment of eddy viscosity models on predicting performance parameters of valves. *Nucl. Eng. Des.* **2019**, *342*, 60–77. [CrossRef]
34. Wilcox, D.C. *Turbulence Modeling for CFD*; DCW Industries Inc.: La Canada, CA, USA, 2006.
35. Launder, B.E.; Reece, G.J.; Rodi, W. Progress in the development of a Reynolds-stress turbulence closure. *J. Fluid Mech.* **1975**, *68*, 537–566. [CrossRef]
36. Wilcox, D.C. Formulation of the k-w Turbulence Model Revisited. *AIAA J.* **2008**, *46*, 2823–2838. [CrossRef]
37. Menter, F.R. Two-equation eddy-viscosity turbulence models for engineering applications. *AIAA J.* **1994**, *32*, 1598–1605. [CrossRef]
38. Menter, F.R. Review of the shear-stress transport turbulence model experience from an industrial perspective. *Int. J. Comput. Fluid Dyn.* **2009**, *23*, 305–316. [CrossRef]
39. Karbon, M.; Sleiti, A.K. Turbulence Modeling Using Z-F Model and RSM for Flow Analysis in Z-SHAPE Ducts. *J. Eng.* **2020**, *2020*, 1–10. [CrossRef]
40. Gibson, M.M.; Launder, B.E. Ground effects on pressure fluctuations in the atmospheric boundary layer. *J. Fluid Mech.* **1978**, *86*, 491–511. [CrossRef]
41. Altaç, Z.; Uğurlubilek, N. Assessment of turbulence models in natural convection from two- and three-dimensional rectangular enclosures. *Int. J. Therm. Sci.* **2016**, *107*, 237–246. [CrossRef]
42. Maly, M.; Slama, J.; Cejpek, O.; Jedelsky, J. Searching for a Numerical Model for Prediction of Pressure-Swirl Atomizer Internal Flow. *Appl. Sci.* **2022**, *12*, 6357. [CrossRef]
43. Van Doormaal, J.; Raithby, G.D. Enhancements of the Simple Method for Predicting Incompressible Fluid Flows. *Numer. Heat Transf. Part. B Fundam.* **1984**, *7*, 147–163. [CrossRef]

water

MDPI

Article

Interaction of Irregular Distribution of Submerged Rigid Vegetation and Flow within a Straight Pool

Kourosh Nosrati [1], Hossein Afzalimehr [1,*] and Jueyi Sui [2]

1 Natural Disasters Prevention Research Center, Faculty of Civil Engineering, Iran University of Science and Technology, Tehran 16846-13114, Iran; koursoh_nosrati@civileng.iust.ac.ir

2 School of Engineering, University of Northern British Columbia, Prince George, BC V2N 4Z9, Canada; jueyi.sui@unbc.ca

* Correspondence: hafzali@iust.ac.ir; Tel.: +98-913-2175524

Abstract: The interaction of bedform and vegetation cover significantly affects the turbulent flow parameters. To investigate this interaction, experiments were carried out in both a gravel-bed river and a laboratory flume. The purpose of field investigations was to find the slopes for both the entrance section and exit section of pools, the grain size of the bed material, and the flow condition. Based on field data, without considering any scaling analysis, a straight pool was constructed in a laboratory flume that was 0.9 m wide, 0.6 m deep, and 14 m long. The entry and exit slopes of the straight pool were 7.4° and 4°, respectively. The straight pool had vertical side walls and a gravel bed with a median grain size of d_{50} = 23.3 mm. Plastic cylinders planted in an irregular pattern in the channel beds were used to model rigid submerged vegetation. The velocity components were recorded by using an ADV at 200 Hz. In this study, the distributions of velocity, Reynolds stress, and TKE were investigated for flows in the presence of submerged rigid vegetation in channel beds with various area densities of vegetation. Results show that the shape of Reynolds stress distribution depends on the entrance and exit slopes of the pool, as well as the irregular distribution pattern of vegetated elements. Inside the pool with the presence of submerged vegetation in the channel bed, the maximum TKE appears above the bed surface with a larger distance depending on the area density of vegetation in the channel bed. However, the momentum exchange and turbulent energy are likely influenced by the secondary circulation of the flows associated with the irregular distribution of vegetated elements in the channel bed. Results of the quadrant analysis show that the momentum between the flow, bedform, and vegetated elements is mostly transferred by sweep and ejection events. The outward event tends to grow toward the water surface, reaching the highest amount near the water surface. At the pool entrance section where the flow is decelerating, the ejection event is dominant near the bed while the sweep event is strong near the water surface. With the decrease in the vegetation density in the pool bed, both the ejection and outward events become dominant.

Keywords: submerged rigid vegetation; 3D bedform; Reynolds stress; quadrant analysis

Citation: Nosrati, K.; Afzalimehr, H.; Sui, J. Interaction of Irregular Distribution of Submerged Rigid Vegetation and Flow within a Straight Pool. *Water* **2022**, *14*, 2036. https://doi.org/10.3390/w14132036

Academic Editors: Anargiros I. Delis and Giuseppe Pezzinga

Received: 4 April 2022
Accepted: 21 June 2022
Published: 25 June 2022

1. Introduction

Vegetation in channel beds and river banks plays a very important role in aquatic systems and benefits the river environment by improving the river water quality by producing oxygen [1], preventing river bank erosion [2], and providing appropriate habitat diversity for fish [3]. Physically, the presence of vegetation patches in rivers modifies the velocity profiles resulting in either positive responses, such as the accumulation of nutrients due to the stress reduction, or negative responses, such as the depletion of nutrients due to the stress enhancement [4].

The effect of vegetation on the velocity profile and turbulence characteristics depends on the type of vegetation (rigid or flexible) and the flow conditions around the vegetation (submerged or emergent). Significant research has been conducted to study the interaction between hydrodynamics and vegetation in recent years [5–9]. Flume experiments have been carried out by using natural vegetation [10–12] or simple elements such as strips or cylinders [13,14]. Field measurements of flow have also been conducted in natural rivers in the presence of submerged vegetation [15–17]. Many theoretical and numerical investigations have been performed, focusing mainly on the evaluation of velocity, shear stress, and turbulence intensity distributions [8,18,19]. For submerged vegetation in channels, Baptist et al. (2007) pointed out that there are four distinct zones in the vertical velocity profile, although the velocity profile is often described by two interacting zones, known as the two-layer approach, namely, the vegetation layer which contains cylindrical elements known as vegetation, and the surface layer above the vegetation (up to the water surface) [20].

For vegetation with the same arrangement pattern and the same height, the characteristics of flow within a group of submerged cylindrical elements are similar to those in the case of emergent vegetation [21]. Above the cylindrical vegetated elements, the flow velocity moves to a higher inflection point. The two co-flowing streams (the upper one and the one between the vegetal elements) increase to generate a Kelvin–Helmholtz instability, which causes the flow to rotate clockwise, and vortices to become larger in the downstream direction, forcing the inflection point to locate in the vegetal zone. For sparse vegetation, the vortex affects the whole vegetated layer, whereas, in the case of dense vegetation, it affects only a layer confined to the top of the vegetated elements [7]. The longitudinal turbulence intensities reach a maximum near the top of the vegetated elements. It has the largest values behind the vegetated elements and decreases in the flow direction. In the free stream region, the longitudinal turbulence intensity reaches the lowest value.

Rigid submerged vegetation in channel beds has attracted the attention of quite a few researchers [7,18,20,22–25]. Some researchers provided the average velocity values in the two layers, while others derived the velocity distribution [7,24]. In the vegetated layer, the streamwise velocity is usually considered constant over the layer depth [13,24], while in the surface layer, various expressions were adopted for the velocity distribution [26], namely, the logarithmic theory [22], the Kolmogorov theory of turbulence [27], the genetic programming [20], and the representative roughness height [28]. Nikora et al. (2013) reported that the flows can be divided into five sub-layers: bed-boundary, uniform, mixing, logarithmic, and wake layers [29]. The bed-boundary layer is usually very thin. The velocity in this layer rapidly increases with the distance from the channel bed. In the uniform layer, the drag force is balanced by the sliding force, causing a relatively uniform velocity distribution. The mixing layer is the most complicated and can be described by a hyperbolic tangent profile [7]. In the logarithmic layer, the velocity follows the law of the wall. In the wake layer, the velocity profile should be modified with a wake term. However, this kind of velocity distribution is complex and hardly occurs in practice. The resistance resulting from vegetated elements reduces the velocity within the canopy of vegetation compared to the overflow, such that the canopy of vegetation interface is a region with strong shear resembling a free-shear layer [30]. Raupach et al. (1996) demonstrated the analogy between terrestrial canopy-shear-layers and free-shear-layers, which includes similarity in the scaling of both mean flow and turbulent statistics [31]. This kind of analogy can be extended to aquatic vegetation [32]. It is important to note that a shear layer is generated only when the momentum absorption by the vegetation canopy is sufficient to produce an inflection point in the velocity profile, which needs to trigger the Kelvin–Helmholtz instability. The turbulent flow type in the horizontal-vertical plan varies with an increase in the area density of vegetation, from the bed-shear flow to the free-shear flow [33]. The bed-shear flow happens because of bed drag near the channel bed, and the presence of vegetation contributes to the bed roughness and increases the bed friction. For the medium dense distribution of vegetation, the flow discontinuity caused by the drag of the vegetation canopy would occur at the top of the vegetation, generating the free-shear flow in a certain vertical region [34].

There are some models proposed to consider the effect of vegetation on flow resistance [24,28]. These models are mostly restricted to vegetated elements with specific arrangements and morphology in channel bed, and thus, they are not easily applied to other vegetation species or morphologies. For example, Yang and Choi (2010) proposed a two-layer model where the momentum balance was applied to each layer and an expression for the mean velocities was proposed. The flow velocity was assumed uniform in the vegetation layer and logarithmic in the upper layer [24]. Cheng (2011) proposed a representative roughness height to describe the resistance caused by vegetation in open-channel flows. He pointed out that the friction factor determined for the surface layer above the vegetation normally increases with the relative roughness (d_{50}/H) [28]. Therefore, the flow characteristics for submerged rigid vegetation should be analyzed to reveal general reasonable results based on laboratory data.

However, to the best of our knowledge, the interaction between randomly vegetal elements and 3D bedforms over a range of vegetation area densities has not been reported. The longitudinal profiles of various gravel-bed rivers are defined as the changes in topographic highs and lows describing macroscale bedforms such as pools and riffles. Pools and riffles are geomorphological features often found in straight, meandering, and braiding river systems [35]. They have been considered essential elements for initiating and developing river bends [36]. Recently, it has been realized that the hydrodynamic variability of the pools-riffles cascade provides a variety of habitat conditions required for different fish species and the same species at different stages of its life [37].

It is well understood that the characteristics of velocity and Reynolds stress distributions depend on a balance between the accelerating force of gravity and the resistance (channel bed and walls) in a uniform flow. In a pool, the flow depth will first increase and then decrease, which inevitably results in a decelerating flow (DF) and an adverse pressure gradient in the streamwise direction, followed by an accelerating flow (AF) and a favorable pressure gradient. The inertial effects say that in the outer flow zone (farther from the bed), streamwise velocity remains relatively high in the DF and relatively low in the AF [38]. The vegetation cover in the channel bed significantly affects the concentration of the suspended load, oxygen level, and water quality. However, the interaction of vegetation and bedforms is an important factor in river morpho-dynamics. In recent decades, many fields and laboratory studies have been conducted to investigate the effect of vegetation on flow hydrodynamics in the presence of bedforms [6,39,40], and also periodic beds [41,42]. For example, in a flume with vegetated walls, Nasiri et al. (2011) found that the Reynolds stress at downstream and upstream locations of a gravel bedform has a convex distribution profile.

To the best of our knowledge, the impact of the irregular distributions of vegetated patches on the characteristics of turbulent flow in the presence of 3D bedforms has hardly been studied. The objectives of the present study are, therefore, to investigate the following questions:

How do the velocity, Reynolds stresses, and turbulence kinetic energy change with the irregularity of vegetation cover over an artificial 3D pool? What are the dominant events in the bursting process when the submerged vegetation is distributed irregularly over a pool?

To answer these questions, laboratory experiments have been conducted in a straight, artificial pool with the presence of submerged rigid vegetation patches with various irregular distribution patterns. Laboratory results were compared to those of a natural gravel river in central Iran which has a similar flow depth, vegetation cover, and grain size.

2. Materials and Methods

Experiments were carried out in both a natural river and a laboratory flume. The purpose of field investigations in a natural river was to find the slopes for both the entrance section and exit section of pools in a gravel-bed river, grain size of bed material, and flow condition. Based on the data collected in the natural gravel-bed river, a pool was constructed in a laboratory flume without conducting any dimensional analysis and scaling. The pool in the laboratory flume had approximately similar conditions to those of the gravel-bed river, including morphology, grain size, and flow depth.

2.1. The Field Study

The Marbor Padena River is located in the south of Isfahan Provence in central Iran. The study site was a straight gravel-bed reach. Figure 1 shows the location of the study river reach where a pool with irregular submerged vegetation is prevalent.

Figure 1. (**a**) Location of the selected reach in the Padena Marbor River; (**b**) topographic surveying.

The geometric dimensions of this pool were determined based on the topographic survey along the selected river reach which is 8.5 m long. The average width of this reach is 4.4 m, with the entrance and exit slopes of 7.4 degrees and 4 degrees, respectively. The flow depth H at the centerline of the pool entrance was 0.45 m, measured with a ruler with an accuracy of 1 mm. The flow velocity was also measured using a Butterfly Current Meter with an accuracy of 0.1 m/s. The time of velocity recordings at each point was 50 s, and the measurements at each point were repeated three times. The flow rate Q was calculated by using the continuity equation $Q = \sum_{i=1}^{n} u_i A_i$ where A_i is the area of each sub-cross-section and u_i is the mean velocity in each sub-cross-section. At the pool entrance, the discharge was Q = 1.79 m/s^3, the average flow velocity U = 1.0 m/s, the Reynolds number $Re = UH/\vartheta_m = 424{,}620$, and the Froud number $Fr = U/(gH)^{0.5} = 0.5$. The values of Re and Fr indicate a fully turbulent and sub-critical flow at the pool entrance.

Figure 2 presents the grain size distribution of bed material in both the selected pool in the river and the artificial pool in the flume, obtained by using the Wolman technique [41]. In fact, gradings were done at the pool entrance, in the pool, and the pool exit for surface layers (as the upper limit of grading) and the lower layer (as the lower limit of grading). The grading used in the laboratory study is the middle one of these grain size distributions. The median grain size of the bed material in the field study ranged from 9.2 mm to 23.4 mm with the geometric standard deviation ($\sigma_g = (d_{84}/d_{16})^{0.5}$) = 2.63, where, d_{84} and d_{16} are 16th and 84th percentile of the particle-size distribution, respectively, and show the grain size distribution is non-monotonous [17].

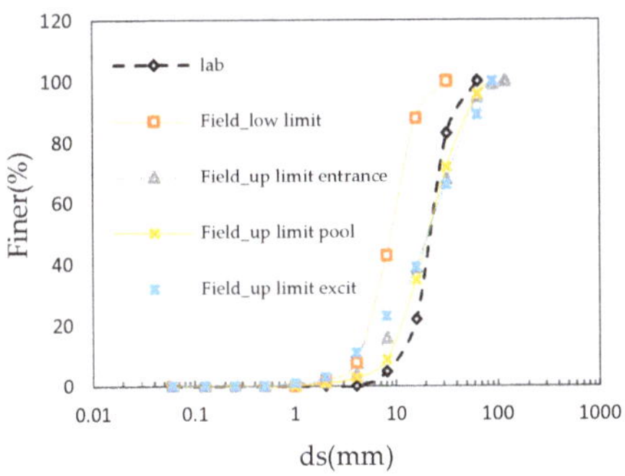

Figure 2. The grain size distribution of bed material in the river and flume.

higher values for the Coles wake parameter have been observed in wind tunnel experiments with larger pressure gradients (e.g., $\Pi_{max} = 6.9$ in a boundary layer close to the separation zone by Krogstad and Skare (1995) [64] and the flow redistribution in a straight artificial pool (e.g., $\Pi_{max} = 4$ [55]). The wake parameter Π value is positive from the upstream portion of the pool to the downstream of the pool regardless of whether the channel bed of the pool is vegetated or not, indicating that the velocity distributions continue to carry the inertial effects from the upstream of the pool into the downstream portion of the pool. Downstream of the pool (X/λ =0.92), the estimated Π value increased again, indicating that the flow is barely recovering toward uniform flow due to the presence of an irregular distribution pattern of vegetated elements for Cases I to III, and due to the variation of bedform in the riffle-pool-riffle for Case IV. However, the minimum of the wake parameter with a negative value of $\Pi = -0.8$ occurring at the downstream limit of the pool has been observed [52]. As shown in Figure 5, for all Cases except Case II in the AF section, the values of u^*_{TKE} are generally less than the values of u^*_{LW} and u^*_{RS}, even if the uniform flow reaches upstream of the pool, implying that the shear velocities calculated by these methods are different, confirming findings of some previous studies [50,65]. In the outer zone, the inertial effects of the flow changed the streamwise velocity profile such that Π increased in the DF section and decreased through the middle of the pool and in the AF section. It also, it increased in the uniform depth section downstream of the pool (Figure 5d).

4.2. The Reynolds Stress along the Channel Centerline

The Reynolds stress $-\overline{u'w'}$ is the most important parameter for estimating the shear stress in a channel bed and investigating the effects of vegetation over the bedform. As shown in Figure 6, similar to the streamwise velocity, the shape of the Reynolds stress $-\overline{u'w'}$ profiles in the pool deviated from the shape of the $-\overline{u'w'}$ profiles in the uniform flow ($X/\lambda = 0.017$, upstream of the pool) due to the AF and DF as well as the presence of an irregular distribution pattern of vegetated elements in channel bed for all Cases. Figure 6 shows the Reynolds stress distribution for all Cases. In the upstream section of the pool (Figure 6a) the flow is normal (e.g., $X/\lambda = 0.5$) and $-\overline{u'w'}$ shows an irregular pattern which is due to a complex interaction of the flow and submerged vegetation. For all Cases, the Reynolds stresses also changed significantly near the bed due to a complex interaction of bedform and an irregular vegetation patch on the bed. Upstream of the pool, very close to the channel bed, the Reynolds stress $-\overline{u'w'}$ decreased toward the bed within a layer approximately 1.3 cm thick. This trend indicates that the bed may not have been utterly flat in the upstream section of the pool. For all Cases inside of the pool, the maximum $-\overline{u'w'}$ occurs at a larger distance above the bed surface depending on the area density of vegetation. For instance, for vegetated Case I, the maximum $-\overline{u'w'}$ occurred at a distance of about 16 cm from the bed, and for vegetated Case III, at a distance of about 19 cm (Figure 6(c-i,c-j)). This effect normally occurs in the DF section (Figure 6(b-i,b-j)) and the AF section (Figure 6(d-i,d-j)) due to the positive and negative streamwise pressure gradients, respectively. In the downstream section of the pool (Figure 6(e-i,e-j)), the distribution of $-\overline{u'w'}$ is different from the distribution observed in the upstream section of the pool due to severe perturbations caused by bedform and the irregular pattern of submerged vegetated elements. The $-\overline{u'w'}$ at some cross-sections (e.g., $X/\lambda = 0.78$ and $X/\lambda = 0.98$) had increased significantly, especially for Case I (Figure 6(d-i,e-i)). It could be explained by flow velocities inducing strong fluctuations in the channel when they met the vegetated elements, especially in these two different zones. In a rectangular channel, the Reynolds stress $-\overline{u'w'}$ can be negative due to a velocity dip caused by secondary currents [66]. A negative Reynolds stress $-\overline{u'w'}$ indicates an upward vertical transport of momentum with negative velocity gradients. Furthermore, negative values of $-\overline{u'w'}$ in zones with positive velocity gradients may occur due to the transfer of turbulent energy from some localized zones to the mean flow [67]. In this study, due to the resistance resulting from vegetation elements, negative values of $-\overline{u'w'}$ can also be observed close to the edge of the vegetation patch (e.g., Figure 6(d-j,e-i,e-j)). However, the irregular fluctuations of stresses are due to a complex flow field caused by the pool and submerged vegetation elements which cannot be easily explained by the momentum equation.

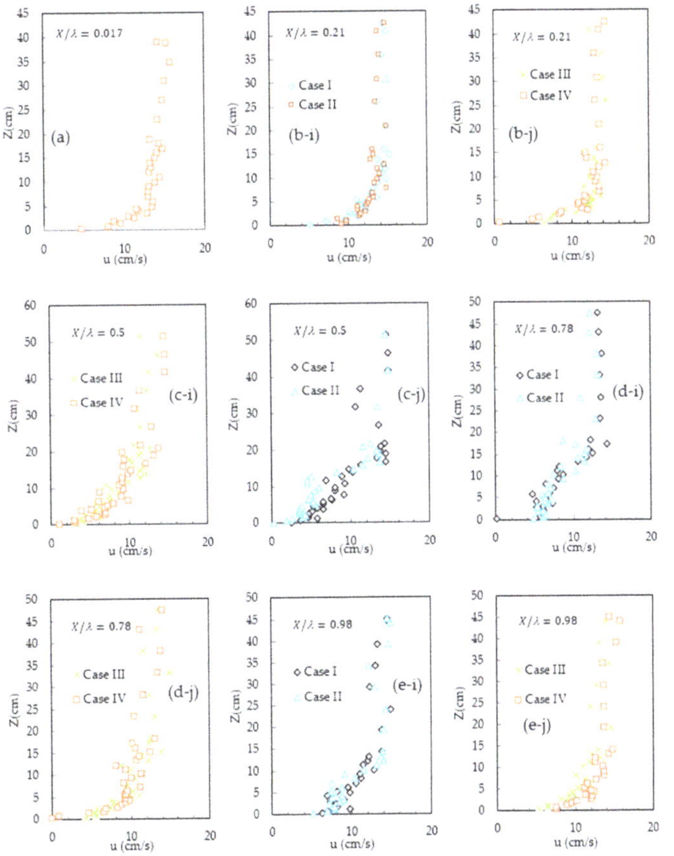

Figure 4. Profiles of streamwise velocity along the channel centerline for all Cases. (**a**) The upstream section of the pool (X/λ = 0.017, (**b-i**,**b-j**) the pool entrance (X/λ = 0.21), (**c-i**,**c-j**) the middle of the pool (X/λ = 0.5), (**d-i**,**d-j**) the pool exit (X/λ = 0.78), and (**e-i**,**e-j**) the downstream section of the pool (X/λ = 0.98).

4.3. Turbulence Kinetic Energy

To examine zones for the turbulence production and dissipation, the flow of turbulent energy can be determined by calculating the Turbulent Kinetic Energy (TKE = 0.5 $\rho\left[\overline{u'^2}+\overline{v'^2}+\overline{w'^2}\right]$), where ρ is the density of water and $\overline{u'^2}$, $\overline{v'^2}$, and $\overline{w'^2}$ are the mean square values of the fluctuations in velocity in the streamwise, lateral, and vertical directions, respectively. As shown in Figure 7, the TKE value is presented against the water depth z. The turbulent kinetic energy profile upstream of the pool (Figure 7a at X/λ = 0.017) shows approximately the same pattern as that for the Reynolds stress in Figure 6a. For all Cases in the DF section, the TKE decreases with the water depth z and its maximum appears approximately at the channel bed. However, for vegetated Case I, the TKE attains its peak at both the channel bed and a certain distance above the bed (z ≅ 1 cm). Similarly, for Case III, the TKE attains its peak at a distance from the bed of z ≅ 2.5 cm from the bed, however, the maximum TKE for Case IV (bare bed) is smaller than that for Case III (Figure 7(b-i,b-j)). For Cases I and

IV (Figure 7(b-i,b-j)) near the bed, the TKE reached its maximum value, indicating that the turbulence generated by the irregular pattern of vegetated elements in the bed plays a key role in the vertical distribution of the TKE. Inside the pool (Figure 7(c-i,c-j)), the maximum TKE occurs at a larger distance from the bed surface, depending on the area density of vegetation. The momentum exchange and turbulent energy are likely influenced by the secondary circulations associated with the irregular distribution of vegetated elements in the channel bed. However, as the area density decreases, the maximum TKE decreases in the AF section, indicating the influence of the AF pool and the submerged vegetation. Downstream of the pool ($X/\lambda = 0.98$), the TKE profile shows a similar pattern as that observed upstream of the pool (Figure 7(e-i,e-j)). The TKE values indicate irregular distribution near the vegetated elements. However, with an increase in the distance from the bed, the irregular distribution reduces (Figure 7(d-i)).

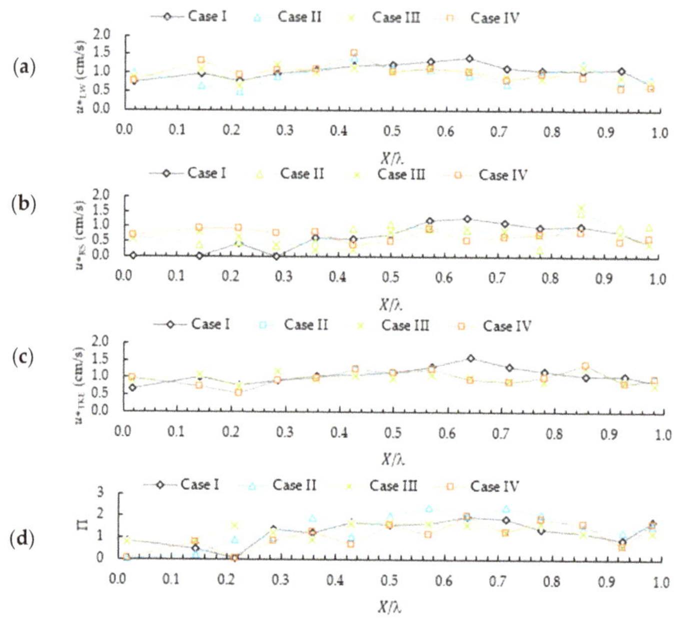

Figure 5. (**a**) Shear velocity according to the law of wall (u^*_{Lw}), (**b**) shear velocity according to the Reynolds stress (u^*_{RS}), (**c**) shear velocity according to the turbulent kinetic energy (u^*_{TKE}), and (**d**) Coles wake parameter (Π) along the channel centerline from upstream to downstream of the bedform.

The differences in TKE values at a certain flow depth might be caused by positive and negative streamwise pressure gradients and irregular patterns of vegetated elements in the AF and DF sections, respectively. These results were similar to those observed by Sukhodolov (2012), where the bias of the TKE profile indicated the cumulative effect of mixing and boundary layers generated by the area density of vegetation [68].

4.4. Quadrant Analysis

In this paper, a quadrant analysis was performed using a bursting cycle detection method [67], as shown in Figure 8. Two quadrants (Q_{ej} and Q_{sw}) contribute positively and the other two quadrants (Q_{out} interactions and Q_{inw} interactions) contribute negatively to the overall principal Reynolds stresses. To determine the contribution of each quadrant to the shear events, a computer program was written in MATLAB. Figure 8 shows the results of the quadrant analysis in the presence of irregular vegetation patterns for all Cases from the upstream section of the pool (Figure 8a at $X/\lambda = 0.017$), the pool entrance (Figure 8b at X/λ = 0.21), the middle of the pool (Figure 8c at X/λ = 0.5), the pool exit (Figure 8d at X/λ = 0.78) and the downstream section of the pool (Figure 8e at X/λ = 0.98), respectively. A quadrant analysis using the bursting cycle detection method at points of specified height, at the entire depth of the flow, was performed from the upstream section of the pool to the downstream of the pool for all Cases. For Case IV, results showed that ejections are dominant the upstream of the pool (X/λ = 0.5) at a distance of about 4 cm from the bed, as expected from the boundary layer development in a uniform flow [54], and the sweeps are stronger at a distance between 4 cm to 11 cm from the bed. Near the water surface, the velocity is higher. Thus, toward the water surface, the outward event tends to grow and reach the highest value. In the FD section (X/λ = 0.21), ejections are dominant near the bed while sweeps are stronger near the water surface for all Cases. With the presence of the irregular distribution of vegetated elements in the channel bed of the pool (X/λ = 0.5), ejections become dominant (for Case I, II, and III) and then the outward event becomes stronger as the area density of vegetation decreases (for Cases IV).

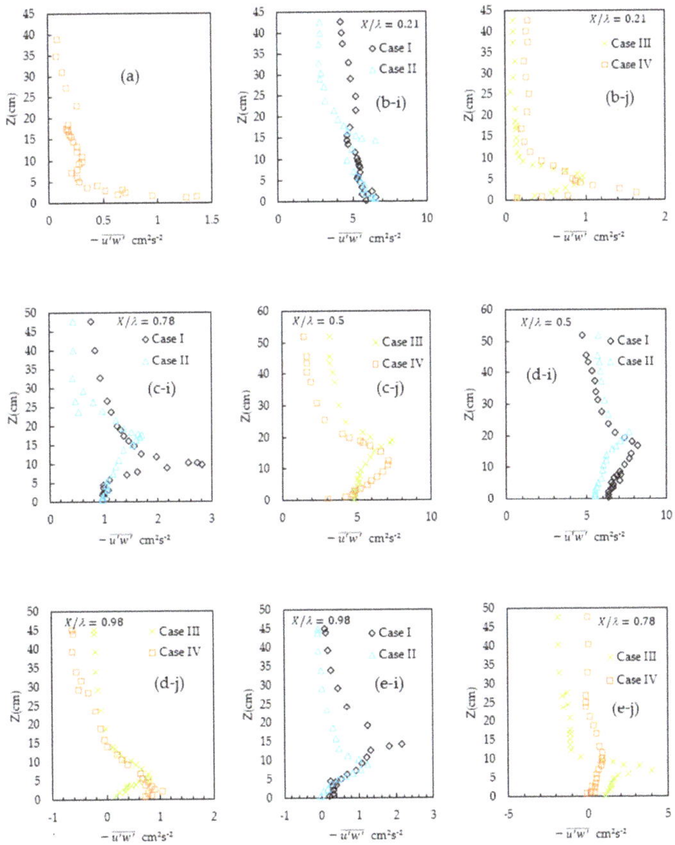

Figure 6. Profiles of the Reynolds stress $-\overline{u'w'}$ along the channel centerline for all Cases. (**a**) The upstream section of the pool (X/λ = 0.017), (**b-i**,**b-j**) the pool entrance (X/λ = 0.21), (**c-i**,**c-j**) the middle of the pool (X/λ = 0.5), (**d-i**,**d-j**) the pool exit (X/λ = 0.78), and (**e-i**,**e-j**) the downstream section of the pool (X/λ = 0.98).

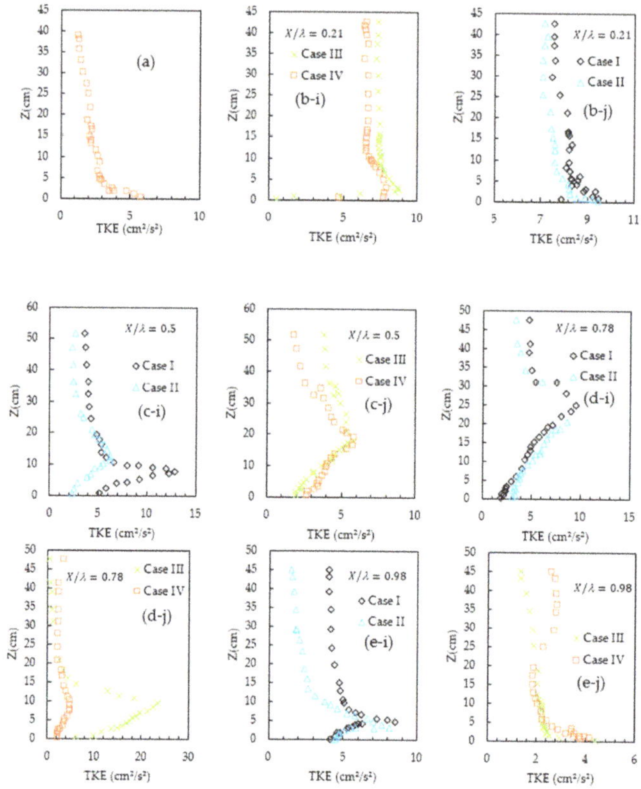

Figure 7. Profiles of the TKE values along the channel centerline for all Cases. (**a**) The upstream section of the pool (X/λ = 0.017), (**b-i**,**b-j**) the pool entrance (X/λ = 0.21), (**c-i**,**c-j**) the middle of the pool (X/λ = 0.5), (**d-i**,**d-j**) the pool exit (X/λ = 0.78), and (**e-i**,**e-j**) the downstream section of the pool (X/λ = 0.98).

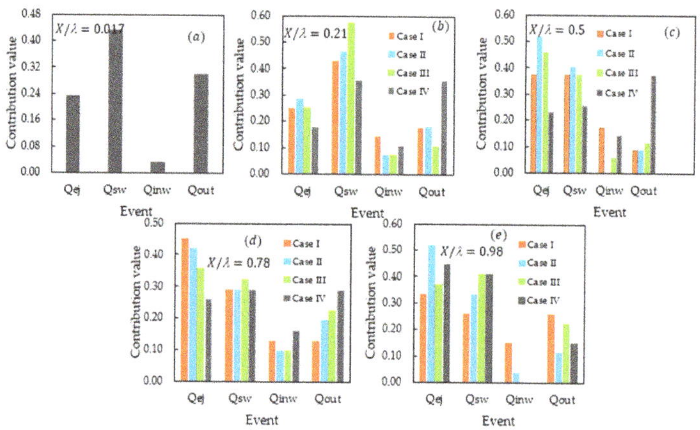

Figure 8. Quadrant analysis using the bursting cycle detection method at (**a**) the upstream section of the pool (X/λ = 0.017), (**b**) the pool entrance (X/λ = 0.21), (**c**) the middle of the pool (X/λ = 0.5), (**d**) the pool exit (X/λ = 0.78), and (**e**) the downstream section of the pool (X/λ = 0.98) for all Cases.

Within the AF section ($X/\lambda = 0.78$), sweeps are again dominant for Case IV and ejections for Cases I, II, and III. In the downstream section of the pool ($X/\lambda = 0.98$), for all Cases, ejections and sweeps are dominant. However, in the presence of an irregular pattern of vegetated elements in the pool, both the ejections and sweeps are stronger events than both inward and outward events.

5. Conclusions

In this paper, the turbulent flow field has been studied based on laboratory experiments with the presence of submerged rigid vegetation elements in the channel bed of a straight pool with different entry and exit slopes. Considering the distribution profiles of velocity, shear stress, turbulence kinetic energy, and quadrant analysis, the following results have been drawn from this study:

- The variations in velocity along the pool section with the submerged vegetation in the channel bed are different from that in the bare pool. In the entrance section of the pool where the flow is decelerating, a regular velocity distribution pattern is observed without any reversal flow even though the pressure gradient is unfavorable. However, in the exit section of the pool where the flow is accelerating and the pressure gradient is favorable, a reverse flow near the bed is observed. This is due to the appearance of a positive pressure gradient developed in this section. This difference in velocity pattern shows the influence of the submerged vegetation on the velocity distribution.
- A suitable method for estimating the shear velocity for the flow with the presence of submerged vegetation in the channel bed is the TKE method (u^*_{TKE}), since all the fluctuation components of turbulence have been used in this method. In fact, the strong lateral component of turbulence fluctuation v'^2 does not appear in the Reynolds stress method. In addition, when the area density of vegetation decreases, the location of the maximum u^*_{TKE} changes along with the pool.
- The shape and the location of the maximum value of the Reynolds stress distribution depend on the slopes of the entrance and the exit section of the pool. In addition, the distribution pattern of vegetated elements in the channel bed also affects the shape and the location of the maximum value of the Reynolds stress distribution. In general, the Reynolds stress distribution in the pool with the presence of vegetation in the bed is irregular and is considerably different from that in the bare pool.
- For all investigated Cases within the pool, the maximum Reynolds stress and TKE occur at a larger distance above the bed surface, depending on the area density of vegetation. The irregular distributions of Reynolds stress and TKE result from the secondary circulations associated with the irregular distribution pattern of vegetated elements in the channel bed.
- Results of the quadrant analysis show that the momentum between flow, bedform and vegetated elements are mostly transferred by sweep and ejection events. Toward the water surface, the outward event becomes the dominant event. At the pool entrance where the flow is decelerating, the ejection event is dominant near the bed while the sweep event is strong near the water surface. In the presence of an irregular distribution of submerged vegetated elements in the pool-bed, ejections become dominant (for Case I, II, and III) and then the outward event becomes stronger as the area density of vegetation decreases (for Cases IV).

Author Contributions: K.N., field works, methodology, software, writing—original draft, preparation; H.A., supervision, writing—review, methodology, validation and editing; J.S., methodology, writing—review and editing. All authors have read and agreed to the published version of the manuscript.

Funding: This research received no external funding.

Institutional Review Board Statement: Not applicable.

Informed Consent Statement: Not applicable.

Data Availability Statement: Data are contained within the article.

Acknowledgments: The authors would like to acknowledge that Ruhollah Nouri Pour, a water engineering graduate from Shiraz University in Iran, helped with the topographic survey of natural rivers.

Conflicts of Interest: The authors declare no conflict of interest.

References

1. Wilcock, R.J.; Nagels, J.W.; Rodda, H.J.; O'Connor, M.B.; Thorrold, B.S.; Barnett, J.W. Water quality of a lowland stream in a New Zealand dairy farming catchment. *N. Z. J. Mar. Freshw. Res.* **1999**, *33*, 683–696. [CrossRef]
2. Yamasaki, T.N.; Jiang, B.; Janzen, J.G.; Nepf, H.M. Feedback between vegetation, flow, and deposition: A study of artificial vegetation patch development. *J. Hydrol.* **2021**, *598*, 126232. [CrossRef]
3. Beltrão, G.d.B.M.; Medeiros, E.S.F.; Ramos, R.T.d.C. Effects of riparian vegetation on the structure of the marginal aquatic habitat and the associated fish assemblage in a tropical Brazilian reservoir. *Biota Neotrop.* **2009**, *9*, 37–43. [CrossRef]
4. Schoelynck, J.; De Groote, T.; Bal, K.; Vandenbruwaene, W.; Meire, P.; Temmerman, S. Self-organised patchiness and scale-dependent bio-geomorphic feedbacks in aquatic river vegetation. *Ecography* **2012**, *35*, 760–768. [CrossRef]
5. Nosrati, K.; Afzalimehr, H.; Sui, J. Drag Coefficient of Submerged Flexible Vegetation Patches in Gravel Bed Rivers. *Water* **2022**, *14*, 743. [CrossRef]
6. Nasiri Dehsorkhi, E.; Afzalimehr, H.; Singh, V.P. Effect of bed forms and vegetated banks on velocity distributions and turbulent flow structure. *J. Hydrol. Eng.* **2011**, *16*, 495–507. [CrossRef]
7. Nepf, H.M. Hydrodynamics of vegetated channels. *J. Hydraul. Res.* **2012**, *50*, 262–279. [CrossRef]
8. Nepf, H.M.; Vivoni, E. Flow structure in depth-limited, vegetated flow. *J. Geophys. Res. Ocean.* **2000**, *105*, 28547–28557. [CrossRef]
9. Kazem, M.; Afzalimehr, H.; Sui, J. Formation of Coherent Flow Structures beyond Vegetation Patches in Channel. *Water* **2021**, *13*, 2812. [CrossRef]
10. Gambi, M.C.; Nowell, A.R.; Jumars, P.A. Flume observations on flow dynamics in Zostera marina (eelgrass) beds. *Mar. Ecol. Prog. Ser.* **1990**, *61*, 159–169. [CrossRef]
11. Nepf, H.; Vivoni, E. Turbulence structure in depth-limited vegetated flow: Transition between emergent and submerged regimes. In Proceedings of the 28th International IAHR Conference, Graz, Austria, 22–27 August 1999.
12. Afzalimehr, H.; Subhasish, D. Influence of bank vegetation and gravel bed on velocity and Reynolds stress distributions. *Int. J. Sediment Res.* **2009**, *24*, 236–246. [CrossRef]
13. Baptist, M.J. A flume experiment on sediment transport with flexible, submerged vegetation. In Proceedings of the International Workshop on Riparian Forest Vegetated Channels: Hydraulic, Morphological and Ecological Aspects, RIPFOR, Trento, Italy, 20–22 February 2003.
14. Afzalimehr, H.; Barahimi, M.; Sui, J. Non-uniform flow over cobble bed with submerged vegetation strip. In Proceedings of the Institution of Civil Engineers-Water Management; Thomas Telford Ltd.: London, UK, 2019; Volume 172, pp. 86–101.
15. Ackerman, J.; Okubo, A. Reduced mixing in a marine macrophyte canopy. *Funct. Ecol.* **1993**, *7*, 305–309. [CrossRef]
16. Lynn, A.L.; Reed, D.J. Hydrodynamics and sediment transport through tidal marsh canopies. *J. Coast. Res.* **2002**, *36*, 459–469. [CrossRef]
17. Klopstra, D.; Barneveld, H.; Van Noortwijk, J.; Van Velzen, E. Analytical model for hydraulic roughness of submerged vegetation. In *Proceedings of Theme A, Managing Water: Coping with Scarcity and Abundance, Proceedings of the 27th Congress of the International Association for Hydraulic Research, San Francisco, CA, USA, 10–15 August 1997*; American Society of Civil Engineers (ASCE): New York, NY, USA, 1997; pp. 775–780.
18. Jahadi, M.; Afzalimehr, H.; Ashrafizaadeh, M.; Kumar, B. A numerical study on hydraulic resistance in flow with vegetation patch. *ISH J. Hydraul. Eng.* **2022**, *28*, 243–250. [CrossRef]
19. Baptist, M.; Babovic, V.; Rodríguez Uthurburu, J.; Keijzer, M.; Uittenbogaard, R.; Mynett, A.; Verwey, A. On inducing equations for vegetation resistance. *J. Hydraul. Res.* **2007**, *45*, 435–450. [CrossRef]
20. Liu, D.; Diplas, P.; Fairbanks, J.; Hodges, C. An experimental study of flow through rigid vegetation. *J. Geophys. Res. Earth Surf.* **2008**, *113*. [CrossRef]
21. Stone, B.M.; Shen, H.T. Hydraulic resistance of flow in channels with cylindrical roughness. *J. Hydraul. Eng.* **2002**, *128*, 500–506. [CrossRef]
22. Defina, A.; Bixio, A.C. Mean flow and turbulence in vegetated open channel flow. *Water Resour. Res.* **2005**, *41*. [CrossRef]
23. Yang, W.; Choi, S.-U. A two-layer approach for depth-limited open-channel flows with submerged vegetation. *J. Hydraul. Res.* **2010**, *48*, 466–475. [CrossRef]
24. Li, S.; Shi, H.; Xiong, Z.; Huai, W.; Cheng, N. New formulation for the effective relative roughness height of open channel flows with submerged vegetation. *Adv. Water Resour.* **2015**, *86*, 46–57. [CrossRef]
25. Pasquino, V.; Gualtieri, P. Flow resistance of submerged rigid vegetation: Focus and validation on two layer approach. In Proceedings of the 37th IAHR World Congress, Kuala Lampur, Malaysia, 13–18 October 2017; pp. 13–18.
26. Huthoff, F.; Augustijn, D.C.; Hulscher, S.J. Analytical solution of the depth-averaged flow velocity in case of submerged rigid cylindrical vegetation. *Water Resour. Res.* **2007**, *43*. [CrossRef]
27. Cheng, N.S. Representative roughness height of submerged vegetation. *Water Resour. Res.* **2011**, *47*. [CrossRef]

28. Nikora, N.; Nikora, V.; O'Donoghue, T. Velocity profiles in vegetated open-channel flows: Combined effects of multiple mechanisms. *J. Hydraul. Eng.* **2013**, *139*, 1021–1032. [CrossRef]
29. Nepf, H.; Ghisalberti, M. Flow and transport in channels with submerged vegetation. *Acta Geophys.* **2008**, *56*, 753–777. [CrossRef]
30. Raupach, M.R.; Finnigan, J.J.; Brunet, Y. Coherent eddies and turbulence in vegetation canopies: The mixing-layer analogy. In *Boundary-Layer Meteorology 25th Anniversary Volume, 1970–1995*; Springer: Berlin/Heidelberg, Germany, 1996; pp. 351–382.
31. Ghisalberti, M.; Nepf, H.M. 3-Mixing layers and coherent structures in vegetated aquatic flows. *J. Geophys. Res.-Part C-Ocean.* **2002**, *107*, 1. [CrossRef]
32. Nepf, H.M. Flow and transport in regions with aquatic vegetation. *Annu. Rev. Fluid Mech.* **2012**, *44*, 123–142. [CrossRef]
33. Zhao, H.; Yan, J.; Yuan, S.; Liu, J.; Zheng, J. Effects of submerged vegetation density on turbulent flow characteristics in an open channel. *Water* **2019**, *11*, 2154. [CrossRef]
34. Rhoads, B.L. Review of River Channels: Environment and Process. *Geogr. Rev.* **1989**, *79*, 119–121. [CrossRef]
35. Rhoads, B.L.; Welford, M.R. Initiation of river meandering. *Prog. Phys. Geogr.* **1991**, *15*, 127–156. [CrossRef]
36. Simon, A.; Bennett, S.J.; Castro, J.M. *Stream Restoration in Dynamic Fluvial Systems: Scientific Approaches, Analyses, and Tools*; John Wiley & Sons: Hoboken, NJ, USA, 2013; Volume 194.
37. Yang, S.-Q.; Chow, A.T. Turbulence structures in non-uniform flows. *Adv. Water Resour.* **2008**, *31*, 1344–1351. [CrossRef]
38. Kabiri, F.; Afzalimehr, H.; Sui, J. Flow structure over a wavy bed with vegetation cover. *Int. J. Sediment Res.* **2017**, *32*, 186–194. [CrossRef]
39. Afzalimehr, H.; Nosrati, K.; Kazem, M. Resistance to Flow in a Cobble-Gravel Bed River with Irregular Vegetation Patches and Pool-Riffle Bedforms (Case study: Padena Marbor River). *Ferdowsi Civ. Eng. (JFCEI)* **2021**, *2*, 35–50.
40. Chakraborty, P.; Sarkar, A. Flow characteristics through submerged rigid vegetation over a sinusoidal perturbed bed. *Int. J. River Basin Manag.* **2016**, *14*, 255–266. [CrossRef]
41. Chakraborty, P.; Sarkar, A. Turbulent flow through a random rigid submerged vegetation over a sinusoidal bed. *J. Appl. Water Eng. Res.* **2021**, *9*, 147–160. [CrossRef]
42. Tubino, M. Growth of alternate bars in unsteady flow. *Water Resour. Res.* **1991**, *27*, 37–52. [CrossRef]
43. Lanzoni, S.; Tubino, M. Grain sorting and bar instability. *J. Fluid Mech.* **1999**, *393*, 149–174. [CrossRef]
44. Buffington, J.M.; Lisle, T.E.; Woodsmith, R.D.; Hilton, S. Controls on the size and occurrence of pools in coarse-grained forest rivers. *River Res. Appl.* **2002**, *18*, 507–531. [CrossRef]
45. Leopold, L.B.; Wolman, M.G.; Miller, J.P.; Wohl, E.; Wohl, E.E. *Fluvial Processes in Geomorphology*; W. H. Freeman & Company: San Francisco, CA, USA, 1964.
46. Carling, P.A.; Orr, H.G. Morphology of riffle–pool sequences in the River Severn, England. *Earth Surf. Processes Landf. J. Br. Geomorphol. Res. Group* **2000**, *25*, 369–384. [CrossRef]
47. Lisle, T.E.; Hilton, S. The Volume of Fine Sediment in Pools: An Index of Sediment Supply in Gravel-Bed Streams 1. *JAWRA J. Am. Water Resour. Assoc.* **1992**, *28*, 371–383. [CrossRef]
48. Kironoto, B.; Graf, W.H.; Reynolds. Turbulence characteristics in rough uniform open-channel flow. *Proc. Inst. Civ. Eng.-Water Marit. Energy* **1994**, *106*, 333–344. [CrossRef]
49. Afzalimehr, H.; Rennie, C.D. Determination of bed shear stress in gravel-bed rivers using boundary-layer parameters. *Hydrol. Sci. J.* **2009**, *54*, 147–159. [CrossRef]
50. Afzalimehr, H.; Anctil, F. Accelerating shear velocity in gravel-bed channels. *Hydrol. Sci. J.* **2000**, *45*, 113–124. [CrossRef]
51. Kironoto, B.; Graf, W.H.; Reynolds. Turbulence characteristics in rough non-uniform open-channel flow. *Proc. Inst. Civ. Eng.-Water Marit. Energy* **1995**, *112*, 336–348. [CrossRef]
52. Song, T.; Chiew, Y. Turbulence measurement in nonuniform open-channel flow using acoustic Doppler velocimeter (ADV). *J. Eng. Mech.* **2001**, *127*, 219–232. [CrossRef]
53. Coles, D. The law of the wake in the turbulent boundary layer. *J. Fluid Mech.* **1956**, *1*, 191–226. [CrossRef]
54. MacVicar, B.J.; Rennie, C.D. Flow and turbulence redistribution in a straight artificial pool. *Water Resour. Res.* **2012**, *48*. [CrossRef]
55. Stapleton, K.; Huntley, D. Seabed stress determinations using the inertial dissipation method and the turbulent kinetic energy method. *Earth Surf. Processes Landf.* **1995**, *20*, 807–815. [CrossRef]
56. Biron, P.M.; Robson, C.; Lapointe, M.F.; Gaskin, S.J. Comparing different methods of bed shear stress estimates in simple and complex flow fields. *Earth Surf. Processes Landf. J. Br. Geomorphol. Res. Group* **2004**, *29*, 1403–1415. [CrossRef]
57. Dodaro, G.; Tafarojnoruz, A.; Stefanucci, F.; Adduce, C.; Calomino, F.; Gaudio, R.; Sciortino, G. An experimental and numerical study on the spatial and temporal evolution of a scour hole downstream of a rigid bed. In Proceedings of the International Conference on Fluvial Hydraulics, River Flow, Lausanne, Switzerland, 3–5 September 2014; pp. 3–5.
58. Dodaro, G.; Tafarojnoruz, A.; Sciortino, G.; Adduce, C.; Calomino, F.; Gaudio, R. Modified Einstein sediment transport method to simulate the local scour evolution downstream of a rigid bed. *J. Hydraul. Eng.* **2016**, *142*, 04016041. [CrossRef]
59. Yang, J.Q.; Kerger, F.; Nepf, H.M. Estimation of the bed shear stress in vegetated and bare channels with smooth beds. *Water Resour. Res.* **2015**, *51*, 3647–3663. [CrossRef]
60. Rowiński, P.M.; Kubrak, J. A mixing-length model for predicting vertical velocity distribution in flows through emergent vegetation. *Hydrol. Sci. J.* **2002**, *47*, 893–904. [CrossRef]
61. Onitsuka, K.; Akiyama, J.; Matsuoka, S. Prediction of velocity profiles and Reynolds stress distributions in turbulent open-channel flows with adverse pressure gradient. *J. Hydraul. Res.* **2009**, *47*, 58–65. [CrossRef]

62. Coleman, H.W.; Moffat, R.J.; Kays, W.M. The accelerated fully rough turbulent boundary layer. *J. Fluid Mech.* **1977**, *82*, 507–528. [CrossRef]
63. Krogstad, P.Å.; Skåre, P.E. Influence of a strong adverse pressure gradient on the turbulent structure in a boundary layer. *Phys. Fluids* **1995**, *7*, 2014–2024. [CrossRef]
64. Kim, S.-C.; Friedrichs, C.; Maa, J.-Y.; Wright, L. Estimating bottom stress in tidal boundary layer from acoustic Doppler velocimeter data. *J. Hydraul. Eng.* **2000**, *126*, 399–406. [CrossRef]
65. Kang, H.; Choi, S.U. Reynolds stress modelling of rectangular open-channel flow. *Int. J. Numer. Methods Fluids* **2006**, *51*, 1319–1334. [CrossRef]
66. Siniscalchi, F.; Nikora, V.I.; Aberle, J. Plant patch hydrodynamics in streams: Mean flow, turbulence, and drag forces. *Water Resour. Res.* **2012**, *48*. [CrossRef]
67. Sukhodolov, A.N.; Sukhodolova, T.A. Vegetated mixing layer around a finite-size patch of submerged plants: Part 2. Turbulence statistics and structures. *Water Resour. Res.* **2012**, *48*. [CrossRef]
68. Willmarth, W.; Lu, S. Structure of the Reynolds stress near the wall. *J. Fluid Mech.* **1972**, *55*, 65–92. [CrossRef]

water

MDPI

Article

Assessment of Annual Erosion and Sediment Yield Using Empirical Methods and Validating with Field Measurements—A Case Study

Ehsan Shahiri Tabarestani [1], Hossein Afzalimehr [1,*] and Jueyi Sui [2]

1 Faculty of Civil Engineering, Iran University of Science and Technology, Tehran 16846-13114, Iran; ehsan_shahiri96@civileng.iust.ac.ir
2 School of Engineering, University of Northern British Columbia, Prince George, BC V2N 4Z9, Canada; jueyi.sui@unbc.ca
* Correspondence: hafzali@iust.ac.ir; Tel.: +0098-913-2175524

Abstract: To implement soil conservation approaches, it is necessary to estimate the amount of annual sediment production from a watershed. The purpose of this study was to determine the erosion intensity and sedimentation rate from a watershed by employing empirical models, including the modified Pacific Southwest Inter-Agency Committee (MPSIAC), the erosion potential method (EPM), and Fournier. Moreover, the accuracy of these empirical models was studied based on field measurements. Field measurements were conducted along two reaches of Babolroud River. Total sediment transport, including suspended load and bed load, was predicted. Bed load transport rate was measured using a Helly–Smith sampler, and suspended load discharge was calculated by a sediment rating curve. The results of this study indicate that the erosion intensity coefficient (Z) of the Babolroud watershed is 0.54, with a deposition rate of 166.469 $m^3/(km^2.year)$. Due to the existence of unusable crops, the highest amount of erosion appeared in the northern region of the watershed. The results using the EPM and MPSIAC models were compared with field measurements and indicated that both models provided good accuracy, with differences of 22.42% and 20.5% from the field results, respectively. Additionally, it could be concluded that the Fournier method is not an efficient method since it is unable to consider the erosion potential.

Keywords: erosion intensity; Babolroud watershed; empirical methods; field measurements; sediment rating curve

Citation: Tabarestani, E.S.; Afzalimehr, H.; Sui, J. Assessment of Annual Erosion and Sediment Yield Using Empirical Methods and Validating with Field Measurements—A Case Study. *Water* **2022**, *14*, 1602. https://doi.org/10.3390/w14101602

Academic Editor: Hung Tao Shen

Received: 26 March 2022
Accepted: 9 May 2022
Published: 17 May 2022

1. Introduction

Soil is one of the most important nonrenewable natural resources in the world. The process of soil erosion involves three distinct stages: the separation of soil particles, the transfer of particles, and the sedimentation of transported particles. Soil erosion is a global concern due to its economic and environmental impacts and has received increasingly more attention from researchers in recent years. Protecting soil erosion is crucial due to its effects on soil fertility, water quality, and flooding prediction. Due to soil erosion, millions of tons of sediments enter reservoirs and cause a decrease in their storage capacity, damage to dams, reduction of the life of dams, changes in water quality, and tremendous economic losses [1,2]. In order to implement programs to control soil erosion and reduce sedimentation, it is necessary to estimate the total volume of sedimentation and the intensity of erosion from a watershed and identify the factors influencing the erosion of the watershed. Identification of these factors will help choose appropriate approaches to control erosion and conserve natural resources [3].

There have been several methods to estimate the sediment yield from a watershed. The earliest empirical method was the universal soil loss equation (USLE) [4]. The USLE method can be used to estimate the average annual erosion rate from a watershed based on rainfall pattern, soil characteristics, topography, and ground cover [5,6]. Some modifications have been suggested to enhance the performance of the USLE model, namely, the revised universal soil loss equation (RUSLE) [7] and modified universal soil loss equation (MUSLE) models [8]. Additional empirical methods have been named, such as Fournier, the Food and Agriculture Organization (FAO), the Pacific Southwest Inter-Agency Committee (PSIAC), the modified PSIAC (MPSIAC), and the erosion potential method (EPM). These methods have been successfully used by researchers in many watersheds [9–18].

Several researchers have estimated the amount of erosion and sediment using both the MPSIAC and EPM models for developing erosion maps in different basins in Iran; however, the Fournier method has not been evaluated in such basins. In some studies, better results have been generated using the MPSIAC model compared to those using the EPM model and vice versa, depending on the watersheds studied and their climatic and geological attributes. The level of soil erosion in a small agricultural watershed in eastern India was evaluated using the USLE model. The results showed that most of the eroded soil was deposited in rice crop check basins before reaching the outlet [12]. In one study, the total amount of soil loss and sediment yield was estimated using the RUSLE model by combining it with a geographic information system (GIS). As reported by the researchers, the sediment delivery ratio of a watershed studied in Ethiopia ranged from 0 to 0.26, and the highest value was reported for the central and eastern parts of the watershed [16].

Geographic information system (GIS) and remote sensing (RS) techniques have been successfully implemented in the assessment of erosion and sediment yield [19–21]. These techniques are cost- and time-effective and, in many cases, result in high accuracy [22–24]. Therefore, these techniques have been used all around the world as tools for the assessment and control of soil erosion and water resources.

Total sediment transport mainly consists of suspended load and bed load. Generally, the ratio of bed load to suspended load of a river is about 5–25% [25]. Due to difficulties of measuring bed load in the field, few studies have been conducted in this regard [26–28]. In one study, changes in runoff and sediment transport in the Middle Reach of the Huai River were studied using 58 years of field data [29]. Bed load sediments were calculated based on data collected using a Helley–Smith sampler in gravel bed rivers, and some of the universal bed load predictors were evaluated with the measured data. The objectives of field measurements were to evaluate the bed load transport of Babolroad River and predict suspended load with the rating curve for two sedimentation stations: Darounkola and Kerikchal Stations.

Accurate identification of eroded areas and the severity of the destruction of soil resources in a watershed will help planners optimize the management of a watershed. In this study, based on data collected in the Babolroud watershed, three empirical models, including EPM, MPSIAC, and Fournier, were evaluated. The accuracy of these models was assessed according to field measurement data at Darounkola and Kerikchal Stations.

The main difference between this study and other studies mentioned in the literature is that the results of field measurements are used for validating the empirical methods. The originality of this study is the application of field measurements for the validation of erosion intensity. To our knowledge, most of the reported studies have evaluated empirical methods by comparing the results with the amount of sedimentation reported in some stations, but they have not assessed the accuracy of field measurements.

2. Materials and Methods

2.1. Study Area

The Babolroud watershed is located between 36°0′2″ and 36°36′35″ north latitude and 52°28′40″ to 52°47′2″ east longitude in Mazandaran Province, Iran (Figure 1). Babolroud River originates from the northern front of Alborz Mountain, and it is bounded by the Talar watershed and Siahroud River in the east and Haraz River in the west. The annual flow discharge of Babolroud River is about 11 m^3/s, and the riverbed mainly consists of gravel and coarse particles. The drainage area of the Babolroud River watershed is about

962 km^2, and this watershed contains five main sub-basins. The area is mostly cold and semi-wet, with an annual precipitation depth of 782 mm and an average temperature of 14.14 centigrade. The maximum and minimum elevations of the watershed are 3677.6 and −14.8 m, respectively. A large part of the middle and southern regions is mountainous and covered by a dense forest of beech, oak, and broadleaf. In the northern region, the land is mainly used for agricultural purposes.

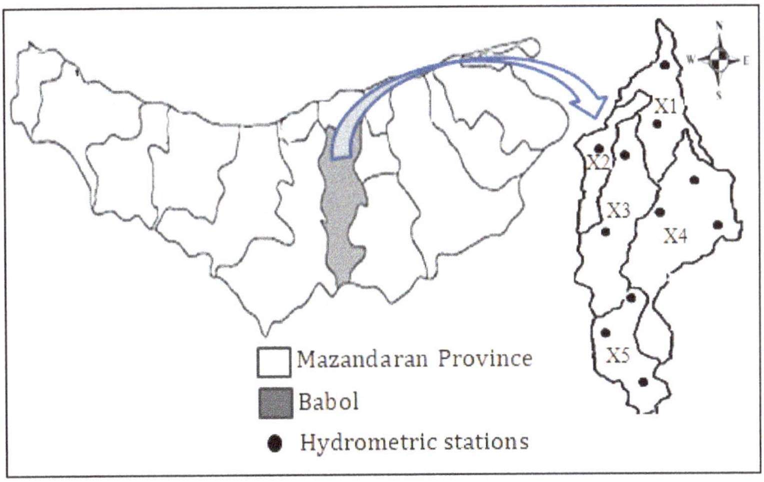

Figure 1. Location of the Babolroud watershed in Iran, Mazandaran Province.

2.2. *MPSIAC Method*

The PSIAC method was proposed in 1968 by the Water Management Committee of the United States to calculate the severity of soil erosion and sediment production in arid and semi-arid regions of the western United States. After modification, it was named modified PSIAC (MPSIAC) [30]. Compared to other existing empirical methods, the MPSIAC method considers more effective erosion factors and reduces the error in estimating the amount of sediment transported [1]. In the MPSIAC method, the impact of 9 effective factors with respect to soil erosion, including geology, soils, climate, runoff, topography, land cover, land use, upland erosion, and channel erosion, has been evaluated. Brief explanations of these factors are presented below.

2.2.1. Surface Geology Factor (f_1)

The surface geology factor is related to the geologic erosion index (Y_1) determined by rock types and their characteristics. Loose rocks are usually easily exposed to erosion and play a key role in sediment yield. Depending on the resistance degree of rocks against erosion, the values of this factor may vary from 0 and 10 [1], which are given in Table 1.

Table 1. Surface geology factor scoring.

Geounit	Description	f_1
Qm	Swamp and marsh	2
Pel	Medium- to thick-bedded limestone	6
Mm,s,l	Marl, calcareous sandstone, sandy limestone, and minor conglomerate	5
TRJs	Dark-gray shale and sandstone	9
K2l2	Thick-bedded to massive limestone	5
Plc	Polymictic conglomerate and sandstone	5
TRe	Bedded dolomite and dolomitic limestone	3

Table 1. *Cont.*

Geounit	Description	f_1
Ktzl	Thick-bedded to massive, white to pinkish orbitolina-bearing limestone	6
Jl	Light-gray, thin-bedded to massive limestone	5
Kbvt	Basaltic volcanic tuff	5
Qft2	Low-level piedmont fan and valley terrace deposits	5

2.2.2. Soil Factor (f_2)

This factor is estimated using $16.67 \times k$, in which k is the soil erodibility factor depending on soil texture and the amount of silt, lime, gravel, and organic matter in soil [30]. The range of changes for this factor is based on soil texture, stability of aggregates, amount of lime, organic matter, ability to spread clay particles, and soil moisture. Table 2 shows the scores allocated to the types of soils in the field.

Table 2. Soil factor scoring.

Type of Soil	f_2	k
Mollisols	6	0.36
Rock Outcrops/Entisols	3	0.18
Alfisols	7.1	0.43
Inceptisols	8	0.48
Mollisols	6	0.36
Inceptisols	8	0.48
Alfisols	7.1	0.43

2.2.3. Climate Factor (f_3)

The amount of runoff from a watershed depends on the amount and intensity of precipitation. The amount and intensity of precipitation influence the amount of erosion. This factor depends on the frequency of precipitation, the intensity of precipitation, and the period of precipitation, snow, ice, and melting. This factor is determined by $0.2 \times P_2$, in which P_2 is the precipitation amount during a period of 6 h with a return period of 2 years (mm).

2.2.4. Runoff Factor (f_4)

To assess the effect of runoff on soil erosion, it is necessary to consider the hydrological characteristics of the watershed, such as the specific flow of floods ($m^3s^{-1}km^{-2}$), the specific flow with different return periods, and the hydrological groups of soils. The runoff factor is estimated by $f_4 = 0.006R + 10Q_P$, in which R is the total average runoff depth (mm) that is interpolated from measurements at the meteorological stations, and Q_P is the peak special discharge ($m^3s^{-1}km^{-2}$) determined from the peak discharge at the hydrological units.

2.2.5. Topography (f_5)

The topography factor is usually determined in accordance with the average slope of a watershed. Erosion usually increases with the slope of a watershed because of the increase in the speed of the runoff generated from a watershed. This factor can be calculated by $0.33 \times S$, in which S is the average slope of a watershed in percentage. The map of the average slope can be generated from the digital elevation model. The topography factor is very important in determining soil erosion from a watershed in the MPSIAC method by considering the score of this factor ranging from 0 to 20 [1].

2.2.6. Ground Cover (f_6)

Vegetation, litter, and rocks are types of ground cover. The presence of any of these three covers can have positive effects on preventing the watershed from soil erosion and sediment yield. The ground cover factor can be determined by $0.2 \times P_b$, in which P_b is the percentage of the bare cover accounted for in a watershed. The value of this factor ranges from −10 to 10 [1].

2.2.7. Land Use (f_7)

To determine this factor, two criteria are usually considered: the first is agricultural activities, and the second is livestock grazing status. If agricultural activities are not common at the basin level, or the watershed area is covered with dense vegetation and is less likely to be domesticated, the role of this factor in soil erosion and the sediment yield from this watershed are negative. This factor can be determined by $20-0.2\ P_C$, in which P_C is the coverage of the plant canopy in percentage. The value of the land use factor ranges from −10 to 10 [1].

2.2.8. Upland Erosion (f_8)

Surface erosion in a watershed is assessed using this factor. This factor is considerably important for determining the sediment yield from a watershed, with a score ranging from 0 to 25 [1]. To assess the surface factor of soil (S.S.F.), seven aspects are considered, including soil mass movement, petiole cover, rock surface cover, pedestalling rock fragments, surface grooves, waterway form, and development of ditch erosion. This factor is estimated by $0.25 \times$ S.S.F., in which S.S.F. is the sum of scores in the BLM method [30].

2.2.9. Channel Erosion (f_9)

Regarding the erosion from channels in a watershed, both erosion from channel banks and sediment transport by the flow are examined. Channel erosion is the result of the destruction of channel banks, which occurs mostly during floods and watery seasons. Some factors that have major effects on the deformation of the channel bed and sediment transport are the average slope of riverbeds, type of rocks along rivers and potential energy of floods. This factor ranges from 0 to 25 [1] and can be calculated by $1.67 \times$ SSF.g, in which SSF.g is the gully erosion in the BLM method [30].

2.2.10. Sediment Flux

Based on the degree of impact of each factor, scores are assigned to each factor Finally, the total score is calculated, and the annual rate of sediment yield (Q_S) is estimated by the following equation.

$$Q_s = 38.77e^{0.0353R} \tag{1}$$

where R is the total sum of factors, and Q_S is the annual rate of sediment yield from each sub-basin in m^3/km^2. In this method, the amount of soil erosion of each unit is called sediment load, which is the sum of suspended load and bed load. According to the amount of sediment produced from a watershed, the sedimentation class of each sub-basin can be obtained from Table 3.

Table 3. Classification of soil erosion in MPSIAC model.

Sediment Production $m^3/(km^2.year)$	Erosion Intensity	Erosion Classification
>1429	Very high	V
476–1429	High	IV
238–476	Moderate	III
95–238	Low	II
<95	Very low	I

2.3. EPM Method

The erosion potential method (EPM) was initially developed based on data collected in Yugoslavia [31]. In this method, four criteria, including watershed erosion coefficient ($\varnothing\phi$), land use coefficient (X_a), coefficient of rock and soil resistance to erosion (Y), and average slope of the watershed (I), are examined. The erosion intensity coefficient (Z-factor) from a sub-basin can be determined using Equation (2):

$$Z = X_a.Y.\left(\phi + I^{0.5}\right) \quad (2)$$

The classification of sub-basins is shown in Table 4 according to the severity of erosion. In this method, the annual rate of sedimentation, q_s in $\frac{m^3}{km^2.year}$, can be calculated by Equation (3):

$$q_s = T.H.\pi.Z^{1.5} \quad (3)$$

where Z is the sediment yield from a sub-basin and can be calculated from Equation (2), H is the mean annual precipitation depth (mm), and T is the temperature coefficient determined by the following equation:

$$T = \left(\frac{t}{10} + 0.1\right)^{0.5} \quad (4)$$

where t is the mean annual air temperature in the watershed (°C)

Table 4. Classification of erosion intensity in EPM method.

Ranges	Erosion Intensity	Erosion Classification
Z > 1	Very high	V
0.71 < Z < 1	High	IV
0.41 < Z < 0.71	Moderate	III
0.2 < Z < 0.71	Low	II
Z < 0.2	Very low	I

2.4. Fournier Method

The Fournier method is a rather simple method for the assessment of erosion intensity because it does not require complex calculations and experimental research. It was initially developed for estimating erosion resulting from rainfall [32]. Fournier proposed two different methods for estimating the annual rate of sediment yield from a watershed.

The first method proposed by Fournier for estimating sediment yield from a watershed is as follows:

$$\text{Log } Q_{S1} = 2.65 \text{ Log}\frac{P_W^2}{P_a} + 0.46 \text{ Log } H\,(\tan S) - 1.56 \quad (5)$$

where Q_S is the annual sediment yield in $\frac{ton}{km^2.year}$, P_W is the average precipitation depth during the rainiest month of each year in the statistical period (mm), P_a is the annual precipitation depth (mm), H is the average height of the watershed (m), and S is the average slope of the watershed (degree).

The second method proposed by Fournier for estimating sediment yield from a watershed is as follows:

$$\text{Log } Q_{S2} = 2.65 \text{ Log}\frac{P_W}{P_a} + 0.46 \text{ Log}\frac{H^2}{A} - 1.56 \quad (6)$$

where A is the drainage area of the watershed (km^2), and other terms are similar to Equation (5). One of the main disadvantages of the Fournier methods is that they do not examine the erosion potential of the basin [32]. Therefore, if two regions are similar in terms related to Equations (5) and (6), but different in terms of geological, soil, and vegetation conditions, the estimated sedimentation using Equations (5) and (6) will be the same.

2.5. Research Data

Estimating the amount of sediment yield from a watershed requires various information such as topography, geology, soil, land use, rainfall, land slope, and temperature of the area. Figure 2a shows the digital elevation map of the Babolroud watershed. The southern part of the watershed is mainly occupied by mountains, but there are plain areas with low elevations in the northern part of the watershed. Figure 2b shows the geological map of the study area. There are a lot of alluvial formations in the northern region, and there is a layer of hard and shallow rocks in the southern part. A large part of the watershed has Alfisol soil, as shown in Figure 2c. This type of soil is often observed in humid and semi-humid areas with the presence of forest. Figure 2d shows the land use map of the Babolroud watershed. One can see from this figure that the watershed is covered by dense forest and mountainous areas, orchards, forest–agricultural areas, and groves. Areas with dense vegetation, which covers a major part of the watershed, play a key role in preventing this watershed from the soil erosion process.

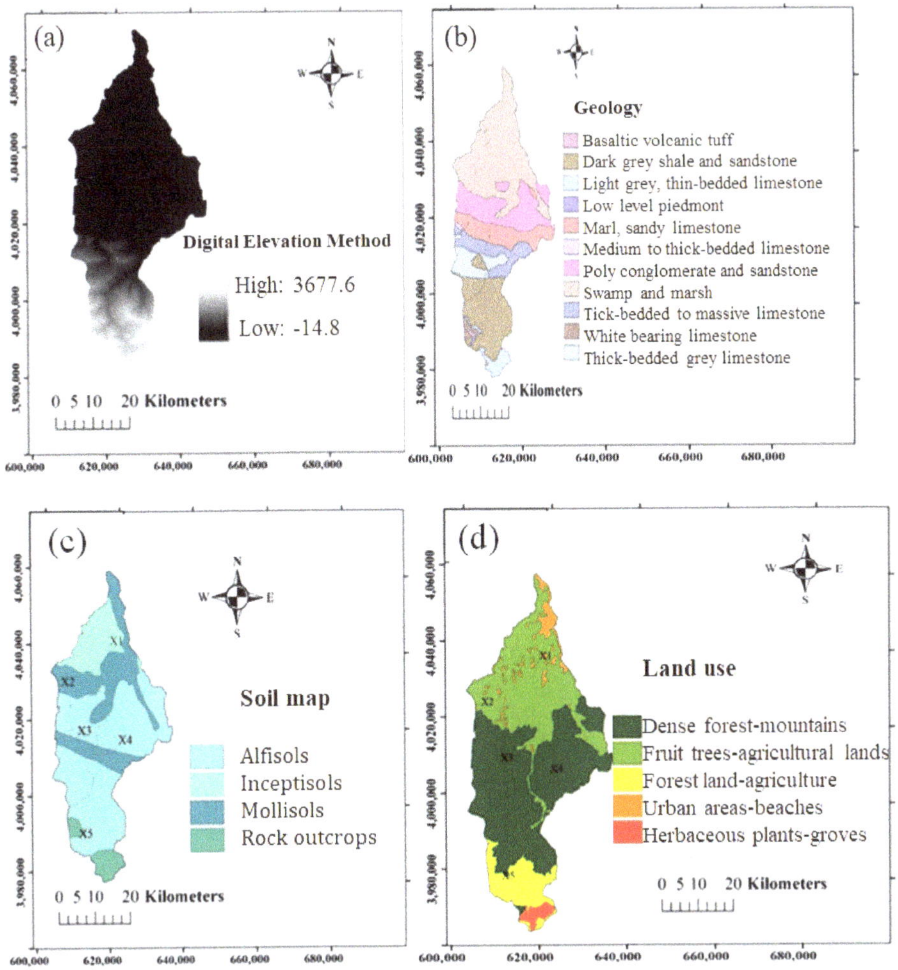

Figure 2. *Cont.*

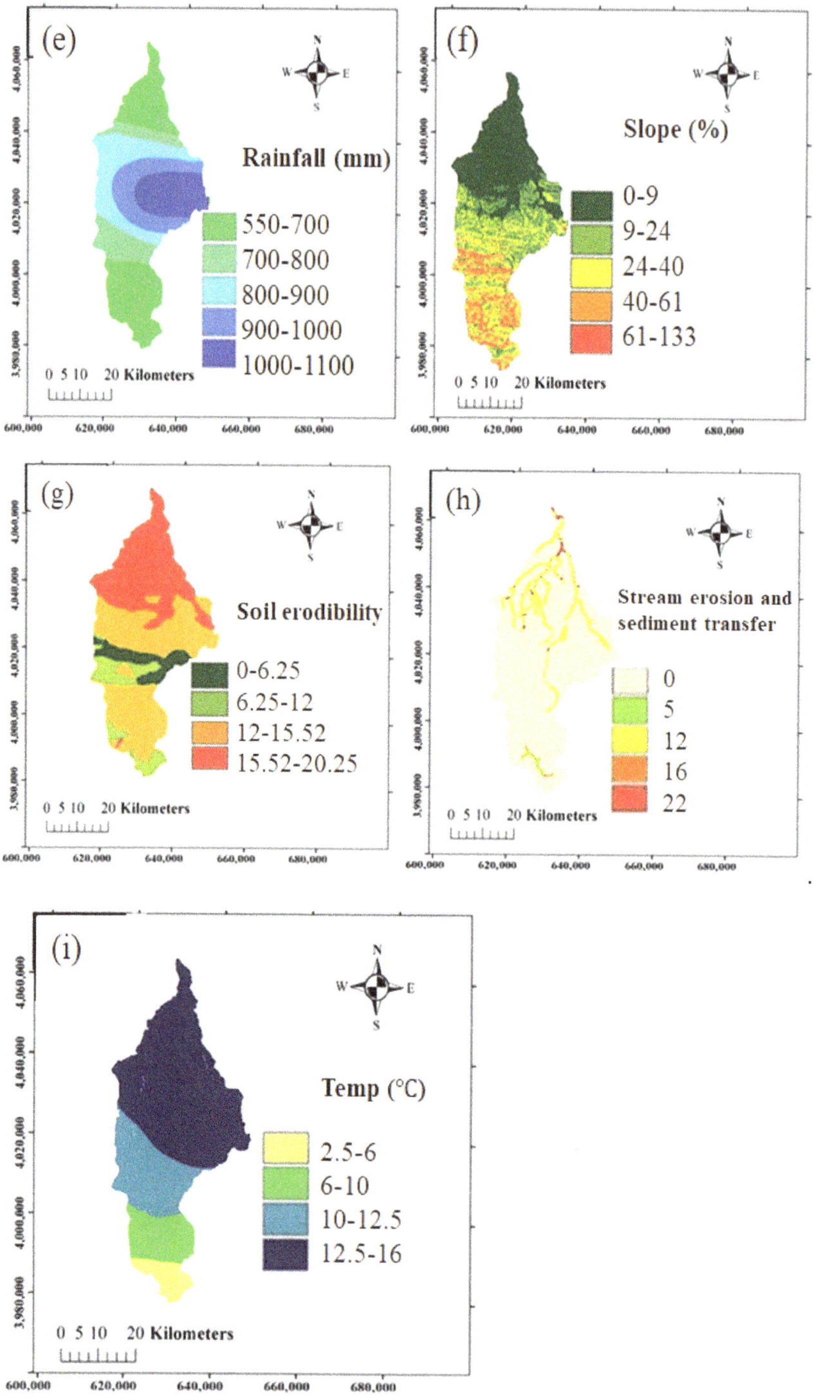

Figure 2. Maps of the study watershed: (**a**) digital elevation model; (**b**) geology; (**c**) soil map; (**d**) land use; (**e**) annual precipitation (mm) (**f**) slope; (**g**) upland erosion; (**h**) channel erosion; (**i**) temperature.

Based on data collected at 11 meteorological stations during the period from 2008 to 2018 in this watershed, statistical analysis was conducted. To develop a spatial map of rainfall intensity, the modified Fournier index was determined based on the meteorological information of 11 hydrometric stations from 2008 to 2018, as shown in Equation (7).

$$MFI = \sum_{i=1}^{i=12} \frac{P_i^2}{P} \quad (7)$$

MFI is the modified Fournier index, Pi is the average monthly rainfall depth, and P is the average annual rainfall depth at the "i" meteorological station. This index shows the sum of the weighted monthly rainfall depth at the "i" station. By means of this technique, the daily rainfall data are converted to monthly rainfall for the purpose of developing a raster rainfall map. The average annual precipitation depth in this watershed ranges from 500 to 1100 mm, as shown in Figure 2e. The maximum precipitation depth occurs in the east part of the watershed and gradually decreases towards the north and south of the watershed. As shown in Figure 2f, the highest slope of the watershed appears in the southern part of the basin due to its mountainous landscape, and the lowest slope in the northern part of the basin where a lot of flat plains appear. Additionally, based on temperature data collected during the period from 2008 to 2018, the temperature map of this watershed was developed. As shown in Figure 2i, in the northern part of the watershed, higher temperatures are observed, as the elevation is lower, and in the southern part of the watershed, lower temperatures are observed, as the elevation is higher. The sources and types of each influential factor are presented in Table 5.

Table 5. Sources and types of applied data.

Dataset	Source	Data Type	Scale of Source Data	Derived Factors
Digital elevation model (DEM)	United States Geological Survey (USGS) site	Raster	1:25,000	Elevation, slope
Rainfall	10-year meteorological data (2009–2019), Iran	Vector	1:25,000	Rainfall map
Geological map	Mazandaran Regional Water Authority, Iran	Vector	1:100,000	Geology, soil type
Land cover	Mazandaran Regional Water Authority, Iran	Vector	1:100,000	Land use

2.6. Field Data

Field data were collected at 7 cross-sections along two straight and relatively stable river reaches of Babolroud River, named the Kelarikola and Darounkola reaches (Figure 3). The Kelarikola reach is located about 8 km downstream of the Darounkola reach. The stations for measurement sediment in the Darounkola and Kerikchal reaches are located at cross-sections D2 and K3, respectively.

All data were collected during the early spring since the maximum sediment transport occurs during the spring season with high flow. At each cross-section, the channel width (from the left bank to the right bank) was divided into equal spacing intervals of one meter. By applying Wolman's method, the median grain size of both surface and subsurface layers in the channel bed was more than 10 mm. Flow depth, flow velocity, bed slope, and bed load transport rate were determined at each cross-section. The flow velocity was measured using a current meter (BFMS-N-002-1678) with an impeller diameter of 40 mm. At each point, flow velocity was measured 3 times, and the average value of these 3 measurements was used to represent the flow velocity at this point. Along each vertical line from the channel bed to the water surface, there were about 12 points for velocity measurements. Based on measured results, the velocity profile along this vertical line was developed. In this study, a Helly–Smith sampler was used for measuring the bed load sediment. This sampler was placed on the channel bed. After a period of 60 s, particles that were larger than the mesh size were collected in a bag. Then, all samples collected were taken to the laboratory, and grain size was obtained for each cross-section. In this study, to conduct the bed load sampling measurements, the flow cross-section was divided into 4 to 5 equal subsections. In each subsection, bed load sampling measurements were carried out and repeated 8 to 10 times at each point. The bed load samples collected at each point were analyzed. The grain size distribution of the bed load sediment for one cross-section (K3) is shown as an example in Figure 4.

Figure 3. Presentation of two selected reaches: (**a**) Kelarikola reach; (**b**) Darounkola reach.

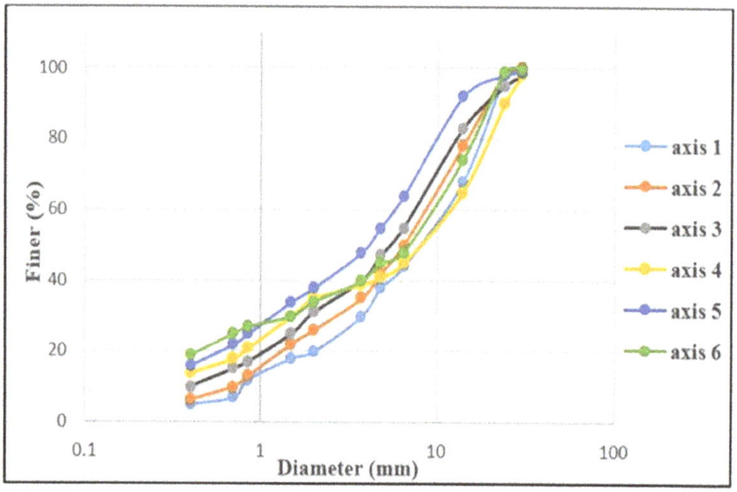

Figure 4. Grain size distribution of bed load at cross-section K3.

Figure 5 shows the bed load discharge at cross-sections D2 and K3. By calculating the area under the curve in Figure 5, the bed load per unit time (gr/s/m) was determined for cross-sections D2 and K3, respectively. In this figure, the bed load rate and the width of the riverbed are shown on the vertical and horizontal axes, respectively.

Bed load discharge is calculated using the following equation [33].

$$Q_b = \frac{1}{2}[(L_1 \times W_{tb1}) + L_2 \times (W_{tb1} + W_{tb2}) + \ldots + L_i \times (W_{i-1} + W_i) \quad (8)$$

where Q_b is the bed load discharge (g/s), L is the length between two points (m), and W_{tb} is the dry weight per unit time and per unit width (g/s/m) and is estimated by the following equation.

$$W_{tb} = \frac{M}{W_s \times n_s \times t_s} \tag{9}$$

where M is the dry mass (gr), W_S is the width of the sampler (m), n_s is the number of repeated samplings, and t_s is the time of the sampling duration (s).

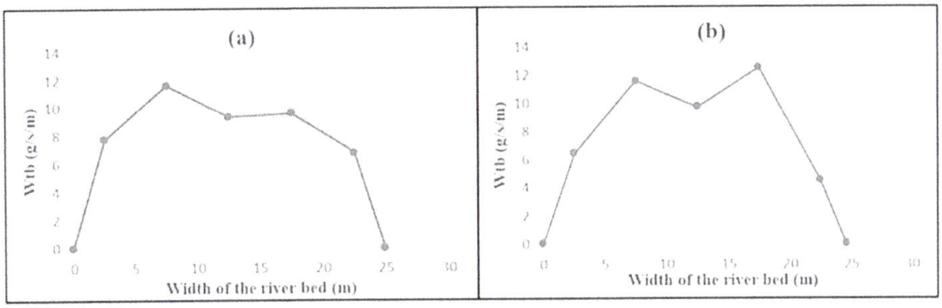

Figure 5. Presentation of bed load discharge at cross-section: (**a**) D2; (**b**) K3.

3. Results

3.1. Determination of Sediment Production and Erosion Class using the MPSIAC Method

As discussed above, in the MPSIAC method, there are nine effective factors impacting the rate of erosion and sediment production. By determining these factors and using Equation (1), the R parameter can be obtained, as shown in Figure 6.

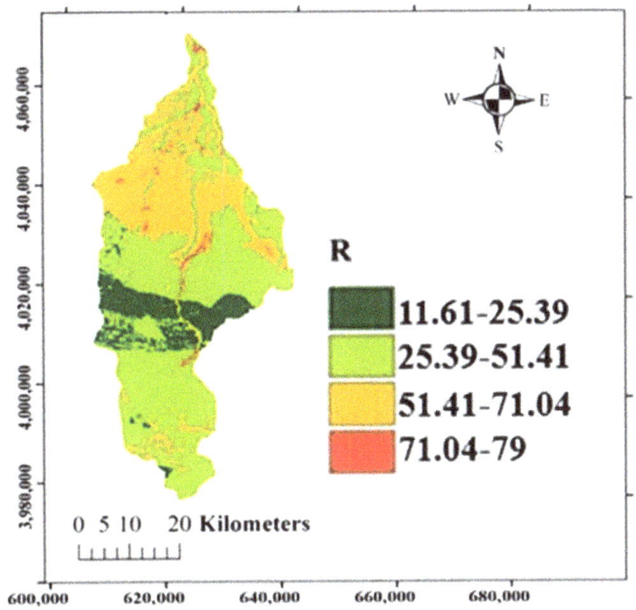

Figure 6. R parameter in the Babolroud watershed.

As shown in Table 6, the amount of sediment production is obtained for the sub-basins. A watershed can be classified into different erosion classes. According to Table 5, sub-basin X_1 located in the northern part of the watershed has a moderate erosion class, and the rest of the sub-basins are located in regions with a low erosion class. The main reason for this is the existence of uncultivated lands in the northern part of the watershed. Furthermore, the whole area with the low erosion class is attributed to the presence of dense pastures and calcareous formations and rocks with medium to high hardness in most areas. Additionally, in the middle belt of the watershed, where the soil is of Mollisol type, R values are placed in the lowest category, which indicates the importance of the type of soil in the sediment yield in this region.

Table 6. Amount of sediment yield and classification of sub-basins of the Babolroud basin.

Region	R	q_s ($\frac{m^3}{km^2 * year}$)	Area (km^2)	Qs ($\frac{m^3}{year}$)	Class
X_1	51.9	238.44	166	39,582.19	III
X_2	45.167	188.38	94	17,708.26	II
X_3	36.327	138.25	226	31,245.38	II
X_4	35.562	134.60	147	19,786.38	II
X_5	40.053	157.51	329	51,821.45	II
basin	41.27	166.469	962	160,143.17	II

3.2. *Calculation of Erosion Intensity Coefficient and Annual Sediment Yield by the EPM Method*

In order to estimate parameter Z, the following information will be used, including the amount of gully and groove erosion, the type of land use, the type of formation and soil, and the slope of the area. For this purpose, the values for this required information for the Babolroud watershed were determined and are summarized in Tables 7–9. Using the values in Tables 7–9, the sub-basins of the Babolroud were classified using the EPM method based on the intensity of erosion and the amount of sediment produced, as shown in Table 10.

Table 7. Estimation of parameter, X_a for the Babolroud watershed.

Herbaceous Plants–Groves	Urban Areas–Beaches	Forest Land–Agriculture	Fruit Trees–Agricultural Lands	Dense Forest–Mountainous Lands
0.4	1	0.3	0.7	0.2

Table 8. Estimation of parameter, Y for the Babolroud watershed.

Geo-unit	Description	Y
Qm	Swamp and marsh	2
Pel	Medium- to thick-bedded limestone	1
Mm,s,l	Marl, calcareous sandstone, sandy limestone, and minor conglomerate	1
TRJs	Dark-gray shale and sandstone	1
K2l2	Thick-bedded to massive limestone	1
Plc	Polymictic conglomerate and sandstone	1.2
TRe	bedded dolomite and dolomitic limestone	1
Ktzl	Thick-bedded to massive, white to pinkish orbitolina-bearing limestone	1
Jl	Light-gray, thin-bedded to massive limestone	1
Kbvt	Basaltic volcanic tuff	1
Qft2	Low-level piedmont fan and valley terrace deposits	2

Table 9. Estimation of parameter, ϕ for the Babolroud watershed.

Urban Areas	Floodplain	Lowlands	Alluvial Plain	Hillside	Plateau	Crop Coverage	Forest Cover
0.3	1	0.6	0.8	0.5	0.2	0.15	0.1

Table 10. Classification of sub-basins using EPM method.

Region	Z	R_u	$q_s\ (\frac{m^3}{km^2.year})$	Area (km^2)	$Q_S\ (\frac{m^3}{year})$	Class
X_1	1.2	0.31	1057.36	166	175,521.76	V
X_2	0.81	0.32	738.13	94	69,384.22	IV
X_3	0.45	0.57	688.35	226	15,5567.1	III
X_4	0.32	0.74	578.87	147	85,093.89	II
X_5	0.23	1.25	236.06	329	77,663.74	II
Basin	0.54	0.79	585.47	962	563,230.71	III

As indicated in Figure 7a, the erosion intensity coefficients in the X_1 and X_2 sub-basins are very high and high, respectively, due to the presence of plains, orchards, and alluvial soils in the northern area of the watershed. In addition, according to the EPM method, the entire watershed is in the category of the moderate erosion intensity, which indicates that the calculated amount of erosion and sediment yield using this model are more than those using the MPSIAC model.

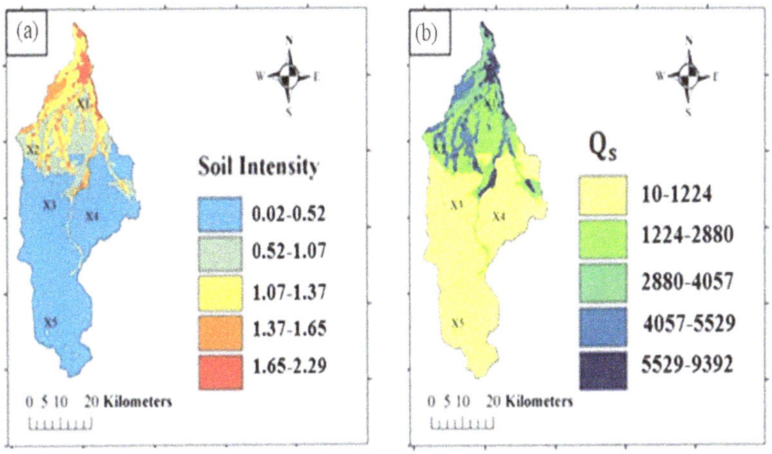

Figure 7. (**a**) Erosion intensity coefficient; (**b**) annual rate of sediment yield in the Babolroud watershed.

3.3. Calculation of Erosion Intensity Coefficient and Annual Rate of Sediment Yield Using the Fournier Method

Table 11 shows the values of the annual rate of sediment yield calculated using the first and second Fournier methods. Contrary to results using the methods discussed above, the lowest erosion rates calculated appeared in sub-basins X_1 and X_2 and the highest erosion rate in sub-basin X_5. Additionally, the values obtained using the first method have a high error, while the values determined using the second method are closer to the results using the other two MPSIAC and EPM methods. The main reason for the obvious difference between the results using the Fournier method and those using the EPM and MPSIAC methods is the lack of erosion potential in the study region. Figure 8 shows the annual sediment yield from sub-basins based on the first and second Fournier methods.

Table 11. Annual sediment yield of the Babolroud watershed based on (**a**) the first Fournier method and (**b**) the second Fournier method.

Region	Area (km^2)	$Qs_{S_1}\left(\frac{Ton}{year}\right)$	$Qs_{S_2}\left(\frac{Ton}{year}\right)$
X_1	166	1.3×10^9	23.24
X_2	94	4.2×10^9	88.36
X_3	226	3.5×10^{10}	537.88
X_4	147	3.3×10^{10}	327.81
X_5	329	8.6×10^{10}	3911.81
Basin	962	6.7×10^{11}	4889.1

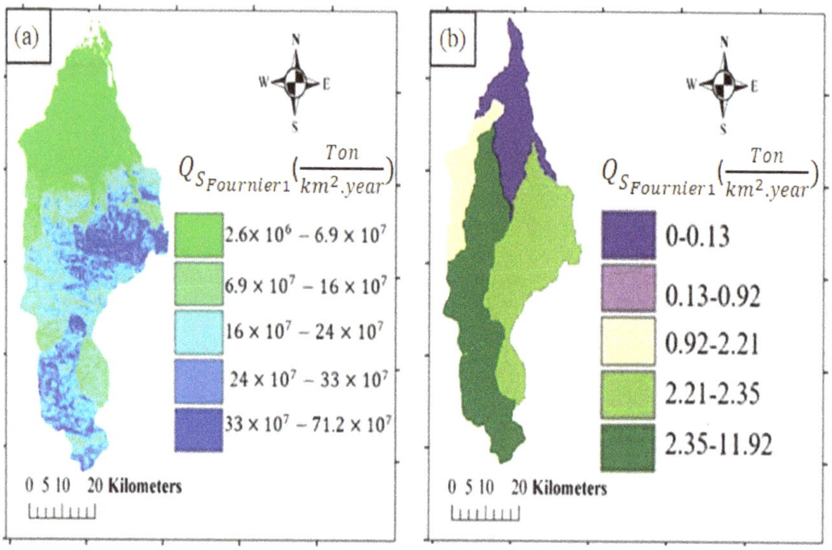

Figure 8. Annual sediment yield based on (**a**) the first Fournier method and (**b**) the second Fournier method.

3.4. Verification of Model with Field Measurements

To verify the results obtained from the models, field data collected at the two sedimentation stations, Darounkola (at D2 cross-section) and Kerikchal (at K3 cross-section), are used. Table 12 summarizes some parameters collected at these two stations in Babolroud River.

Table 12. Measured parameters in Babolroud River.

Cross-Section	Slope S (m/m)	Width W (m)	Hydraulic Depth h (m)	Mean Flow Velocity U_{eq} (m/s)	Bed Load Transport Rate q_b (ton/day)	Discharge q (m^2/s)
D1	0.0071	23.3	0.395	0.989	0.634	0.391
D2	0.0077	25	0.391	1.094	0.717	0.428
D3	0.0056	24.7	0.432	0.965	0.702	0.417
K1	0.0009	28	0.385	1.093	0.580	0.421
K2	0.0007	25.2	0.521	0.95	0.762	0.496
K3	0.0058	24.6	0.570	0.926	0.736	0.528
K4	0.0078	25.4	0.561	0.862	0.612	0.484

On the other hand, the hydrometric data set, including the suspended sediment concentration (C), water discharge (Q), and average rate of suspended load (milligrams/liter), at the four gauging stations in Babolroud River were collected by the Mazandaran Regional

Water Authority. As reported by other researchers, based on long-term field measurement data, the relationship between discharge and sediment transport can be employed to determine the features of sediment transport and assess the changes in runoff and sediment yield from watersheds [34–36].

Suspended sediment concentration samples at these stations in this river were collected several times a year, including a data set for the period from 2008 to 2018. With this data set, the sediment rating curve can be plotted (Figure 9). Similar to the results of other researchers [34–36], a regression analysis was performed to obtain a power function between suspended sediment discharge (Q_S) and water discharge (Q) as follows:

$$Q_s = 23.539Q^{1.1436} \quad (10)$$

$$R^2 = 0.678 \quad (11)$$

where Q_S is the suspended sediment discharge (ton/day), and Q is water discharge (m^3/s).

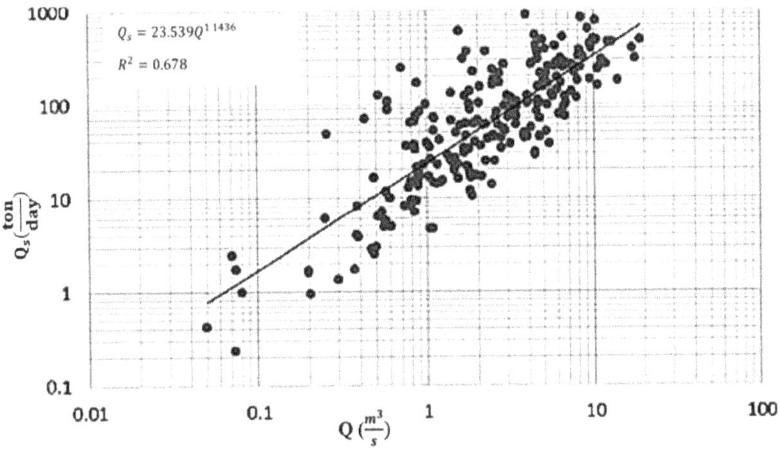

Figure 9. Sediment rating curve for Babolroud River.

Applying the sediment rating curve for Babolroud River, the discharge values of suspended sediment at both Darounkola and Kerikchal Stations were calculated. Table 13 shows results of flow discharge, suspended sediment discharge (Q_S), and bed load sediment discharge (Q_b) at all cross-sections along these two river reaches.

Table 13. Flow and sediment discharge at all cross-sections.

Cross-Section	Q (m^3/s)	Q_s ($\frac{ton}{day}$) Calculated	Q_b ($\frac{ton}{day}$) Measured
D1	9.1	294.136	14.772
D2	10.7	353.990	17.925
D3	10.3	338.897	17.339
K1	11.8	395.906	16.240
K2	12.5	422.877	19.202
K3	13	442.276	18.106
K4	12.3	415.148	15.545

Table 14 shows a comparison of the results calculated using the EPM and MPSIAC methods compared to those of field measurements at both Darounkola and Kerikchal Stations. Due to the large differences between the results using the Fournier method and those using the other two methods (EPM and MPSIAC methods), the results of the Fournier model were refused in this study.

Table 14. Results predicted by MPSIAC and EPM methods compared to the field measurements.

	Field Measurements				MPSIAC	EPM
Station	Q $(\frac{m^3}{s})$	Qs Suspended $(\frac{ton}{day})$	Qs Bed $(\frac{ton}{day})$	Qs Total $(\frac{ton}{day})$	Qs Total $(\frac{ton}{day})$	Qs Total $(\frac{ton}{day})$
Darounkola	10.7	353.99	9.86	371.915	287.38	248.272
Kerikchal	13	442.276	18.106	460.382	376.24	520.72

As shown in Figure 10, at Darounkola Station, results calculated using the MPSIAC model are closer to the results of field measurements compared to results using the EPM. At Kerikchal Station, however, results calculated using the EPM are slightly better than results calculated using the MPSIAC model. Overall, the calculation error using the EPM method for these two sedimentation stations is 22.42%, and the MPSIAC model is 20.5%. Thus, compared to the results using the EPM method, the MPSIAC method can be used to predict sediment yield closer to results from rating curves generated based on field measurements at these two stations, indicating the better performance of the MPSIAC model in the Babolroud watershed.

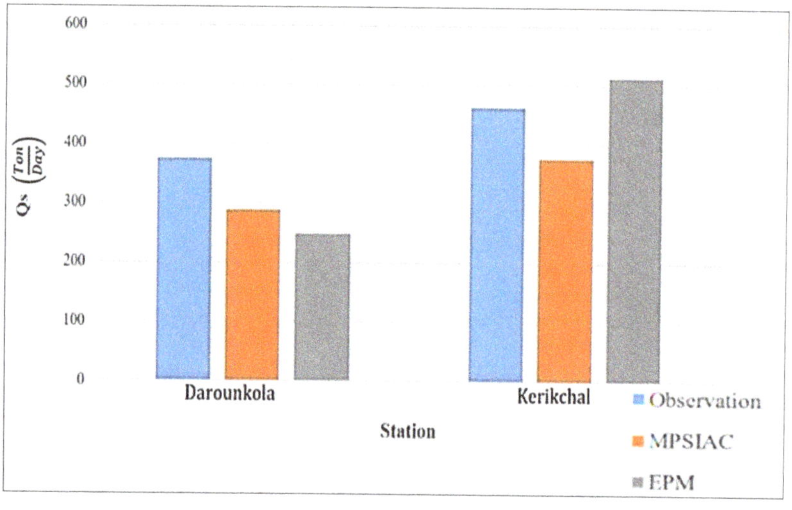

Figure 10. Comparison between the results predicted by both MPSIAC and EPM models to those of field measurement at Darounkola and Kerikchal Stations.

4. Discussion

Empirical methods for assessing soil erosion can be worked out with available inputs to estimate the sediment yield for areas exposed to high erosion risk. This research assessed three empirical methods for soil erosion estimation, including MPSIAC, EPM, and Fournier methods integrated with a GIS 10.5 model, compared to the results of field measurements in the Babolroud watershed, Iran.

The results of this research show that the sediment yield estimated by the MPSIAC method is closer to the results of field measurements at the two sedimentation stations, compared to the results by using the EPM method. This finding indicates that the MPSIAC model should be preferred to apply in the Babolroud watershed. As also reported by other researchers, the MPSIAC method has better performance in comparison with the EPM method in the Talar watershed, which is also located in Mazandaran Province and has similar geographic characteristics to the watershed studied [15]. Nevertheless, the results of a study in the DEZ watershed showed that the EPM model generates better results than those using the MPSIAC model [10]. It should be noted that this watershed is located in the south of Iran and has different climatic and geological features in comparison to those located in Mazandaran Province. In another study in Khorasan Province in Iran, it was proved that the MPSIAC method underestimates the sediment yield [13], which is in line with the findings of the current research. Moreover, it can be concluded that the Fournier model should not be considered for areas with the same characteristics as those of the Babolroud watershed, as shown in a previous study, where this method had no efficiency in predicting erosion intensity [37]. The results of another study that was conducted in the mountainous region with a semi-arid climate suggested that the MPSIAC model is suitable for predicting the annual average sediment yield of Iranian watersheds under similar conditions [38]. In another area in the east of Iran that has a semi-arid climate, the application of the PSIAC and MPSIAC models was evaluated. The results showed that the calculated annual sediment yields in most parts using both models agree well with those of field observations [39].

The application of field measurements in determining the bed load of fluvial rivers was proved in several studies [40–43]. In this regard, bed load sediments have been estimated with reasonable accuracy. The sediment rating curve is also an accurate method for calculating suspended load in rivers [44–46]. Resultantly, field measurement can be a precise method for the validation of employed numerical methods.

It is suggested to apply these empirical models in further research for determining the annual sediment yields from other watersheds with different geographic features and compare the results with those of the present study.

5. Conclusions

Accurate estimation of the amount of erosion and sedimentation is impossible in all areas due to technical, protective, and economic reasons. Therefore, the most appropriate method is to estimate the amount of erosion and sediment yield, which requires knowledge of the erosion mechanisms and the factors affecting them. On the other hand, choosing the appropriate model for each region requires the evaluation of the accuracy of different methods by comparing their results with the measured values in a watershed. In the present study, the EPM, MPSIAC, and Fournier methods were used to estimate the amount of erosion and sediment yield. Field measurements were also carried out at seven cross-sections along two reaches of Babolroad River, and the bed load transport rate was calculated. Suspended sediment discharge was calculated by applying the sediment rating curves. The total sediment load was determined based on field measurement data. The results of empirical methods showed that the erosion status of the area is in the moderate erosion class, and it is necessary to carry out watershed management and soil protection in this area. The highest erosion intensity is in the X_1 sub-basin, mainly due to the lack of uncultivated land in the northern part of the watershed. The southern areas were less exposed to erosion due to the layer covered by hard and shallow rocks, forest, and mountain coverings. A comparison of results by using both empirical methods and field measurements at the two sedimentation stations on Babolroud River showed that both the EPM and MPSIAC methods can better predict the intensity of erosion and sediment production from the Babolroud watershed compared to the Fournier method. The total sedimentation of Darounkola Station was 371.915, 287.38, and 248.272 ton/day for field measurement, MPSIAC, and EPM, respectively. Additionally, the values for Kerikchal Station were 460.382, 376.24, and 520.72 ton/day for field measurement, MPSIAC, and EPM, respectively. The calculation error for these two sedimentation stations was 22.42% and 20.5% for the EPM and MPSIAC methods, respectively, indicating the better performance of the MPSIAC model in the Babolroud watershed.

Author Contributions: E.S.T.; conceptualization, methodology, software, measurements, and writing, H.A.; supervision, methodology, and validation, J.S.; methodology and writing reviewing. All authors have read and agreed to the published version of the manuscript.

Funding: This research received no external funding.

Institutional Review Board Statement: Not applicable.

Informed Consent Statement: Not applicable.

Data Availability Statement: Data will be available upon request to the authors.

Acknowledgments: Special thanks to Mazandaran Regional Water Authority for providing data and information of this article.

Conflicts of Interest: The authors declare no conflict of interest.

References

1. Refahi, H.G. *Water Erosion and Conservatio*; University of Tehran Publication: Tehran, Iran, 2015.
2. Wang, G.; Gertner, G.; Fang, S.; Anderson, A.B. Mapping Multiple Variables for Predicting Soil Loss by Geostatistical Methods with TM Images and a Slope Map. *Photogramm. Eng. Remote Sens.* **2003**, *69*, 889–898. [CrossRef]
3. Alizadeh, A. *Soil Erosion and Conservation*; Astan Qods Razavi Publication: Mashhad, Iran, 1990.
4. Wischmeier, W.H.; Smith, D.D. Predicting rainfall erosion losses. In *USDA Agricultural Research Services Handbook*; USDA: Washington, DC, USA, 1978.
5. Jain, M.K.; Das, D. Estimation of sediment yield and areas of soil erosion and deposition for watershed prioritization using GIS and remote sensing. *Water Resour. Manag.* **2010**, *24*, 2091–2112. [CrossRef]
6. Jemai, S.; Kallel, A.; Agoubi, B.; Abida, H. Soil Erosion Estimation in Arid Area by USLE Model Applying GIS and RS: Case of Oued El Hamma Catchment, South-Eastern Tunisia. *J. Indian Soc. Remote Sens.* **2021**, *49*, 1293–1305. [CrossRef]
7. Renard, K.G.; Foster, G.R.; Weesies, G.A.; Porter, J.P. RUSLE: Revised universal soil loss equation. *J. Soil Water Conserv.* **1991**, *46*, 30–33.
8. Williams, J.R. Sediment routing for agricultural watersheds. *JAWRA J. Am. Water Resour. Assoc.* **1975**, *11*, 965–974. [CrossRef]
9. Bagherzadeh, A.; Daneshvar, M.R.M. Sediment yield assessment by EPM and PSIAC models using GIS data in semi-arid region. *Front. Earth Sci.* **2011**, *5*, 207–216. [CrossRef]
10. Noori, H.; Siadatmousavi, S.M.; Mojaradi, B. Assessment of sediment yield using RS and GIS at two sub-basins of Dez Watershed, Iran. *Int. Soil Water Conserv. Res.* **2016**, *4*, 199–206. [CrossRef]
11. Ganasri, B.P.; Ramesh, H. Assessment of soil erosion by RUSLE model using remote sensing and GIS—A case study of Nethravathi Basin. *Geosci. Front.* **2016**, *7*, 953–961. [CrossRef]
12. Singh, G.; Panda, R.K. Grid-cell based assessment of soil erosion potential for identification of critical erosion prone areas using USLE, GIS and remote sensing: A case study in the Kapgari watershed, India. *Int. Soil Water Conserv. Res.* **2017**, *5*, 202–211. [CrossRef]
13. Pourkarimi, M.; Mahmoudi, S.; Masihabadi, M.; Pazira, E.; Moeini, A. Use of MPSIAC and EPM to estimate sediment yield and erosion-a case study of a watershed of the second urban phase, Mashhad, Khorasan Province. *Agric. For.* **2017**, *63*, 201–213. [CrossRef]
14. Batista, P.V.G.; Silva, M.L.N.; Silva, B.P.C.; Curi, N.; Bueno, I.T.; Acérbi Júnior, F.W.; Davies, J.; Quinton, J. Modelling spatially distributed soil losses and sediment yield in the upper Grande River Basin—Brazil. *Catena* **2017**, *157*, 139–150. [CrossRef]
15. Mirakhorlo, M.S.; Rahimzadegan, M. Application of sediment rating curves to evaluate efficiency of EPM and MPSIAC using RS and GIS. *Environ. Earth Sci.* **2018**, *77*, 723. [CrossRef]
16. Kidane, M.; Bezie, A.; Kesete, N.; Tolessa, T. The impact of land use and land cover (LULC) dynamics on soil erosion and sediment yield in Ethiopia. *Heliyon* **2019**, *5*, e02981. [CrossRef] [PubMed]
17. Rajbanshi, J.; Bhattacharya, S. Assessment of soil erosion, sediment yield and basin specific controlling factors using RUSLE-SDR and PLSR approach in Konar river basin, India. *J. Hydrol.* **2020**, *587*, 124935. [CrossRef]
18. Pijl, A.; Reuter, L.E.H.; Quarella, E.; Vogel, T.A.; Tarolli, P. GIS-based soil erosion modelling under various steep-slope vineyard practices. *Catena* **2020**, *193*, 104604. [CrossRef]
19. Mezosi, G.; Mucsi, L. Soil erosion assessment with the help of remote sensing methods. In Proceedings of the International Symposium of Operationalization of Remote Sensing, Enschede, The Netherlands, 19–23 April 1993; pp. 19–23.
20. Tangestani, M.H. Integrating geographic information systems in erosion and sediment yield applications using the erosion potential method (EPM). In Proceedings of the GIS Research UK, Ninth Annual Conference, Glamorgan, Wales, 18–20 April 2001; pp. 18–20.
21. Lin, C.Y.; Lin, W.T.; Chou, W.C. Soil erosion prediction and sediment yield estimation: The Taiwan experience. *Soil Tillage Res.* **2002**, *68*, 143–152. [CrossRef]
22. Shrimali, S.S.; Aggarwal, S.P.; Samra, J.S. Prioritizing erosion-prone areas in hills using remote sensing and GIS—A case study of the Sukhna Lake catchment, Northern India. *Int. J. Appl. Earth Obs. Geoinf.* **2001**, *3*, 54–60. [CrossRef]

23. Shahiri Tabarestani, E.; Afzalimehr, H. Artificial neural network and multi-criteria decision-making models for flood simulation in GIS: Mazandaran Province, Iran. *Stoch. Environ. Res. Risk Assess.* **2021**, *35*, 2439–2457. [CrossRef]
24. Hadian, S.; Tabarestani, E.S.; Pham, Q.B. Multi attributive ideal-real comparative analysis (MAIRCA) method for evaluating flood susceptibility in a temperate Mediterranean climate. *Hydrol. Sci. J.* **2022**, *67*, 401–418. [CrossRef]
25. Yang, C.T. *Sediment Transport: Theory and Practice*; McGraw-Hill: New York, NY, USA, 1996; ISBN 0070723109.
26. Haddadchi, A.; Mohammad, H.O.; Amir, A.D.; Haddadchi, A.; Mohammad, H.O.; Amir, A.D. Assessment of Bed-Load Predictors Based on Sampling in a Gravel Bed River. *J. Hydrodyn.* **2012**, *24*, 145–151. [CrossRef]
27. López, R.; Vericat, D.; Batalla, R.J. Evaluation of bed load transport formulae in a large regulated gravel bed river: The lower Ebro (NE Iberian Peninsula). *J. Hydrol.* **2014**, *510*, 164–181. [CrossRef]
28. Shahiri Tabarestani, E.; Afzalimehr, H.; Pham, Q.B. Flow structure investigation over a pool-riffle sequence in a variable width river. *Acta Geophys.* **2022**, *1*, 713–727. [CrossRef]
29. Yu, B.Y.; Wu, P.; Sui, J.; Ni, J.; Whitcombe, T. Variation of runoff and sediment transport in the huai river—A case study. *J. Environ. Inform.* **2020**, *35*, 138–147. [CrossRef]
30. Jhonson, C.W.; Gembhart, A.C. Predicting sediment yield from sagerbrush range lands. *Agric. Rev. Man.* **1982**, *26*, 145–156.
31. Gavrilovic, Z. The Use of an Empirical Method for Calculating Sediment Production and Transport in Unsuited or Torrential Streams. *Open J. Geol.* **1988**, *6*, 411–422.
32. Costea, M. Using the Fournier indexes in estimating rainfall erosivity. Case study-the Secasul Mare Basin. *Aerul Si Apa. Compon. ale Mediu.* **2012**, *2012*, 313–320. Available online: http://aerapa.conference.ubbcluj.ro/ (accessed on 25 March 2022).
33. Song, T.; Chiew, Y.M.; Chin, C.O. Effect of Bed-Load Movement on Flow Friction Factor. *J. Hydraul. Eng.* **1998**, *124*, 165–175. [CrossRef]
34. Sui, J.; He, Y.; Liu, C. Changes in sediment transport in the Kuye River in the Loess Plateau in China. *Int. J. Sediment Res.* **2009**, *24*, 201–213. [CrossRef]
35. Liu, C.; Sui, J.; Wang, Z.Y. Changes in runoff and sediment yield along the Yellow River during the period from 1950 to 2006. *J. Environ. Inform.* **2008**, *12*, 129–139. [CrossRef]
36. Liu, C.; Sui, J.; Wang, Z.Y. Sediment load reduction in Chinese rivers. *Int. J. Sediment Res.* **2008**, *23*, 44–55. [CrossRef]
37. Kavian, A.; Safari, A. Determining the appropriate model for estimating sedimentation using statistical methods. *J. Appl. Res. Geogr. Sci.* **2013**, *30*, 111–130.
38. Najm, Z.; Keyhani, N.; Rezaei, K.; Nezamabad, A.N.; Vaziri, S.H. Sediment yield and soil erosion assessment by using an empirical model of MPSIAC for Afjeh & Lavarak sub-watersheds, Iran. *Earth* **2013**, *2*, 14–22.
39. Daneshvar, M.; Bagherzadeh, A. Evaluation of sediment yield in PSIAC and MPSIAC models by using GIS at Toroq Watershed, Northeast of Iran. *Front. Earth Sci.* **2012**, *6*, 83–94. [CrossRef]
40. Núñez-González, F.; Rovira, A.; Ibàñez, C. Bed load transport and incipient motion below a large gravel bed river bend. *Adv. Water Resour.* **2018**, *120*, 83–97. [CrossRef]
41. Afzalimehr, H.; Hadian, S.; Shahiri Tabarestani, E.; Mohammadi, M. Influence of Suspended Sediment Load on Roughness Coefficient and Intensity of Flow Turbulence (Case study: Haraz, Rostamabad and Beheshtabad Rivers). *Environ. Water Eng.* **2020**, *6*, 459–472. [CrossRef]
42. Shahiri Tabarestani, E.; Afzalimehr, H.; Pham, Q.B. Validation of double averaged velocity method in a variable width river. *Earth Sci. Inform.* **2021**, *14*, 2265–2278. [CrossRef]
43. Yang, C.T.; Huang, C. Applicability of sediment transport formulas. *Int. J. Sediment Res.* **2001**, *16*, 335–353.
44. Bravo-Espinosa, M.; Osterkamp, W.R.; Lopes, V.L. Bedload Transport in Alluvial Channels. *J. Hydraul. Eng.* **2003**, *129*, 783–795. [CrossRef]
45. Yu, B.Y.; Wu, P.; Sui, J.; Yang, X.; Ni, J. Fluvial geomorphology of the Middle Reach of the Huai River. *Int. J. Sediment Res.* **2014**, *29*, 24–33. [CrossRef]
46. Sui, J.; He, Y.; Karney, B.W. Flow and high sediment yield from the Huangfuchuan watershed. *Int. J. Environ. Sci. Tech* **2008**, *5*, 149–160. [CrossRef]

MDPI
St. Alban-Anlage 66
4052 Basel
Switzerland
www.mdpi.com

Water Editorial Office
E-mail: water@mdpi.com
www.mdpi.com/journal/water

www.ingramcontent.com/pod-product-compliance
Lightning Source LLC
Chambersburg PA
CBHW040754250926
51850CB00007BB/1491
9783036587455